The Conversational Interface

Michael McTear · Zoraida Callejas
David Griol

The Conversational Interface

Talking to Smart Devices

 Springer

Michael McTear
School of Computing and Mathematics
Ulster University
Northern Ireland
UK

David Griol
Department of Computer Science
Universidad Carlos III de Madrid
Madrid
Spain

Zoraida Callejas
ETSI Informática y Telecomunicación
University of Granada
Granada
Spain

ISBN 978-3-319-32965-9   ISBN 978-3-319-32967-3   (eBook)
DOI 10.1007/978-3-319-32967-3

Library of Congress Control Number: 2016937351

Printed on acid-free paper

This Springer imprint is published by Springer Nature
The registered company is Springer International Publishing AG Switzerland

Foreword

Some of us who have been in the field of "computers that understand speech" for many years have experienced firsthand a tremendous evolution of all the technologies that are required for computers to talk and understand speech and language. These technologies, including automatic speech recognition, natural language understanding, language generation, and text to speech are extremely complex and have required decades for scientists, researchers, and practitioners around the world to create algorithms and systems that would allow us to communicate with machines using speech, which is the preferred and most effective channel for humans.

Even though we are not there yet, in the sense that computers do not yet have a mastering of speech and language comparable to that of humans, we have made great strides toward that goal. The introduction of Interactive Voice Response (IVR) systems in the 1990s created programs that could automatically handle simple requests on the phone and allow large corporations to scale up their customer care services at a reasonable cost. With the evolution of speech recognition and natural language technologies, IVR systems rapidly became more sophisticated and enabled the creation of complex dialog systems that could handle natural language queries and many turns of interaction. That success prompted the industry to create standards such as VoiceXML that contributed to making the task of developers easier, and IVR applications became the test bed and catalyzer for the evolution of technologies related to automatically understanding and producing human language.

Today, while IVRs are still in use to serve millions of people, a new technology that penetrated the market half a decade ago is becoming more important in everyday life: that of "Virtual Personal Assistants." Apple's Siri, Google Now, and Microsoft's Cortana allow whoever has a smartphone to ask virtually unlimited requests and interact with applications in the cloud or on the device. The embodiment of virtual personal assistants into connected devices such as Amazon's Echo and multimodal social robots such as Jibo are on the verge of enriching the human–computer communication experience and defining new ways to interact with the vast knowledge repositories on the Web, and eventually with the Internet of Things. Virtual Personal Assistants are being integrated into cars to make it safer to interact with onboard entertainment and navigation systems, phones, and the Internet. We can imagine how all of these technologies will be able to enhance the usability of self-driving cars when they will become a reality. This is what the conversational interface is about, today and in the near future.

The possibilities of conversational interfaces are endless and the science of computers that understand speech, once the prerogative of a few scientists who had access to complex math and sophisticated computers, is today accessible to many who are interested and want to understand it, use it, and perhaps contribute to its progress.

Michael McTear, Zoraida Callejas, and David Griol have produced an excellent book that is the first to fill a gap in the literature for researchers and practitioners who want to take on the challenge of building a conversational machine with available tools. This is an opportune time for a book like this, filled with the depth of understanding that Michael McTear has accumulated in a career dedicated to technologies such as speech recognition, natural language understanding, language generation and dialog, and the inspiration he has brought to the field.

In fact, until now, there was no book available for those who want to understand the challenges and the solutions, and to build a conversational interface by using available modern open source software. The authors do an excellent job in setting the stage by explaining the technology behind each module of a conversational system. They describe the different approaches in accessible language and propose solutions using available software, giving appropriate examples and questions to stimulate the reader to further research.

The Conversational Interface is a must read for students, researchers, interaction designers, and practitioners who want to be part of the revolution brought by "computers that understand speech."

San Francisco   Roberto Pieraccini
February 2016   Director of Advanced Conversational Technologies
                                                                                               Jibo, Inc.

Preface

When we first started planning to write a book on how people would be able to talk in a natural way to their smartphones, devices and robots, we could not have anticipated that the conversational interface would become such a hot topic.

During the course of writing the book we have kept touch with the most recent advances in technology and applications. The technologies required for conversational interfaces have improved dramatically in the past few years, making what was once a dream of visionaries and researchers into a commercial reality. New applications that make use of conversational interfaces are now appearing on an almost weekly basis.

All of this has made writing the book exciting and challenging. We have been guided by the comments of Deborah Dahl, Wolfgang Minker, and Roberto Pieraccini on our initial book proposal. Roberto also read the finished manuscript and kindly agreed to write the foreword to the book.

We have received ongoing support from the team at Springer, in particular Mary James, Senior Editor for Applied Science, who first encouraged us to consider writing the book, as well as Brian Halm and Zoe Kennedy, who answered our many questions during the final stages of writing. Special thanks to Ms. Shalini Selvam and her team at Scientific Publishing Services (SPS) for their meticulous editing and for transforming our typescript into the published version of the book.

We have come to this book with a background in spoken dialog systems, having all worked for many years in this field. During this time we have received advice, feedback, and encouragement from many colleagues, including: Jim Larson, who has kept us abreast of developments in the commercial world of conversational interfaces; Richard Wallace, who encouraged us to explore the world of chatbot technology; Ramón López-Cózar, Emilio Sanchis, Encarna Segarra and Lluís F. Hurtado, who introduced us to spoken dialog research; and José Manuel Molina and Araceli Sanchis for the opportunity to continue working in this area. We would also like to thank Wolfgang Minker, Sebastian Möller, Jan Nouza, Catherine Pelachaud, Giusseppe Riccardi, Huiru (Jane) Zheng, and their respective teams for welcoming us to their labs and sharing their knowledge with us.

Writing a book like this would not be possible without the support of family and friends. Michael would like to acknowledge the encouragement and patience of his wife Sandra who has supported and encouraged him throughout. Zoraida and David would like to thank their parents, Francisco Callejas, Francisca Carrión, Amadeo Griol, and Francisca Barres for being supportive and inspirational, with a heartfelt and loving memory of Amadeo, who will always be a model of honesty and passion in everything they do.

Contents

About the Authors

Michael McTear is Emeritus Professor at the University of Ulster with a special research interest in spoken language technologies. He graduated in German Language and Literature from Queens University Belfast in 1965, was awarded MA in Linguistics at University of Essex in 1975, and a Ph.D. at the University of Ulster in 1981. He has been a Visiting Professor at the University of Hawaii (1986–1987), the University of Koblenz, Germany (1994–1995), and University of Granada, Spain (2006–2010). He has been researching in the field of spoken dialog systems for more than 15 years and is the author of the widely used textbook *Spoken dialogue technology: toward the conversational user interface* (Springer, 2004). He also is a coauthor (with Kristiina Jokinen) of the book *Spoken Dialogue Systems*, (Morgan and Claypool, 2010), and (with Zoraida Callejas) of the book *Voice Application Development for Android* (Packt Publishing, 2013).

Zoraida Callejas is Associate Professor at the University of Granada, Spain, and member of the CITIC-UGR Research Institute. She graduated in Computer Science in 2005, and was awarded a Ph.D. in 2008, also at the University of Granada. Her research is focused on dialog systems, especially on affective human–machine conversational interaction, and she has made more than 160 contributions to books, journals, and conferences. She has participated in several research projects related to these topics and contributes to scientific committees and societies in the area. She has been a Visiting Professor at the Technical University of Liberec, Czech Republic (2007–2013), University of Trento, Italy (2008), University of Ulster, Northern Ireland (2009), Technical University of Berlin, Germany (2010), University of Ulm, Germany (2012, 2014) and ParisTech, France (2013). She is coauthor (with Michael McTear) of the book *Voice Application Development for Android* (Packt Publishing, 2013).

David Griol is Professor at the Department of Computer Science in the Carlos III University of Madrid (Spain). He obtained his Ph.D. degree in Computer Science from the Technical University of Valencia (Spain) in 2007. He also has a B.S. in Telecommunication Science from this University. He has participated in several European and Spanish projects related to natural language processing and conversational interfaces. His main research activities are mostly related to the study of statistical methodologies for dialog management, dialog simulation, user modeling, adaptation and evaluation of dialog systems, mobile technologies, and multimodal interfaces. His research results have been applied to several application domains including Education, Healthcare, Virtual Environments, Augmented Reality, Ambient Intelligence and Smart Environments, and he has published in a number of international journals and conferences. He has been a visiting researcher at the University of Ulster (Belfast, UK), Technical University of Liberec (Liberec, Czech Republic), University of Trento (Trento, Italy), Technical University of Berlin (Berlin, Germany), and ParisTech University (Paris, France). He is a member of several research associations for Artificial Intelligence, Speech Processing, and Human–Computer Interaction.

Abbreviations

ABNF	Augmented Backus-Naur Form
AI	Artificial intelligence
AIML	Artificial Intelligence Markup Language
AM	Application manager
AmI	Ambient intelligence
API	Application programming interface
ASR	Automatic speech recognition
ATIS	Airline Travel Information Service
AU	Action unit
AuBT	Augsburg Biosignal Toolbox
AuDB	Augsburg Database of Biosignals
BDI	Belief, desire, intention
BML	Behavior Markup Language
BP	Blood pressure
BVP	Blood volume pulse
CA	Conversation analysis
CAS	Common answer specification
CCXML	Call Control eXtensible Markup Language
CFG	Context-free grammar
CITIA	Conversational Interaction Technology Innovation Alliance
CLI	Command line interface
COCOSDA	International Committee for Co-ordination and Standardisation of Speech Databases
CRF	Conditional random field
CTI	Computer–telephone integration
CVG	Compositional vector grammar
DBN	Dynamic Bayesian networks
DG	Dependency grammar
DiAML	Dialog Act Markup Language
DIT	Dynamic Interpretation Theory
DM	Dialog management

DME	Dialog move engine
DNN	Deep neural network
DR	Dialog register
DST	Dialog state tracking
DSTC	Dialog state tracking challenge
DTMF	Dual-tone multifrequency
ECA	Embodied Conversational Agent
EEG	Electroencephalography
ELRA	European Language Resources Association
EM	Expectation maximization
EMG	Electromyography
EMMA	Extensible MultiModal Annotation
EmotionML	Emotion Markup Language
ESS	Expressive speech synthesis
EU	European Union
FACS	Facial action coding system
FAQ	Frequently asked questions
FIA	Form Interpretation Algorithm
FML	Functional Markup Language
FOPC	First-order predicate calculus
GMM	Gaussian Mixture Model
GPU	Graphics processing unit
GSR	Galvanic skin response
GUI	Graphical user interface
HCI	Human–computer interaction
HMM	Hidden Markov model
HRV	Heart rate variability
IoT	Internet of Things
IQ	Interaction quality
ISCA	International Speech Communication Association
ISU	Information State Update
ITU-T	International Telecommunication Union
IVR	Interactive voice response
JSGF	Java Speech Grammar Format
JSON	JavaScript Object Notation
LGPL	Lesser General Public License
LPC	Linear predictive coding
LPCC	Linear prediction cepstral coefficient
LSA	Latent semantic analysis
LUIS	Language Understanding Intelligent Service (Microsoft)
LVCSR	Large vocabulary continuous speech recognition
MDP	Markov decision process
MFCC	Mel frequency cepstral coefficients
MLE	Maximum likelihood estimation
MLP	Multilayer perceptron

MURML	Multimodal Utterance Representation Markup Language
NARS	Negative Attitudes toward Robots Scale
NER	Named entity recognition
NIST	National Institute of Standards and Technology
NLG	Natural language generation
NLP	Natural language processing
NLTK	Natural Language Toolkit
NP	Noun phrase
OCC	Ortony, Clore, and Collins
PAD	Pleasure, arousal, and dominance
PARADISE	PARAdigm for DIalogue Evaluation System
PLS	Pronunciation Lexicon Specification
POMDP	Partially observable Markov decision process
POS	Part of speech
PP	Perplexity
PP	Prepositional phrase
PPG	Photoplethysmograph
PSTN	Public switched telephone network
QA	Question answering
RAS	Robot Anxiety Scale
RG	Response generation
RL	Reinforcement learning
RNN	Recurrent (or Recursive) neural network
ROS	Robot Operating System
RSA	Respiratory sinus arrythmia
RVDK	Robotic Voice Development Kit
SAIBA	Situation, Agent, Intention, Behavior, Animation
SASSI	Subjective Assessment of Speech System Interfaces
SC	Skin conductance
SCXML	State Chart XML: State Machine Notation for Control Abstraction
SDK	Software development kit
SDS	Spoken dialog system
SIGGEN	Special Interest Group on Natural Language Generation
SISR	Semantic Interpretation for Speech Recognition
SLU	Spoken language understanding
SMIL	Synchronized Multimedia Integration Language
SMILE	Speech and Music Interpretation by Large-space feature Extraction
SRGS	Speech Recognition Grammar Specification
SSI	Social Signal Interpretation
SSLU	Statistical spoken language understanding
SSML	Speech Synthesis Markup Language
SURF	Speeded Up Robust Features
SVM	Support vector machine
TAM	Technology acceptance model

TIPI	Ten-Item Personality Inventory
TTS	Text-to-speech synthesis
UI	User interface
UMLS	Unified Medical Language System
URI	Uniform resource identifier
VoiceXML	Voice Extensible Markup Language
VP	Verb phrase
VPA	Virtual personal assistant
VUI	Voice user interface
W3C	World Wide Web Consortium
WA	Word accuracy
Weka	Waikato Environment for Knowledge Analysis
WER	Word error rate
WOZ	Wizard of OZ
XML	Extensible Markup Language
ZCR	Zero crossing rate

Chapter 1
Introducing the Conversational Interface

Abstract Conversational interfaces enable people to interact with smart devices using conversational spoken language. This book describes the technologies behind the conversational interface. Following a brief introduction, we describe the intended readership of the book and how the book is organized. The final section lists the apps and code that have been developed to illustrate the technology of conversational interfaces and to enable readers to gain hands-on experience using open-source software.

1.1 Introduction

The idea of being able to hold a conversation with a computer has fascinated people for a long time and has featured in many science fiction books and movies. With recent advances in spoken language technology, artificial intelligence, and conversational interface design, coupled with the emergence of smart devices, it is now possible to use voice to perform many tasks on a device—for example, sending a text message, updating the calendar, or setting an alarm. Often these tasks would require multiple steps to complete using touch, scrolling, and text input, but they can now be achieved with a single spoken command. Indeed, voice input is often the most appropriate mode of interaction, especially on small devices where the physical limitations of the real estate of the device make typing and tapping more difficult.

The topic of this book is the conversational interface. Conversational interfaces enable people to interact with smart devices using spoken language in a natural way—just like engaging in a conversation with a person. The book provides an introduction and reference source for students and developers who wish to understand this rapidly emerging field and to implement practical and useful applications. The book aims at a middle ground between reviews and blogs in online technology magazines on the one hand and, on the other, technical books, journal articles, and conference presentations intended for specialists in the field.

M. McTear et al., *The Conversational Interface*,
DOI 10.1007/978-3-319-32967-3_1

Currently, there is no comparable book that brings together information on conversational interfaces comprehensively and in a readable style.

1.2   Who Should Read the Book?

This book is intended as an introduction to conversational interfaces for two types of reader:

- Final-year undergraduate and graduate students with some background in computer science and/or linguistics who are also taking courses in topics such as spoken language technology, mobile applications, artificial intelligence, Internet technology, and human–computer interaction.
- Computing professionals interested in or working in the fields of spoken language technology, mobile applications, and artificial intelligence.

The book provides a comprehensive introduction to conversational interfaces at a level that is suitable for students and computing professionals who are not specialists in spoken language technology and conversational interface development.

In order to follow the programming exercises, readers should have some knowledge of Java and of Android app development. Completing the exercises will enable readers to put their ideas into practice by developing and deploying apps on their own devices using open-source tools and software.

1.3   A Road Map for the Book

The book is organized into four parts. The chapters in Part I present the background to conversational interfaces, looking at how they have emerged recently and why, the mechanisms involved in engaging in conversation, and past and present work on spoken language interaction with computers that provides a basis for the next generation of conversational interfaces. Part II covers the various technologies that are required to build a conversational interface with speech for input and output, along with practical chapters and exercises using open-source tools. Part III extends the conversational interface to look at interactions with smart devices, wearables, and robots, as well as the ability to recognize and express emotion and personality. Part IV examines methods for evaluating conversational interfaces and looks at current trends and future directions. Figure 1.1 shows a road map for the book.

The following is a brief summary of the contents of each chapter.

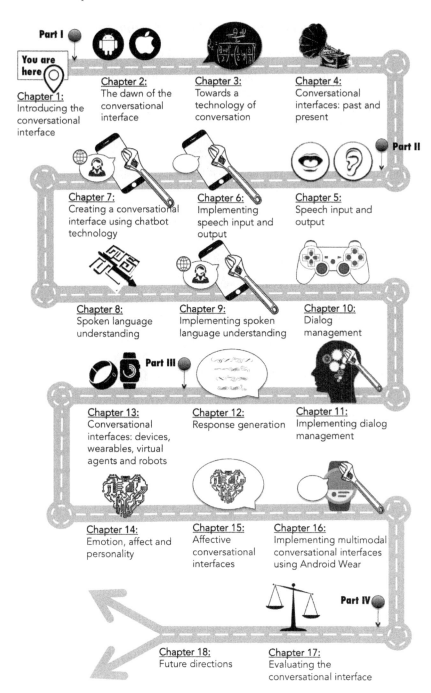

Fig. 1.1 A road map for the book

1.3.1 *Part I: Conversational Interfaces: Preliminaries*

Chapter 2: The dawn of the conversational interface

This chapter presents some examples of conversational interfaces on smartphones and reviews the technological advances that have made conversational interfaces possible. This is followed by a brief overview of the technologies that make up a conversational interface.

Chapter 3: Toward a technology of conversation

Although conversation is natural and intuitive for most people, it is not so obvious how conversation works. This chapter describes the various mechanisms that come into play when we engage in conversation. An understanding of these mechanisms is essential for developers wishing to design and implement effective conversational interfaces.

Chapter 4: Conversational interfaces: past and present

This chapter presents an overview of several different approaches to the modeling of conversational interaction with computers and of the achievements from these approaches that can contribute to the next generation of conversational interfaces.

1.3.2 *Part II: Developing a Speech-Based Conversational Interface*

Chapter 5: Speech input and output

In this chapter, the technologies that allow computers to recognize and produce speech are reviewed. These technologies have seen enormous advances during the past decade. Recent developments are outlined.

Chapter 6: Implementing speech input and output

Building on the overview of speech recognition and text to speech synthesis in Chap. 5, this chapter presents two open-source tools—the HTML5 Web Speech API and the Android Speech API—along with a series of practical exercises illustrating their usage.

Chapter 7: Creating a conversational interface using chatbot technology

This chapter shows how a conversational interface can be created using fairly simple techniques that have been applied widely in the development of chatbots. Some working examples of conversational interfaces using the Pandorabots platform are presented.

Chapter 8: Spoken language understanding

As conversational interfaces become more complex, more sophisticated techniques are required to interpret the user's spoken input and extract a representation of its meaning. This chapter describes and illustrates a range of technologies that are used for spoken language understanding.

Chapter 9: Implementing spoken language understanding

This chapter provides a tutorial on the use of the Api.ai platform to extract a semantic representation from the user's spoken utterances. The chapter also reviews some similar tools provided by Wit.ai, Amazon Alexa, and Microsoft LUIS, and looks briefly at other tools that have been widely used in natural language processing and that are potentially relevant for conversational interfaces.

Chapter 10: Dialog management

This chapter describes the dialog management component of a conversational interface that controls how the system should respond to the user's input. Two main approaches are reviewed—handcrafted dialog strategies that are used widely in industry and an emerging trend in which dialog strategies are learnt automatically using statistical models trained on corpora of real conversations.

Chapter 11: Implementing dialog management

Building on Chap. 10, this chapter provides practical exercises in rule-based and statistical dialog management, using VoiceXML for the rule-based approach and a corpus of dialogs to illustrate statistical dialog management.

Chapter 12: Response generation

This chapter reviews several approaches that have been used to create the system's response to the user's input, looking at canned text, the technology of natural language generation, and recent statistical methods. Methods for creating responses from unstructured and structured content on the Web are reviewed.

1.3.3   Part III: Conversational Interfaces and Devices

Chapter 13: Conversational interfaces: devices, wearables, virtual agents, and robots

Although conversational interfaces have been used mainly to interact with virtual personal assistants on smartphones, increasingly sensors and other devices provide input and support conversational interaction. This chapter reviews the peculiarities of conversational interaction with wearables, robots, and other smart devices.

Chapter 14: Emotion, affect, and personality

Affect is a key factor in human conversation, and a conversational interface that is to be believable will have to be able to recognize and express emotion and personality. This chapter reviews the issues involved in endowing conversational interfaces with emotion and personality.

Chapter 15: Affective conversational interfaces

This chapter explains how emotion can be recognized from physiological signals, acoustics, text, facial expressions and gestures, and how emotion synthesis is managed through expressive speech and multimodal embodied agents. The main open-source tools and databases are reviewed that are available for developers wishing to incorporate emotion into their conversational interfaces.

Chapter 16: Implementing multimodal conversational interfaces using Android Wear

This chapter discusses the challenges that arise when implementing conversational interfaces for a variety of devices with different input and output capabilities. There is a series of exercises showing how to develop libraries for multimodal interfaces using Android Wear.

1.3.4   Part IV: Evaluation and Future Prospects

Chapter 17: Evaluating the conversational interface

As conversational interfaces become more complex, their evaluation has become multifaceted. It is not only necessary to assess whether they operate correctly and are usable, but there are also novel aspects like their ability to engage in social communication. This chapter discusses the main measures that are employed for evaluating conversational interfaces from a variety of perspectives.

Chapter 18: Future directions

This chapter concludes the book by discussing future directions, including developments in technology and a number of application areas that will benefit from conversational interfaces, including smart environments, health care, care of the elderly, and conversational toys and educational assistants for children. The chapter also reviews issues related to the digital divide for under-resourced languages.

Apps and Code In several chapters, we provide tutorials showing how to develop examples and apps that illustrate different kinds of conversational interface at different levels of complexity. The following apps are provided:

- Web SAPI (Chap. 6). These examples show how to use the Web Speech API to provide spoken input and output on Web pages (you can try it using Google Chrome).

- SimpleTTS, RichTTS, SimpleASR, RichASR, TalkBack (Chap. 6). These apps show with increasing complexity how to build Android apps with speech output and input.
- TalkBot (Chap. 7). Shows how to create a chatbot with Pandorabots and use it with an Android app.
- Understand (Chap. 9). Shows the use of the Api.ai platform to extract semantic representations from the user's spoken utterances.
- PizzaRules (Chap. 11). Shows how to use VoiceXML to build a rule-based dialog manager to manage a pizza service.
- PizzaStat (Chap. 11). Shows how to build a statistical dialog manager to manage a pizza service.
- MorningCoffee, CookingNotifications, WriteBack (Chap. 16). These apps show how to build conversational interfaces for smartwatches with Android Wear, from predefined system voice actions to rich speech input processing.

The code for the examples is in GitHub, in the ConversationalInterface[1] repository.

---

[1]http://zoraidacallejas.github.io/ConversationalInterface.

Part I
Conversational Interfaces: Preliminaries

Chapter 2
The Dawn of the Conversational Interface

Good morning, Theodore. You have a meeting in five minutes.
Do you want to try getting out of bed?
Samantha in the movie Her: Official Trailer

Abstract With a conversational interface, people can speak to their smartphones and other smart devices in a natural way in order to obtain information, access Web services, issue commands, and engage in general chat. This chapter presents some examples of conversational interfaces and reviews technological advances that have made conversational interfaces possible. Following this, there is an overview of the technologies that make up a conversational interface.

2.1 Introduction

In the movie *Her* (2013), Theodore Twombly acquires Samantha, described as "the world's first intelligent operating system." Samantha is constantly available, just like a personal assistant, not only monitoring Theodore's calendar and answering his questions but also providing guidance and support in his personal life. They develop a relationship, and Theodore confesses to a friend that he is "dating" Samantha.

In the real world, being able to talk with a personal assistant in this way on a smartphone or other smart device has almost become a reality. Personal assistants, known by various names such as virtual personal assistants (VPAs), intelligent personal assistants, digital personal assistants, mobile assistants, or voice assistants, have become mainstream. Examples include Apple's Siri, Google Now, Microsoft Cortana, Amazon Alexa, Samsung S Voice, Facebook's M, and Nuance Dragon. Indeed, a search for "personal assistants" on Google Play toward the end of December 2015 returned 100 entries. Many of these VPAs help users to perform a variety of tasks on their smartphones, such as obtaining information using voice search, finding local restaurants, getting directions, setting the alarm, updating the calendar, and engaging in general conversation. Others provide more specialized functions, such as fitness monitoring, personalized preparation of drinks, and recipe planning.

We use the term *conversational interface* to refer to the technology that supports conversational interaction with these VPAs by means of speech and other modalities. To set the scene, we begin with some examples of the sorts of interactions that can be performed using such conversational interfaces.

M. McTear et al., *The Conversational Interface*,
DOI 10.1007/978-3-319-32967-3_2

11

Say "Ok Google"

Fig. 2.1 Google Search box on a Nexus 5 smartphone. Google and the Google logo are registered trademarks of Google Inc., used with permission

2.2   Interacting with a Conversational Interface

The following examples are taken from interactions with the Google Now personal assistant, which is available for Android devices as well as for iPhones and iPads. Google Now can be activated by tapping on the microphone icon in the Google Search box, as shown in Fig. 2.1 and also on more recent Android devices (Android 4.4 onwards) by saying "OK Google". This activates speech recognition, leading to a screen displaying the instruction say "Ok Google".

There is a Google support page that provides information on how to turn on "OK Google" voice search along with examples of what you can say in a large variety of languages.[1] The following example shows the output from a query about the weather.

> User (spoken input): What's the weather in Belfast?
> Google Now (spoken output): It's 7 degrees and partly cloudy in Belfast.

In addition to the spoken response, there is also a visual display of the recognized question, a visual representation of the weather forecast, and, on scrolling down, the addresses of some relevant Web pages (Fig. 2.2).

The next example is a general knowledge query.

> User (spoken input): When was Belfast City Hall built?
> Google Now (spoken output): Belfast City Hall is 110 years old.

The response from Google Now displays the recognized question, the answer, a picture showing Belfast City Hall, when it was built, a map of its location, and, on scrolling down, links to some additional information (Fig. 2.3).

Our final example illustrates access to operations on the device, in this case to set an alarm.

---

[1] https://support.google.com/websearch/answer/2940021?hl=en. Accessed February 19, 2016.

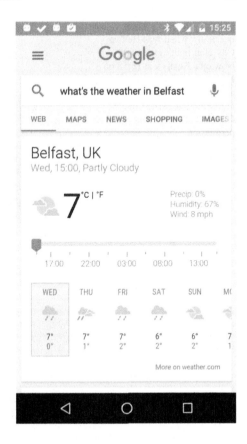

Fig. 2.2 Response to the query "What's the weather in Belfast" to Google Now on a Nexus 5 at 15:25 on February 10, 2016. Google and the Google logo are registered trademarks of Google Inc., used with permission

User (spoken input): Set the alarm for 9 o'clock tomorrow morning;
Google (spoken output): OK, 9 am, setting your alarm.

In addition to the spoken response, Google Now also presents the information shown in Fig. 2.4, which displays the recognized question, and a message confirming that the alarm has been set.

A wide range of device operations can be performed using voice, such as placing a call to a contact, sending a text, or launching an app. In many cases, the use of voice commands enables users to carry out these operations in fewer steps compared with traditional input methods. For example, the following steps would be required to set an alarm manually on a Nexus 5:

1. Tap on the clock icon (e.g. from the home screen).
2. Find the icon representing the alarm and tap on it.
3. Tap on the time displayed.
4. Adjust the hours and minutes to the required time.
5. Tap on "Done" to finish.

Fig. 2.3 Response to the
question "When was Belfast
City Hall built" to Google
Now on a Nexus 5 at 15:58 on
February 10, 2016. Google
and the Google logo are
registered trademarks of
Google Inc., used with
permission

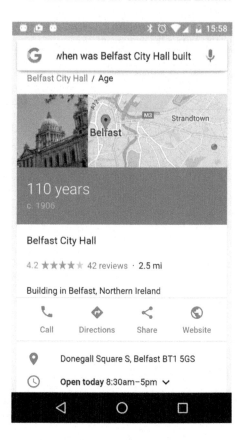

Fig. 2.4 Response to the
command "Set the alarm for 9
o'clock tomorrow morning"
to Google Now on a Nexus 5
on February 10, 2016. Google
and the Google logo are
registered trademarks of
Google Inc., used with
permission

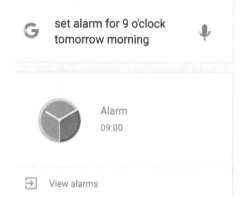

2.3   Conversational Interfaces for Smart Watches and Other Devices

Although conversational interfaces first appeared on smartphones, they are now also being deployed on various other devices such as smart watches, social robots, and devices such as Amazon Echo. In the future, we might expect conversational interfaces to be an integral part of the so-called Internet of Things (IoT), a massive network of connected objects, sensors, and devices that "talk" to each other and in some cases also communicate with humans.

A smart watch is a wearable device that provides many of the functions of a smartphone, such as notifications of incoming messages and emails, navigation, and voice search. Some smart watches also provide fitness tracking and heart rate monitoring. Users interact with a smart watch by speaking, by tapping on items on the screen, and by swiping the displayed cards. Some smart watches also have speakers that provide spoken output. Smart watches do not usually have an Internet connection so they have to be connected to a smartphone via Bluetooth pairing using software such as Android Wear—a dedicated software development kit (SDK) for wearables that is compatible with Android as well as iOS devices. One of the advantages of a smart watch is that actions such as responding to an email or text message can be performed on the fly using voice without having to take the phone from a pocket or handbag, find the required app, and tap in a reply.

Social robots allow users to perform tasks similar to those provided by a virtual personal assistant on a smartphone. However, because of their physical embodiment and because they often possess social qualities such as the ability to recognize and display emotions, they also provide social functions such as companionship for the elderly.

Conversational interfaces on devices such as smart watches and social robots provide many of the same capabilities that are already available on smartphones, although there may be differences in terms of the type of interface they provide. For example, the display on a smart watch is smaller than that on a typical smartphone, while some devices do not have a display for information but rely solely on voice for input and output. Chapter 13 discusses various devices, wearables, and robots; Chaps. 14 and 15 explore emotion and personality in conversational interfaces; and Chap. 16 provides a tutorial on how to develop apps for smart watches using the Android Wear SDK.

2.4   Explaining the Rise of the Conversational Interface

The conversational interface has for a long time been a vision of researchers in speech technology and artificial intelligence (AI), but until recently, this vision has only been realized in science fiction books and movies, such as 2001: A Space Odyssey, Star Wars, and many others. In 1987, Apple released a concept video depicting a device called the Knowledge Navigator, a software agent that provided services similar to those of a present-day VPA and that possessed advanced communication capabilities, including excellent text-to-speech and perfect speech understanding. In 2001, Berners-Lee and colleagues put forward a vision for a Semantic Web in which Semantic Web agents would perform tasks such as checking calendars, making appointments, and finding locations (Berners-Lee et al. 2001). But it was not until 2011 that these visions were realized with the launch of Siri, generally recognized as the first voice enabled VPA.

Siri and similar conversational systems have been made possible as a result of developments in technology and of increasing user acceptance and adoption, as explained further in the following sections.

2.4.1 Technological Developments

Various technological advances have contributed to the recent rise of conversational interfaces.

The renaissance of Artificial Intelligence. Since the mid-1950s, researchers in artificial intelligence (AI) have wrestled with the challenge of creating computers that are capable of intelligent behavior. AI has gone through cycles of euphoria and rejection with some initial successes followed by some spectacular failures. At first, it was believed that intelligent behavior could be reproduced using models of symbolic reasoning based on rules of formal logic. This was known as the *knowledge-based approach*, in which the focus was on problems that are difficult for humans but easy for computers—for example, decision-making and playing chess. Knowledge-based systems (also known as expert systems) were developed in the 1970s and 1980s to assist with decision-making in complex problems such as medical diagnosis, while IBM's chess playing computer Deep Blue defeated a human world champion in 1996. However, it became evident that various aspects of intelligent behavior that are easy for humans but difficult for computers, such as speech recognition and image recognition, could not be solved using these symbolic approaches but required processes such as the extraction of patterns from data and learning from experience. As a result, *subsymbolic approaches* using neural networks and statistical learning methods have come to dominate the field. Several factors have contributed to the recent success of subsymbolic approaches: developments in graphics processing units (GPUs) that have enabled the massive parallel computations required to run neural networks; the availability of vast amounts of data (known as *big data*) that enable AI systems to learn and become increasingly more intelligent; and the development of new algorithms (known as *deep learning*) that run on GPUs and process these vast amounts of data.[2] A sign of the promise of this new AI is that many major companies such as Google, Microsoft, Amazon,

---

[2]http://www.wired.com/2014/10/future-of-artificial-intelligence/. Accessed February 19, 2016.

Facebook, and Baidu—China's leading internet-search company—have recruited the world's leading experts in deep learning to support their research and development work in areas such as search, learning, natural language understanding, and personal assistant technology.

Advances in language technologies. Language technologies have benefitted from the new AI. Speech recognition accuracy has improved dramatically since around 2012 following the adoption of deep learning technologies. There have also been major advances in spoken language understanding. Machine learning approaches to dialog management have brought improved performance compared with traditional handcrafted approaches by enabling systems to learn optimal dialog strategies from data. Furthermore, grand challenges in various areas of speech and language technology, including speech recognition, text-to-speech synthesis, spoken dialog management, and natural language learning, have promoted the exploration and evaluation of different systems and techniques using shared tasks and data, leading to technological advances and wider cooperation within the research communities (Black and Eskenazi 2009).

The emergence of the Semantic Web. The vision of the Semantic Web is that all of the content on the Web should be structured and machine-readable, so that search using the traditional approach of keywords as input has been replaced by semantic search based on the meaning of the input. Semantically tagged pages marked up using encodings such as Resource Description Framework in Attributes (RDFa) and large structured knowledge bases such as Google's Knowledge Graph have enabled search engines to better interpret the semantics of a user's intent, to return structured answers to queries, and, for virtual personal assistants such as Google Now, to support a question/answer type of interaction. Examples of the more complex types of question that can now be answered by the Google app are described here.[3]

Device technologies. Smartphones and other intelligent devices have become more powerful than the large personal computers of only a few years ago. Indeed, in one comparison, it was stated that a single Apple iPhone5 has 2.7 times the processing power of a 1985 Cray-2 supercomputer.[4] Moreover, since smartphones have access to a wide range of contextual information, such as the user's location, time and date, contacts, and calendar, the integration of this contextual information into conversational interfaces enables VPAs to provide help and support that is relevant and personalized to the individual user.

Increased connectivity. Faster wireless speeds, the almost ubiquitous availability of WiFi, more powerful processors in mobile devices, and the advent of cloud computing mean that resource-intensive operations such as speech

---

[3]http://insidesearch.blogspot.com.es/2015/11/the-google-app-now-understands-you.html. Accessed February 19, 2016.

[4]http://www.phonearena.com/news/A-modern-smartphone-or-a-vintage-supercomputer-which-is-more-powerful_id57149. Accessed February 19, 2016.

recognition and search can be performed in the cloud using large banks of powerful computers.

The interest of major technology companies in conversational interfaces. While previously interest in conversational interfaces for VPAs was limited to relatively small niche companies and to enthusiastic evangelists of the AI dream, now many of the largest companies in the world are competing to create their own VPAs, for example, Apple's Siri, Google's Google Now, Amazon's Alexa, Microsoft's Cortana, Facebook's M, and Baidu's Duer. These VPAs enable companies to more accurately profile the users of their VPAs, enabling them to promote their e-commerce services and thus gain a competitive advantage.

Notwithstanding these advances, there is still more work to be done before conversational interfaces achieve a level of performance similar to that of humans. For example, in looking for a possible way forward, Moore (2013) suggests that it is necessary to go beyond the domain of speech technology and draw inspiration from other fields of research that inform communicative interaction, such as the neurobiology of living systems in general.

2.4.2   User Acceptance and Adoption

Even if a product is technologically advanced, it will not succeed unless it is accepted and adopted by users. Until recently, it seemed that users stopped using their VPAs after an initial stage of experimentation. In some cases, they encountered problems such as speech recognition errors and so reverted to more accustomed and more accurate modes of input. Some users found amusement by saying "silly" things to their VPA to see what sort of response they would get. Furthermore, the proliferation of so many virtual personal assistants makes it difficult to select and adopt one particular VPA for regular use.

Evaluations of VPAs have so far been fairly informal, taking the form either of showdowns or of surveys. In a showdown, a large bank of questions is submitted to selected VPAs and the responses are analyzed. In one showdown, Google Now was compared with Microsoft Cortana and Soundhound's VPA Hound,[5] while in another, Google Now was compared also with Microsoft Cortana as well as with Siri.[6]

Conversational interfaces are appealing to users who wish to engage with Web services when on the go. Given the processing power and speed of modern smartphones as well as ongoing Internet connectivity, users no longer need to be located at a desktop PC to search for information or access Web services. Also, with devices becoming smaller to aid portability input is easier using a conversational

---

[5]http://www.greenbot.com/article/2985727/google-apps/android-virtual-assistant-showdown-google-now-vs-cortana-vs-hound.html. Accessed February 19, 2016.

[6]https://www.stonetemple.com/great-knowledge-box-showdown/. Accessed February 19, 2016.

interface compared with tapping on the soft keyboards of smartphones. In any case, some devices will not have keyboards but only microphones for voice input. This is likely to be the case as more and more devices become linked in the Internet of Things, where many of the devices will rely exclusively on voice for input and output.

Young people are also more likely to use conversational interfaces. In a recent study of the use of voice search, it was reported that teenagers talk to their phones more than the average adult and more than half of teenagers between 13 and 18 use voice search daily.[7] Voice search is also widely used in China in VPAs such as Baidu's Duer as it is more difficult to input text in Chinese and so speech is a more convenient input mode.

2.4.3 Enterprise and Specialized VPAs

Enterprise and specialized VPAs provide assistance in specific domains and for specific users. Enterprise and specialized VPAs can assist professionals in their work—for example, helping doctors to manage their workload, schedules, messages, and calls, and to obtain up-to-date and reliable information to assist with diagnosis. They can also assist customers to get help and information about a company's products.

IBM Watson for oncology is an example of a specialized VPA that helps oncologists to make evidence-based treatment decisions based on an analysis of an individual patient's medical records and a search for treatment options in a vast corpus of information from journals, textbooks, and millions of pages of text.[8] The Ask Anna VPA, developed by Artificial Solutions[9] to provide help to customers searching for information about products on the IKEA Web site, is an example of a customer-facing VPA.[10] Other examples include JetStar's Ask Jess virtual assistant, developed on the Nuance Nina platform,[11] that answers customers' queries about bookings, baggage, and seating,[12] and Next IT's Alme, a multimodal, multichannel, and multilanguage platform for customer service in domains such as health care, travel, insurance, finance, and retail.[13]

---

[7]https://googleblog.blogspot.fr/2014/10/omg-mobile-voice-survey-reveals-teens.html. Accessed February 19, 2016.

[8]http://www.ibm.com/smarterplanet/us/en/ibmwatson/watson-oncology.html. Accessed February 19, 2016.

[9]http://www.artificial-solutions.com/. Accessed February19, 2016.

[10]https://www.chatbots.org/virtual_assistant/anna3/. Accessed February 19, 2016.

[11]http://www.nuance.com/company/news-room/press-releases/Jetstar.docx. Accessed February 19, 2016.

[12]http://www.jetstar.com/au/en/customer-service. Accessed February 19, 2016.

[13]http://www.nextit.com/alme/. Accessed February 19, 2016.

VPAs can provide a commercial advantage for companies in the generation of advertising revenues and referral fees by directing users to specific services and Web sites that have been "chosen" by the assistant. Furthermore, as Meisel (2013) points out, they can promote a company's brand and services in a similar way to the company's Web site, but with the added value of a more personalized and more enjoyable interaction.

2.4.4   The Cycle of Increasing Returns

It has been predicted in a number of studies that the global market for VPAs will increase dramatically in the next few years. One factor in addition to those discussed above is the so-called cycle of increasing returns. User acceptance and adoption interact with developments in technology to produce a cycle of increasing returns. As performance improves, more people will use conversational interfaces. With more usage, there will be more data that the systems can use to learn and improve. And the more they improve, the more people will want to use them. Given this cycle, it can be expected that conversational interfaces will see a large uptake for some time to come and that this uptake will be accompanied by enhanced functionalities and performance.

2.5   The Technologies that Make up a Conversational Interface

In this book, we describe the various technologies that make up a conversational interface and that enable users to engage in a conversation with a device using spoken language and other modalities. In Part 2, we will focus on spoken language technologies, as these are the core components of the majority of current conversational interfaces, while in Part 3, we will describe additional aspects of the input and output such as the recognition and display of emotion and personality.

Looking first at conversational interfaces that make use of spoken language technologies, Fig. 2.5 shows the typical components of such a system and the information flow between the components.

Typically, such a conversational interface operates as follows. On receiving spoken input from the user, the system has to:

- Recognize the words that were spoken by the user (speech recognition).
- Interpret the words, i.e., discover what the user meant and intended by speaking these words (spoken language understanding).
- Formulate a response, or if the message was unclear or incomplete, interact with user to seek clarification and elicit the required information (dialog management).

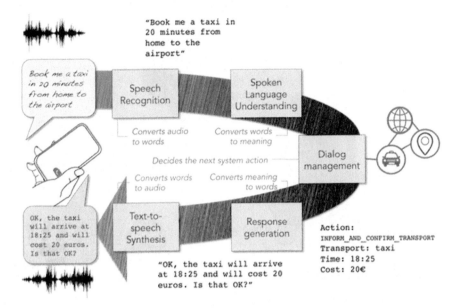

Fig. 2.5 The components of a spoken language conversational interface

- Construct the response, which may be in the form of words or, as in the examples above, accompanied by visual and other types of information (response generation).
- Speak and display the response (text-to-speech synthesis).

However, when people engage in natural conversational interaction, they convey much more than just the meanings of the words spoken. Their speech also conveys their emotional state and aspects of their personality. Additionally, in face-to-face interaction, their nonverbal behaviors, such as their facial expression, gestures, and body posture, also convey meaning. Other information may also be transmitted when speaking to a conversational interface. For example, smartphones and other smart devices have built-in sensors and actuators that gather data about the user and the environment, including location, motion, orientation, and biosignals such as heart rate. Figure 2.6 shows these additional inputs to the conversational interface. We look at these in more detail in Part 3.

As shown in these figures, once the conversational interface has interpreted the input from the user, it constructs queries to Web services and knowledge sources in order to perform tasks and retrieve information to be output by the response generation component. Our focus in this book will be on the conversational interface, and we will not explore how these Web services and knowledge sources are accessed and how information to be presented to the user is retrieved, although in some of the laboratory sessions, we will show how to connect to some existing services such as Google Search and Google Maps.

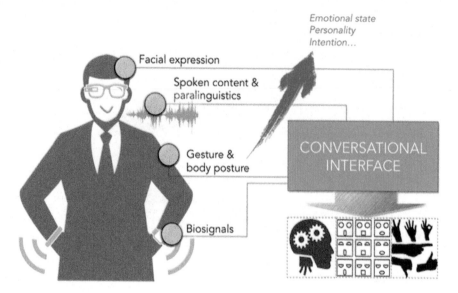

Fig. 2.6 Additional inputs to the conversational interface

2.6 Summary

A conversational interface allows people to talk to their devices in an intuitive and natural way. In this chapter, we have presented some examples of conversational interfaces that are available on smartphones and other devices.

Conversational interfaces have been made possible by recent advances in technology, in particular:

- A renaissance in AI, in which deep learning has brought about dramatic improvements in speech recognition accuracy and more recently in spoken language understanding and dialog management.
- The development of powerful processors that support the massively parallel computations required for deep learning algorithms and that provide the processing power on small devices such as smartphones that were available only to supercomputers a few years ago.
- Advances in the technologies of the Semantic Web that enable almost instantaneous access to the vast stores of unstructured as well as structured knowledge on the Internet.

As a result of these technological advances, user acceptance of technologies such as the conversational interface has increased, leading to increased adoption and consequently producing more data from which systems can learn, in turn resulting in further improvements in the technology.

Before we go on to explore the various technologies that make up the conversational interface, we need to understand what we mean by "conversation." In the next chapter, we look at the technology of conversation, in which we provide an overview of the contributions made by researchers in a variety of fields, including linguistics, psychology, sociology, and AI, to our understanding of the processes of conversational interaction.

Further Reading

There are a number of books and other resources that cover the topics of virtual personal assistants as well as related developments in AI and speech and language technologies. Trappl (2013) explores what is required for a personalized virtual butler and how such a virtual butler might be useful for an aging population. Stork (1998) is a collection of papers that examines the extent to which science fiction's HAL, the computer in the movie Space Odyssey 2011, has become technologically feasible—a bit dated now with recent advances in AI, but still worth reading. Meisel's online book The Software Society discusses the technology of virtual personal assistants,[14] while the short online book by Bouzid and Ma (2013) provides an introduction to the principles and techniques behind effective voice user interface design. An article by Ron Kaplan, Head of Nuance Communications' Natural Language Understanding R&D Lab, argues for conversational user interfaces as a replacement for the traditional graphical user interface.[15] The conversational interface is also the subject of a blog by John M Smart.[16] Pieraccini (2012) provides an accessible overview of speech technology and its history, and discusses the emergence of virtual personal assistants in the final chapter.

There are a number of blogs and other online resources that regularly discuss virtual personal assistants, including Amy Stapleton's Virtual Agent Chat,[17] Bill Meisel's Speech Strategy News,[18] and the Conversational Interaction Technology news board.[19] An online article by Buzzanga discusses recent advances in search that go beyond the use of keywords.[20]

Exercises

1. Watch the video "Behind the mic: the science of talking with computers."[21] This short video, produced by Google Creative Lab, features some of the key researchers in the field talking about speech recognition, language understanding, neural networks, and the use of speech to interact with smart devices.

---

[14]http://thesoftwaresociety.com/. Accessed February 19, 2016.

[15]http://www.wired.com/2013/03/conversational-user-interface/. Accessed February 19, 2016.

[16]http://www.accelerationwatch.com/lui.html. Accessed February 19, 2016.

[17]http://virtualagentchat.com/. Accessed February 19, 2016.

[18]http://www.tmaa.com/speechstrategynews.html. Accessed February 19, 2016.

[19]http://citia.lt-innovate.eu/. Accessed February 19, 2016.

[20]https://www.sla.org/wp-content/uploads/2015/06/2015_Buzzanga.pdf. Accessed February 19, 2016.

[21]https://youtu.be/yxxRAHVtafI. Accessed 2 March 2016.

2. Go to YouTube and search for "virtual personal assistants." You will find a large number of videos. Watch some videos that show examples of different VPAs and take note of the sorts of questions and commands that they can handle.
3. Look for videos on YouTube that show shootouts (or comparisons) of different VPAs, noting the questions and commands that are used in the tests. You can also find several interesting videos and demos created in the Interaction Lab of Heriot Watt University, Edinburgh.[22]
4. Go to Google Play[23] and search for "virtual personal assistants." A large number of VPAs will be returned. Select and download two that you can use for a comparison test. For example, you could download Assistant and Call Mom, as these are featured later in the book. Using the questions and commands noted in exercise 2, test and compare the two VPAs, noting where they succeed and where they fail. For inputs that fail, distinguish between failures due to speech recognition errors and failures due to the back-end components of the app (e.g., if it is not able to make a correct search for the item you requested).

References

Berners-Lee T, Hendler J, Lassila O (2001) The semantic web. Sci Am 284:34–43. doi:10.1038/scientificamerican0501-34

Black AW, Eskenazi M (2009) The spoken dialog challenge. In: Proceedings of the 10th annual meeting of the special interest group in discourse and dialog (SIGDIAL 2009), Queen Mary University of London, September 2009, pp 337–340. doi:10.3115/1708376.1708426

Bouzid A, Ma W (2013) Don't make me tap: a common sense approach to voice usability. ISBN 1-492195-1-7

Meisel W (2013) The personal-assistant model: unifying the technology experience. In: Neustein A, Markowitz JA (eds) Mobile speech and advanced natural language solutions. Springer, New York, pp 35–45. doi:10.1007/978-1-4614-6018-3_3

Moore RK (2013) Spoken language processing: where do we go from here? In: Trappl R (ed) Your virtual butler: the making of. Springer, Berlin, pp 119–133. doi:10.1007/978-3-642-37346-6_10

Pieraccini R (2012) The voice in the machine: building computers that understand speech. MIT Press, Cambridge

Stork D (ed) (1998) HAL's legacy: 2001's computer as dream and reality. MIT press, Cambridge

Trappl R (ed) (2013) Your virtual butler: the making of. Springer, Berlin. doi:10.1007/978-3-642-37346-6

---

[22]https://sites.google.com/site/hwinteractionlab/demos. Accessed March 2, 2016.

[23]https://play.google.com/store. Accessed March 2, 2016.

Chapter 3
Toward a Technology of Conversation

Abstract Conversation is a natural and intuitive mode of interaction. As humans, we engage all the time in conversation without having to think about how conversation actually works. In this chapter, we examine the key features of conversational interaction that will inform us as we develop conversational interfaces for a range of smart devices. In particular, we describe how utterances in a conversation can be viewed as actions that are performed in the pursuit of a goal; how conversation is structured; how participants in conversation collaborate to make conversation work; what the language of conversation looks like; and the implications for developers of applications that engage in conversational interaction with humans.

3.1 Introduction

The main purpose of a conversational interface is to support conversational inter-action between humans and machines. But what do we mean by conversational interaction? Generally, the term *conversation* is used to describe an informal spoken interaction in which news and views are exchanged and in which one of the main purposes is the development and maintenance of social relationships. This form of interaction is also called *small talk*. Conversation can be contrasted with dialog, a term that is generally used to describe an interaction with a more transactional purpose. For example, a user interacts with a *spoken dialog system* in order to perform a specific task, such as requesting flight information, booking a flight, or troubleshooting a faulty device.

The term *conversational* has also been used to describe systems that display more human-like characteristics and that support the use of spontaneous natural language, in contrast to those systems that require a more restricted form of input from the user, such as single words or short phrases. However, since users are likely to want to speak to conversational interfaces for a variety of purposes, sometimes to perform transactions, other times to ask questions, and yet other times simply to

© Springer International Publishing Switzerland 2016
M. McTear et al., *The Conversational Interface*,
DOI 10.1007/978-3-319-32967-3_3

chat, we will disregard these distinctions and use the term *conversation* to cover all types of spoken interaction with smart devices.

We distinguish our approach to the analysis of conversational interaction from what is covered in popular accounts of *the art of conversation*. Typically, writers describing the art of conversation provide advice and tips on how to conduct a successful conversation, for example, by applying strategies such as establishing shared interests, keeping a conversation going, and successfully bringing a conversation to a close. In contrast, in our approach, we will focus on what has been described as a *technology of conversation* (Sacks 1984), drawing on insights from studies of conversational interaction across a range of disciplines, including conversation analysis, psychology, linguistics, and artificial intelligence. Taking this approach will enable us to identify those key characteristics of human–human conversation that we might find useful to include in a computational model of advanced conversational interfaces.

3.2 Conversation as Action

When people engage in conversation, they do not just produce utterances, they perform actions—for example, they ask questions, make promises, pay compliments, and so on. The notion that speakers perform actions with their utterances was first proposed by Wittgenstein (1958) and developed further by Austin (1962) and Searle (1969) as a theory of *speech acts*. The term *dialog act* was introduced by Bunt (1979), while other commonly used terms are *communicative act* (Allwood 1976), *conversation act* (Traum and Hinkelmann 1992), *conversational move* (Carletta et al. 1997), and *dialog move* (Cooper et al. 1999).

One important insight from speech act theory is that the performance of a speech act requires that certain conditions be fulfilled. For example, for an utterance to be intended as a command by a speaker and understood as such by an addressee, the following conditions are required (based on Searle 1969):

1. The utterance is concerned with some future act that the hearer should perform.
2. The hearer is able to do the act, and the speaker believes that the hearer can do the act.
3. It is not obvious to the speaker and hearer that the hearer will do the act in the normal course of events.
4. The speaker wants the hearer to do the act.

These conditions incorporate the intuitions that people normally only ask others to do actions that they want to have done and that they believe that the other person is able to carry out the act and would not otherwise have done so without being asked. Thus, performing a speech act such as a request involves a dialog agent in reasoning about their own and their addressee's beliefs, desires, and intentions (or

Table 3.1 Action schema for a request

REQUEST (S,H,A)	
Constraints	Speaker(S) ∧ Hearer(H) ∧ ACT(A) ∧ H is agent of ACT (S is speaker and H is hearer, and A is an act and H is the agent of the act)
Precondition	Want(S,ACT(H)) (speaker wants hearer to do the action)
Effect	Want(H,ACT(H)) (the hearer wants to do the act)
Body	Believe(H,Want(S,ACT(H))) (the hearer believes that the speaker wants the hearer to do the act)

mental states). In the plan-based model of dialog that became prominent in the 1980s, a speech act such as a request was formalized in predicate calculus as an action schema similar to that used in AI models of planning that specified the preconditions, effects, and body of the action (Allen 1995). Table 3.1 shows an action schema for a request.

Using action schemas such as this along with various processes of inference, an agent could interpret and produce dialog acts in a conversation (see Jurafsky and Martin 2009: 887–881 for a more detailed example). However, one of the problems of the plan-based approach was that it was computationally complex and in the worst case intractable. Also, it has been argued that humans do not necessarily go through a complicated process of inference when engaging in dialog. Nevertheless, this early work on speech acts gave rise to an influential approach to dialog theory known as the *BDI (belief, desire, intention)* approach that is still current in some areas of dialog systems research (see further Chap. 4).

A number of different taxonomies of dialog acts have been proposed to label the functions of utterances in conversation. Traum (2000) compared eight schemes, showing how there are differences in the distribution of dialog act types across different domains, schemes, and corpora. For example, the Verbmobil tagset, which included acts such as suggest, accept, and reject, was intended for a two-party scheduling domain in which the speakers were involved in planning a meeting (Alexandersson et al. 1997).

One of the most widely used dialog act tagsets in computational linguistics is Dialog Act Markup in Several Layers (DAMSL) (Allen and Core 1997). DAMSL is intended as a more general, domain-independent tagset. In DAMSL, four main categories of tag are distinguished:

1. Communicative status—whether the utterance is intelligible and whether it was successfully completed.
2. Information level—the semantic content of the utterance.
3. Forward looking function—how the utterance constrains the future beliefs and actions of the participants and affects the discourse.
4. Backward looking function—how the utterance relates to the previous discourse.

The lists of acts with forward and backward looking functions are similar to acts proposed in other dialog act taxonomies. Forward looking functions describe the functions of utterances mainly in terms of the speaker's intentions and the obligations of the speaker and hearer. For example, requests for action (labeled as Influencing-addressee-future-action) obligate the hearer either to perform the action or at least to acknowledge the request. There are two subcategories: open option and action directive. Open option suggests an action without obligating the hearer, for example:

| A: There's a train at 7. | Open-Option |
| B: Let's take a later one. | Action-Dir |

Action directive, on the other hand, obligates the hearer to perform the action or else to communicate a refusal or inability to perform it, as in the following example:

A: Let's take the train at 7.	Action-Dir
B: No.	Reject (utt1)
B: Let's take a later one.	Action-Dir

In order to assist with labeling, a decision tree was provided to help distinguish between different types of act. There is also a tool that allows the tags to be applied though menus, buttons, and text fields.[1]

Backward looking functions indicate how the current utterance relates to the previous discourse. An utterance can answer, accept, reject, or try to correct some previous utterance or utterances. Four dimensions are involved: agreement, signaling understanding, answering, and relating to preceding discourse in terms of informational content. Backward looking acts are coded in terms of the type of act as well as the previous elements of the discourse that the acts refer to, as in the reject act in the preceding example. The following example shows the different types of response that could be made to an offer:

---

[1]http://www.cs.rochester.edu/research/trains/annotation.

1.	A: Would you like a beer with your sandwich?	Offer
2a.	B: Yes, please.	Accept (utt1)
2b.	B: Just the sandwich, thanks.	Accept-part (utt1)
2c.	B: I'm not sure, maybe later.	Maybe (utt1)
2d.	B: I don't want a drink just now.	Reject-part (utt1)
2e.	B: Actually I have to be getting on, it's late.	Reject (utt1)

A more elaborate version of the DAMSL coding scheme, Switchboard Shallow-Discourse Function Annotation (SWBD-DAMSL), was used to annotate the switchboard corpus, a collection of spontaneous conversations (Jurafsky et al. 1997). A set of about 60 tags was used, many of which provided more subtle distinctions than the original DAMSL categories.

Another widely used coding scheme is the HCRC Map Coding Scheme, developed at the Human Communication Research Centre at the University of Edinburgh. In the map task, two speakers discussed routes on maps. Three levels of dialog unit were encoded—conversational moves (similar to dialog acts), conversational games, and transactions (Carletta et al. 1997). Conversational moves included various types of initiation and response, corresponding roughly to the forward and backward looking functions in DAMSL. A decision tree was provided to assist in the annotation. Conversational games and transactions described larger segments of conversation (see further below).

These two schemes differ in that the HCRC scheme uses surface-level definitions for its categories, whereas in DAMSL the definitions are more intention-based. What these and other similar schemes have in common is that they extend the basic categories of speech act theory by including conversational phenomena, such as grounding and various types of response, along with an indication of how utterances relate to each other in a conversation. In another scheme, Traum and Hinkelmann (1992) suggested that conversational acts could be viewed in terms of a hierarchy consisting of four levels: turn taking, grounding, core speech acts, and argumentation. Examples of these different act types are shown in Table 3.2.

These act types are realized by increasingly larger chunks of conversation as we move down the list, with turn-taking acts often realized by sublexical items and argumentation acts often spanning whole conversations.

Table 3.2 Levels of conversation acts (adapted from Traum and Hinkelmann 1992)

Act type	Sample acts
Turn taking	take-turn keep-turn release-turn assign-turn
Grounding	Initiate Continue Acknowledge Repair RequestRepair RequestAcknowledgement Cancel
Core speech acts	Inform WH-Question YesNo-Question Accept Request Reject Suggest Evaluate RequestPermission Offer Promise
Argumentation	Elaborate Summarize Clarify Question&Answer Convince Find-Plan

The most elaborate approach to dialog acts is the Dialog Act Markup Language (DiAML) annotation framework, which originated in Bunt's dynamic interpretation theory (DIT) (Bunt 1995, 2009). DiAML has been accepted as an ISO standard 24627-2 "Semantic annotation framework, Part 2: Dialog Acts" (Bunt et al. 2012a). The framework is application independent and can be used to annotate typed, spoken, and multimodal dialog.

A key insight of DIT is that utterances in dialog are generally multifunctional, so that utterances often need to be annotated with more than one function. The following example, taken from Bunt (2011: 223), illustrates:

1.   A: What time is the next train to Amersfoort?
2.   B: Let me see ... The next one will be at eleven twenty-five.
3.   A: Eleven twenty-five. There's an earlier train to Amersfoort than that?
4.   B: Amersfoort? I'm sorry, I thought you said Apeldoorn.

As Bunt explains, B's utterance 2 begins with the words "let me see," in which he stalls for time while he finds the required information rather than simply waiting until the information is available. In so doing, he performs a turn-taking act and at the same time signals that he needs time to find the answer to A's question. More generally, it can be shown how utterances in a dialog often do not correspond to a single speech act but to several speech acts performed simultaneously.

DIT is a multidimensional dialog act annotation scheme. Table 3.3 shows the dimensions identified in DIT.

The process of annotating dialogs using DiAML is supported by a facility in the ANVIL annotation tool (Kipp 2012) that produces XML-based output in the DiAML format (Bunt et al. 2012b).

In summary, there are a number of well-established schemes for the annotation of transcripts of conversations in terms of the dialog acts performed by the utterances of the speakers. Recognizing what dialog act is being performed by an utterance is relevant for conversational interfaces to help determine the intentions behind a speaker's utterance and in order to be able to respond appropriately. For example, if the user says "is there a Chinese restaurant near here?", the system has to decide if the speaker is asking a question, to which the answer might be "yes" or "no," or if the speaker is indirectly asking to be given information about local Chinese restaurants, perhaps with directions of how to get there. Determining the speaker's intention has become an important topic in computational models of conversational interaction, particularly as people do not always express their intentions directly and so the system requires mechanisms to be able to infer their intentions from whatever evidence might be available and then provide a helpful response. We discuss methods for the automatic recognition of dialog acts in Chap. 8.

Table 3.3 The dimensions specified in DIT (based on Bunt 2011: 229)

Dimension	Description
1. Task/activity	Dialog acts whose performance contributes to performing the task or activity underlying the dialog
2. Autofeedback	Dialog acts that provide information about the speaker's processing of the previous utterance(s)
3. Allo-feedback	Dialog acts used by the speaker to express opinions about the addressee's processing of the previous utterance(s), or that solicit information about that processing
4. Contact management	Dialog acts for establishing and maintaining contact
5. Turn management	Dialog acts concerned with grabbing, keeping, giving, or accepting the sender role
6. Time management	Dialog acts signaling that the speaker needs a little time to formulate his contribution to the dialog
7. Discourse structuring	Dialog acts for explicitly structuring the conversation, e.g., announcing the next dialog act, or proposing a change of topic
8. Own communication management	Dialog acts where the speaker edits the contribution to the dialog that he is currently producing
9. Partner communication management	The agent who performs these dialog acts does not have the speaker role and assists or corrects the speaker in formulating a contribution to the dialog
10. Social obligation management	Dialog acts that take care of social conventions such as greetings, apologies, thanking, and saying good-bye

Knowing how to respond to a dialog act requires knowledge of what is an appropriate response. In the next section, we show how utterances in a conversation combine in recognizable structures and then discuss how these structures can help a conversational agent select appropriate responses to the user's utterances.

3.3 The Structure of Conversation

Utterances in a conversation do not occur in isolation but relate to each other in a number of ways. Various terms have been used to describe these co-occurring structural units, including adjacency pairs (Schegloff and Sacks 1973), exchanges (Sinclair and Coulthard 1975), discourse segments (Grosz and Sidner 1986), and conversational games (Kowtko et al. 1993). Here, we will use the term *adjacency pair,* which was first introduced within the framework of conversation analysis (CA) and is now used widely to describe sequences of related dialog acts. Typical examples of adjacency pairs are question–answer, greeting–greeting, and offer–acceptance, for example:

A: What's your name?	Question
A: What's your name?	Question
B: Walter.	Answer
A: Hi.	Greeting
B: Hi.	Greeting
A: Would you like a beer?	Offer
B: Cheers.	Acceptance

In CA terminology, the first item in the adjacency pair is called a *first pair part* and the second a *second pair part*. In other schemes, the terms *initiation and response* are used.

However, some first pair parts can have alternative second pair parts. For example, an offer can be followed by a refusal, as in:

A: Would you like a beer?	Offer
B: No thanks, I'm driving.	Refusal

In this example, the refusal is also accompanied by an explanation.

Another observation is that the pairs are not always adjacent, as in the following example:

A: Would you like a beer?	Offer
B: Have you non-alcoholic?	Question
A: Yes.	Answer
B: Sure.	Accept

There are also presequences in which the preconditions of an act such as an invitation can be checked:

A: Are you doing anything tonight?	Check
B: No	Answer Check
A: How about going out for a drink?	Invitation
B: Sure.	Accept

Schemes such as these have proved useful as a basis for annotating transcripts of conversations and analyzing the structural relationships between utterances. However, conversation analysts have gone beyond annotation and have looked at sequences of adjacency pairs to provide explanations for various interactional phenomena that occur in conversation and to suggest how the participants interpret these phenomena.

To take a simple example, as Schegloff (1968) notes, after one participant has produced a first pair part, there is an expectation that the second participant will produce a response that displays an understanding of that first pair part. In this way, each turn provides an opportunity for the participants to monitor the conversation and display their understanding. Schegloff (1968) coined the term *conditional relevance* to describe how the absence of an expected second pair part is noticeable and accountable. Evidence of this sequencing principle can be seen in the following example, discussed in detail in Levinson (1983: 320–321), where a delay in responding is treated as noticeable:

> A: So I was wondering would you be in your office on Monday by any chance?
>   (2.0)
> A: Probably not.

In this example, the two second pause after A's question is taken as indicating a negative answer to the question. This inference is explained in terms of the conversational principle of preference organization, which states that alternative responses to first pair parts are not equivalent alternatives, but that they differ in terms of their interactional implications. For example, a positive response to an invitation is likely to occur promptly, whereas a negative response is marked in some way, for example, by a period of silence or an explanation for the refusal, as illustrated in the following:

> A: uh if you'd care to come and visit a little while this morning I'll give you a cup of coffee
>
> B: heh, well (delay) that's awfully sweet of you (statement of appreciation)
>
> B: I don't think I can make it this morning. I'm running an ad in the paper (explanation of refusal)

This phenomenon is relevant for human–machine conversations as a means of explaining the inferences made by human participants when there is a delay in the machine's response (see Sect. 3.4.1 on turn taking). For example, a brief period of silence might lead the user to infer that the system has not understood or is unable to generate a response. An advanced conversational agent that emulates the processes of human conversation would also have to be able to model conversational phenomena such as these in order to engage naturally in conversation. An example of how this might be done is provided in the TRIPS system where the generation manager, which is responsible for the system's output, can use basic dialog rules to acknowledge the user's utterance while it is still in the process of constructing its next output (Allen et al. 2001; Stent 2002).

Conversational structure is also important for determining what sort of response the system should make to a user's utterance. As mentioned earlier, it is important to know whether an utterance such as "is there a petrol station near here?" is intended as a question or as a request for directions. Having determined the dialog act of the user's utterance, the system can select an appropriate type of response. Recent work in statistical dialog management has shown how such a dialog strategy can be learned from data. For example, in an approach described in Griol et al. (2008, 2014), dialogs are represented as a sequence of dialog acts by the system and the user, and the objective of the dialog manager is to find the best response to the user's previous utterance. A similar approach has been proposed using reinforcement learning (Frampton and Lemon 2005) (for further detail on statistical dialog management and reinforcement learning, see Chap. 10).

3.3.1   Dealing with Longer Sequences

Generally, in current conversations with a machine, short interactions consisting of a single adjacency pair are involved—for example, where the human user asks a question or issues a command and the machine responds. This is known as a one-shot conversation. However, in many cases, longer interactions are required. For example, conversations involving complex transactions such as flight reservations are often broken down into subtasks, such as getting the flight details, which itself breaks down into asking for departure and arrival information, before making the reservation. In commercially deployed systems, the progression through the different subtasks is usually strictly controlled, while some research systems have explored more flexible mechanisms for handling shifts between subtasks. For example, in the CMU Communicator spoken dialog system, the structure of an activity such as creating an itinerary was not predetermined. Although the system had an agenda with a default ordering for the traversal of the itinerary tree, the user could change the focus or introduce a new topic (Rudnicky and Wu 1999).

Conversational transactions tend to be less structured, although nevertheless in conversations in which several topics are discussed the participants need to keep track of items being referenced at different stages in the conversation, monitoring topic shifts, and returning to previous topics where necessary. The following example illustrates how keeping track of topic shifts in a conversation helps in the interpretation of the pronoun *it* in utterance 7:

1. A: Click on the "install" icon to install the program.
2   B: OK
3   B: By the way, did you hear about Bill?
4   A: No, what's up?
5   B: He took his car to be fixed and they've found all sorts of problems.
6   A: Poor Bill. He's always in trouble.
7   A: OK. Is it ready yet?

The referent of "it" in A's last utterance (7) is not Bill's car, which is the most recently mentioned element that matches the pronoun syntactically, but the program that B is installing. In this case, the pronoun and its referent are separated by several turns. However, the use of discourse cues helps the participants keep track of topic shifts. In the example presented here, the main topic is the installation of a program. The intervening turns are part of an unrelated topic (or digression) about Bill and his car. However, the main topic remains open, and it is possible to refer to elements that belong to the main topic later in the dialog using anaphoric devices such as pronouns and definite descriptions. In this example, the beginning of the digression is signaled by the phrase "by the way" and its end is signaled by the word "OK."

Until recently, spoken queries to assistants such as Google Now only supported one-shot dialogs in which the system either answered or did not answer the user's question. In more recent versions of Google Now, it is possible to ask follow-up questions that include pronouns. For example:

Where is the Eiffel tower located?
How high is it?
When was it built?

Pronoun resolution and discourse structure in general are likely to become an important element of the conversational interface as systems become more conversational and more human-like. For recent surveys of computational approaches to discourse structure, see Jurafsky and Martin (2009, Chap. 21) and Webber et al. (2012).

3.4 Conversation as a Joint Activity

When people engage in a conversation, they collaborate to make the conversation work (Clark 1996). This does not necessarily mean that the participants have to agree with each other, but even in conversations where there is disagreement, the participants take turns according to general conventions, they take measures to ensure mutual understanding (grounding), and they work together to resolve misunderstandings and communicative breakdowns.

3.4.1 Turn Taking in Conversation

Participants in conversation take turns at talking. A simple view of conversational turn taking is that one participant talks and the other participant waits until that turn has been completed and then begins to talk. In reality, turn taking is more complex than this and conversational turn taking has been described as an orderly phenomenon governed by rules (in the form of mutually accepted conventions) that determine how turns are allocated and how conversational participants can take the floor (Sacks et al. 1974). Sacks and colleagues made detailed analyses of transcripts of naturally occurring conversations in which overlaps and periods of silence were meticulously annotated. One of their observations was that:

> Transitions from one turn to a next with no gap and no overlap between them are common. Together with transitions characterized by slight gap or slight overlap, they make up the vast majority of transitions.

We have already seen an example where the next speaker does not take the floor, resulting in a period of silence that gives rise to an inference based on a *dispreferred* response. Examples where there is a slight overlap are also interesting as they shed light on how participants manage turn taking in conversation. As Sacks et al. indicate, turn taking is locally managed on a turn-by-turn basis. The size of a turn is not predetermined so that, in order to estimate when a turn is complete, the next speaker has to estimate a point at which it is potentially complete, i.e., what they call a transition-relevance place. The following is a simple example where the heavier shading indicates the point at which the utterances overlap:

A: That's an interesting house	isn't it?
B:	do you like it?

In this example, at the point where A has said the word "house," the utterance is potentially complete and B begins to talk, but as A continues with "isn't it," the result is a brief period of overlap. The interesting conclusion that can be drawn from examples such as these is that participants continually monitor the ongoing talk to predict transition-relevance places where they can attempt to take the floor.

Turn taking in human–machine conversation takes a different form due to the requirements of speech recognition technology. From the perspective of the machine, it is necessary to determine when the user has begun to speak and when the utterance is complete in order to reduce processing load and to increase recognition accuracy. The system generally begins listening at a point where the user presses an icon, such as a microphone, to activate the speech recognizer, or alternatively following a signal such as a beep. An end-pointing algorithm is used to detect the beginning and end of the portion of the acoustic signal to be analyzed as speech, where detecting the end of an utterance usually involves detecting a period of silence at which point it is assumed that the speaker has finished speaking. Problems can occur if the user continues to speak following the brief period of silence as the recognizer may not be able to analyze the additional input as part of a complete utterance, or may have begun producing output that results in overlap.

From the perspective of the user, detecting the end of the machine's utterance is usually obvious at the point where the machine has stopped talking, although a brief period of silence followed by a continuation of the output could cause the user to begin speaking, resulting in overlap. Some systems allow the user to cut in on the machine's output at any point so that an experienced user can cut short a lengthy prompt when they already know its content. This technique is known as *barge-in*. For barge-in to work successfully, the machine's output has to be terminated promptly on the detection of input from the user; otherwise, the recognizer may receive as input a signal containing portions of the machine's speech along with the speech of the user. Another problem is that the user, on continuing to hear output from the machine, will begin to speak louder, or repeat elements of the utterance that were overlapped. In either case, the speech signal will be distorted and is likely to be more difficult to recognize accurately, if at all.

In current human–machine conversation, turn taking is more carefully regulated in order to take account of the requirements of speech recognition technology. Nevertheless, designers have to take account of the expectations that humans bring to these interactions—for example, that a pause in the machine's output signals that the output is complete. Considerable effort has gone into determining the duration and timing of pauses, for example, between menu items or between the different elements of an extended prompt. For further discussion and an example of an interaction containing turn-taking problems due to inappropriate timing, see Lewis (2011: 210–221). Recent work on incremental processing in dialog, described in more detail in Chap. 4, is aimed at producing more human-like conversational turn-taking behaviors by modeling how humans monitor an ongoing utterance to determine when to take the turn (Skantze and Hjalmarsson 2013).

3.4.2 Grounding

Participants in conversation cannot be sure that their utterances have been mutually understood, and so they must engage in a process of joint activity called *grounding* (Clark 1996; Clark and Schaefer 1989; Clark and Brennan 1991). The following is a slightly simplified version of an example discussed in Clark and Brennan (1991: 129):

A: Now, - um do you and your husband have a car?
B: Have a car?
A: Yeah.
B: No.

In this example, A knows that he has asked B a question, but as B indicates that she has not understood the question and seeks clarification, it is only after A's response "Yeah" that the question has been grounded. Following B's subsequent response "No," the knowledge that B and her husband do not have a car becomes mutually understood, i.e., it becomes part of the common ground between A and B.

Clark and colleagues maintain that grounding is a continuous process that participants in conversation engage in on a turn-by-turn basis and that grounding can only be achieved through joint activity. In their model of grounding, a contribution in a conversation is not viewed as a single communicative act; rather, it is viewed as a collaboratively produced element consisting of two parts and involving two participants A and B:

- A *presentation* phase, in which A presents an utterance to B with the expectation that B will provide some evidence to A that the utterance has been understood.
- An *acceptance* phase in which B provides evidence of understanding, on the assumption that this evidence will enable A to believe that B understood A's utterance.

The acceptance phase may take several turns, including sequences of clarification requests and repairs. Once both phases are complete, it is argued that it will be common ground between A and B that B has understood what A meant.

Grounding is often achieved through the use of acknowledgements such as "uh huh" or "mm," also known as back-channel responses (Schegloff 1982), as in the following example (Clark and Brennan 1991: 131):

B: And I went to some second year seminars where there are only about half a dozen people.
A: mm.
B: and they discussed what a word was.

Here, A's "mm," which occurs within the turn of speaker B and overlaps with B's *and*, provides feedback that what is being said has been understood without actually taking the floor. Another mechanism is to initiate a relevant next turn that, following Schegloff's principle of conditional relevance discussed earlier, provides evidence to a first speaker A that the next speaker B has understood what A said. For more detail on Clark's model of grounding, see Clark (1996: 221–252).

In human–machine conversation, grounding is generally implemented using confirmations (also known as verifications), for example:

> System: where do you want to fly to?
> User: London.
> System: Flying to London, ...

See Chap. 10 for a discussion of different confirmation strategies employed in spoken dialog systems and voice user interfaces.

Clark's model of grounding has provided a basis for more advanced computational models of conversation developed. Traum (1994) presents a protocol that determines on a turn-by-turn basis whether utterances have been grounded and what actions are required to ground them. The protocol is implemented in a natural language dialog system in which the system reasons about the state of grounding and performs appropriate actions. These ideas were developed further in information state update theory (see Chap. 4) where common ground is one of the central informational components. A distinction is made between information that is assumed to be private to a participant and information that is shared, i.e., common ground. As a dialog progresses, the common ground is updated—for example, as a result of a grounding act such as an acknowledgement. Table 3.4 shows a simplified example of an update rule triggered by the conversational move `assert` (see Matheson et al. 2001 for a more formal account).

Modeling grounding in terms of mental states and conversational obligations has been further developed within recent versions of Information State Update Theory but has not been taken up within commercially deployed systems. Nevertheless, it could be argued that advanced conversational agents, acting, for example, as conversational companions or social robots, will require a model of common ground including shared beliefs, obligations, and commitments in order to demonstrate human-like conversational capabilities.

Table 3.4 An update rule

Name	Assert
Condition	G (common ground) contains an assert K addressed by A to B
Update	Add to G that if A asserts K, then A is trying to get B to believe K and that if B believes K, then B is also conversationally committed to K

In other words, when A makes an assertion K, A's intention is to get B to believe K. Furthermore, if B believes K, then B is also committed to that belief. TrindiKit, the dialog move engine toolkit, provides support for the development of dialog systems incorporating grounding processes (Larsson and Traum 2000)

3.4.3 *Conversational Repair*

Miscommunication is a frequent occurrence in conversation, and there is a wide range of studies of miscommunication in different domains, including differences in the use of language by males and females (Tannen 2001) and differences between people from different cultural and linguistic backgrounds (Gumperz 1978). Coupland et al. (1991) is a collection of papers investigating miscommunication in terms of gender, generational, and cultural differences, in clinical and organizational contexts, and in person–machine communication. Clark (1996) has conducted numerous detailed studies of how humans communicate (and miscommunicate) with one another, while Nass and Brave (2004) report a range of studies of how people interact with spoken dialog systems, looking at issues such as how the gender of the system's voice has a bearing on whether the system is perceived as being competent and knowledgeable.

It is useful to distinguish between failures to understand, where the hearer is unable to interpret what the speaker said, and misunderstandings, where the hearer's interpretation is different from what the speaker intended to convey. Failure to understand is generally easier to handle and in the simplest case can involve the hearer asking the speaker to repeat or reformulate. Misunderstandings are more difficult as it may not be immediately obvious that a misunderstanding has occurred. Note that detecting and dealing with miscommunication is a requirement for both participants as they collaborate to achieve mutual understanding.

Researchers in conversation analysis have produced a detailed account of how conversational repair is organized (Schegloff et al. 1977). Four types of repair were distinguished, depending on who initiated the repair and who carried it out. The speaker whose utterance contains a repairable element is described as *self*, while the recipient of the utterance is referred to as *other*. This gives rise to a four-way distinction:

- Self-initiated self-repair (in own turn, turn 1);
- Self-initiated self-repair (in turn transition);
- Other-initiated self-repair (in turn 2);
- Other-initiated other-repair (in turn 2).

For example, the current speaker may detect a potential problem in the utterance that they are currently producing and initiate a self-repair, as in:

> A: I was talking to Joan Wat- I mean Joan Wilson yesterday.

Another possibility is that the hearer B detects a problem and initiates a clarification request that invites the speaker A to initiate a repair:

> A: I was talking to Joan Watson yesterday.
> B: Who?
> A: I mean Joan Wilson.

It is also possible for the hearer to initiate and perform the repair, as in:

> A: I was talking to Joan Watson yesterday.
> B: You mean Joan Wilson.

Some interesting findings arose from this analysis. On the one hand, the opportunities for repair are sequentially ordered in that opportunities for self-initiation of repair occur before those for other-initiation of repair, as indicated above. Similarly, repair by self occurs before repair by other. These findings led to a more general conclusion that repairs are organized in terms of a system of *preference* in which self-initiation of repair is preferred to other-initiation, while self-repair is preferred to other-repair. In this analysis, the term *preference* is used to describe the feature of *markedness*. In other words, items that are preferred are unmarked, while those that are *dispreferred* are marked. With reference to repair, this means that dialog acts such as corrections of the other person's utterance are normally delayed and marked in some way—for example, with indicators of uncertainty or hesitation. In this way, the speaker of the utterance containing a potential error is given the opportunity to self-correct, and corrections by the recipient of the utterance are modulated so that they appear less direct.

Generally, these detailed insights on conversational repair have not been addressed in implementations of human–machine conversation where the focus has been mainly on dealing with miscommunication caused by speech recognition errors. However, with new research aiming to produce more human-like conversational interfaces, some of these insights are likely to become more relevant. For example, if the system detects a potential error, one way to rectify it would be to immediately correct the error, as happens usually in current systems. However, in terms of the properties of naturally occurring conversation, this would be a dispreferred strategy and a more human-like system might use either a more mitigated form of correction or a request for clarification, or even decide that in this particular context, it is reasonable to ignore the miscommunication and continue with the conversation. See McTear (2008) for a review of approaches to miscommunication in several different disciplines and of methods used in different interactional contexts to repair (and sometimes ignore) miscommunication.

However, designing repair strategies by hand would be a complicated and time-consuming task, and the question arises whether it is possible for a system to learn repair strategies. Frampton (2009) describes a learned strategy for repairs of automatic speech recognition (ASR) and spoken language understanding (SLU) errors using reinforcement learning that enabled the system to get the dialog back on track following miscommunication (see Chap. 10 for a detailed description of reinforcement learning in spoken dialog systems). Other relevant work includes the dissertations of Skantze (2007) and Bohus (2007) in which data-driven methods and machine learning were used to find optimal strategies for error handling in dialog. Skantze investigated the strategies used by humans to deal with errors and applied a method in which the costs associated with different grounding strategies could be measured and the strategy with the minimum expected cost selected. Bohus examined different recovery strategies and, like Skantze, found that strategies such as MoveOn, where the system advances the task by moving on to a different question rather than asking the user to repeat or rephrase, turned out to be more productive, given that asking the user to repeat or rephrase often resulted in an escalation of the error instead of its resolution. For more detail on error handling in spoken dialog systems, see Jokinen and McTear (2010, Chap. 3).

3.5 The Language of Conversation

The written language that appears in books and newspapers is usually carefully edited and corrected, making it amenable to analysis using grammars that contain rules for well-formed sentences. In contrast, the language of conversation is less regular as it often contains disfluencies, such as hesitation markers, fillers (e.g., "and so on," "you know," and "if you see what I mean"), fragments, and self-corrections that reflect the spontaneous production of speech but that make the processing of the text using methods applicable to written language more problematic. The following is an example from the Air Travel Information Service (ATIS) corpus, cited in Jurafsky and Martin (2009: 391):

> does United offer any one-way flights uh, I mean one way fares for 160 dollars?

In this example, the speaker changes the object noun phrase "one-way flights" to "one-way fares," marking the change by "uh" and "I mean." The resulting string of words could not be parsed using a grammar designed for the analysis of well-formed written text.

There are different ways in which this problem has been addressed. One approach is to edit the text and attempt to make it more regular. Since some of the features of spontaneous spoken language are sufficiently predictable, they can be described using special rules or strategies to filter out the irregularities and produce sentences that can be parsed using a conventional grammar. For example, utterances including false starts have a typical structure as illustrated in the following example:

The meeting will be on Mon-	uh	on Tuesday
reparandum	*editing term*	*alteration*

The *reparandum* is the item that is to be corrected or replaced. The *editing term*, often indicated by a disruption in the prosodic contour, by a word that has been cut off, or by a hesitation marker such as "uh," signals that a self-repair is occurring. Finally, the *alteration* is the corrected version of the *reparandum*. Frequently, there is some similarity between the *reparandum* and the *alteration* in terms of the words used as well as their syntactic structures. For example, a word in the *alteration* that replaces a word in the *reparandum* will often be of a similar word class and have a similar meaning. Given these features, it is possible to devise methods for detecting and correcting self-repairs and other types of disfluency in naturally occurring speech (for further details, see Heeman and Allen 1994).

A second approach, which we will discuss further in Chap. 8, is not to attempt to parse all of the input but rather to extract key words and phrases that indicate the entities being referenced in the input. In this way, the irregularities in the input can be ignored.

3.5.1 Prosodic, Paralinguistic, and Nonverbal Behaviors

While extracting key words and phrases is often sufficient for task-oriented dialogs, where the main objective is to determine the user's requirements, as in a flight booking, in other forms of conversation, such as social chat, there are additional streams of information over and above the literal content of the words in the utterance. These additional streams of information include prosodic and paralinguistic features as well as nonverbal behaviors.

Prosody refers to features such as the following:

- Overall pitch contour—this can determine the dialog act that is being performed, for example, "OK" with a rising tone indicates a checking act, whereas a falling tone indicates acceptance or confirmation.
- Accentuation—the item that receives prominence is generally being marked by the speaker as being new to the discourse as opposed to the other items that are treated as being given.
- Phrasing—the grouping of an utterance into meaningful chunks. For example, "call the ticket office in Belfast" is taken to refer to a ticket office that is in Belfast, whereas "call the ticket office | in Belfast," with a pause between "office" and "Belfast," would convey the meaning of calling the ticket office while the hearer is in Belfast.

Prosodic information can support the text-to-speech synthesis component of a conversational interface by using the correct prosodic forms to distinguish between otherwise ambiguous dialog acts, to indicate what information is new, and to group the utterance into meaningful chunks that will assist the hearer to more easily understand the utterance (Hirschberg 2002). Similarly, dialog understanding can be improved by the use of prosodic information that enables the system to track the dialog structure, to segment utterances correctly, and to predict and interpret dialog acts (Nöth et al. 2002; Hastie et al. 2002).

Paralinguistics refers to properties of the speech signal that can be used, either consciously or subconsciously, to modify meaning and convey emotion. Put simply, it is the study of *how* something is said as opposed to *what* is said. Examples of paralinguistic features include those that accompany and overlay the content of an utterance and modify its meaning, such as pitch, speech rate, voice quality, and loudness, as well as other vocal behaviors, such as sighs, gasps, and laughter. Paralinguistic properties of speech are important in human conversation as they can affect how a listener perceives an utterance. These effects are particularly marked in interactions between people from different cultures. For example, a speaker speaking with a volume that would be normal in his or her own culture could be perceived as being loud and aggressive by a listener from a different culture. Similarly, people can be perceived as being sarcastic, dismissive, or disinterested on account of properties of their voice, even where these effects were not intended.

Nonverbal behavior includes gestures, gaze, body orientation, and posture that accompany speech in face-to-face conversations between humans. We can distinguish between those behaviors that convey information unintentionally and those that have a communicative function. Clear cases of the former would include dress and gait, which often indicate a person's affiliations and attitudes, though not necessarily intentionally. Pointing as a means of indicating something or nodding to convey assent are clear cases of intentional nonverbal behaviors that function as surrogates or accompaniments of speech.

3.5.2 Implications for the Conversational Interface

Prosody, paralinguistics, and nonverbal behavior have a long history of study within linguistics and psychology but have only recently become hot topics in human-machine interaction with the emergence of socially aware computing, where the aim is to bring social intelligence to computers. One aspect of this new area of research is the study of behavioral cues and social signals and their relevance for systems that aim to simulate human social intelligence by being able to recognize and produce human-like social and behavioral cues. Until recently, attention was focused on how to make speech recognition and spoken language understanding more accurate through the analysis of this additional information. Now, the emphasis is on making machines more human-like, for example, in order to develop embodied conversational agents (ECAs) and social robots. This new research thrust has brought together researchers in the social sciences and engineering in a new field known variously as social signal processing (Pentland 2007; Vinciarelli et al.

2009), behavioral signal processing (Narayanan and Georgiou 2013), or affective computing (Schröder et al. 2012). We discuss these topics in greater detail in Chaps. 14 and 15 and provide a laboratory with exercises in Chap. 16.

3.6 Summary

In this chapter, we have reviewed the key features of conversation that will inform the design of conversational interfaces. In particular, we have covered the following aspects:

- Conversation as action—how utterances produced in conversation should be viewed as actions that speakers carry out in order to achieve their goals and how addressees interpret these actions. Recognizing what dialog act is being performed by an utterance is relevant for conversational interfaces to help determine the intentions behind a speaker's utterance and in order to be able to respond appropriately. The view of conversation as action originated in speech act theory and has been further developed in terms of dialog acts and the belief, desire, intention (BDI) model of dialog. Various dialog act tagsets were reviewed, including DAMSL and DiAML.
- The structure of conversation—how utterances in a conversation relate to each other. Various structural units have been used to describe sequences of dialog acts, including adjacency pairs, exchanges, discourse segments, and conversational games. In addition to analyzing conversational structure, conversation analysts have used these structures to explain conversational phenomena, such as how participants interpret the absence of an expected response.
- Conversation as a joint activity—how participants in conversation collaborate in taking turns and in reducing the risk of miscommunication through the process of grounding.
- Conversational repair—the mechanisms for repairing breakdowns in conversation through various types of repair either by the speaker or by the addressee. As shown by conversation analysts, types of repair differ in terms of their sequential properties and preference. Opportunities for self-repairs (repairs by the current speaker) occur sequentially before opportunities for other-repair (repairs by the addressee), and other-repairs are dispreferred (or marked)—for example, with indicators of uncertainty or hesitation.
- The language of conversation—how the language of conversation differs from the language of written text and the challenges that these differences pose for the speech recognition and spoken language understanding components. The language of conversation also includes prosodic, paralinguistic, and nonverbal behaviors that enhance conversational interaction and that convey additional information such as affect and emotional state.

In the next chapter, we provide an overview of the origins of conversational interfaces and highlight a number of key insights and technological achievements that are relevant for developers of conversational interfaces.

Further Reading

Speech act theory has been influential in the computational modeling of dialog. For a recent collection of chapters on speech act theory and pragmatics, see Searle (2013). Bunt and Black (2000) is a collection of chapters on computational pragmatics, while Geis (2006) examines the relationship between speech acts and conversational interaction. Fernández (2014) provides a comprehensive overview of the various dialog phenomena described in this chapter.

There have been a number of recent books on conversation analysis, including introductions to the main findings, methods, and techniques (Sidnell 2010; Hutchby and Wooffitt 2008), and a collection of papers by international experts in the areas of conversation analysis, discourse analysis, linguistics, anthropology, communication, psychology, and sociology (Sidnell and Stivers 2014). Eggins and Slade (2005) compare a number of approaches to the analysis of casual conversation, including conversation analysis, discourse analysis, and the ethnography of speaking. Hayashi et al. (2013) is a collection of papers by researchers in anthropology, communication, linguistics, and sociology on conversational repair and human understanding.

Schuller and Batliner (2013) present a detailed survey of computational approaches to paralinguistics, examining the methods, tools, and techniques used to automatically recognize affect, emotion, and personality in human speech.

For an overview of computational approaches to conversation, see Allen (1995, Chap. 17, Defining a conversational agent) and Jurafsky and Martin (2009, Chap. 24, Dialog and conversational agents).

Exercise

Although conversation is a natural and intuitive behavior for most people, understanding the mechanisms of conversational interaction in order to be able to model them in a conversational interface requires an understanding of the "technology of conversation."

There is an excellent online introductory tutorial on the methods of conversation analysis that provides basic instructions on how to transcribe and analyze conversations (Antaki 2002). Some audio/video clips of conversational interactions are provided along with transcriptions. Following the tutorial will allow you to appreciate some of the complex issues involved in analyzing and modeling conversation, including what to transcribe (e.g., just the words or other aspects such as paralinguistic features and gestures), basic transcription notation conventions, and how to analyze the data.

References

Allen JF (1995) Natural language understanding, 2nd edn. Benjamin Cummings Publishing Company Inc., Redwood

Allen JF, Core M (1997) Draft of DAMSL: dialog act markup in several layers. The Multiparty Discourse Group. University of Rochester, Rochester, USA. http://www.cs.rochester.edu/ research/cisd/resources/damsl/RevisedManual/. Accessed 20 Jan 2016

Allen JF, Ferguson G, Stent A (2001) An architecture for more realistic conversational systems. In: Proceedings of intelligent user interfaces 2001 (IUI-01), Santa Fe, NM, 14–17 Jan 2001. doi:10.1145/359784.359822

Alexandersson J, Buschbeck-Wolf B, Fujinami T, Maier E, Reithinger N, Schmitz B, Siegel M (1997) Dialog acts in VERBMOBIL-2. Verbmobil report 204, May 1997, DFKI GmbH, Saarbrücken Germany

Allwood J (1976) Linguistic communication as action and cooperation. Gothenburg monographs in linguistics 2. University of Göteborg, Department of Linguistics

Antaki C (2002) An introductory tutorial in conversation analysis. http://www-staff.lboro.ac.uk/ ~sscal/sitemenu.htm. Accessed on 26 Jan 2016

Austin JL (1962) How to do things with words. Oxford University Press, Oxford

Bohus D (2007) Error awareness and recovery in conversational spoken language interfaces. Ph.D. dissertation, Carnegie Mellon University, Pittsburgh, PA

Bunt HC (1979) Conversational principles in question-answer dialogs. In: Krallmann D (ed) Zur Theorie der Frage. Narr Verlag, Essen, pp 119–141

Bunt HC (1995) DIT – dynamic interpretation and dialog theory. In: Taylor MM, Neel F, Bouwhuis DG (eds) Proceedings from the second Venaco workshop on multimodal dialog. Benjamins, Amsterdam, pp 139–166

Bunt HC (2009) The DIT++ taxonomy for functional dialog markup. In: Heylen D, Pelachaud C, Catizone R, Traum DR (eds) Proceedings of the AMAAS 2009 workshop towards a standard markup language for embodied dialog acts. Budapest, May 2009, pp 13–24

Bunt HC (2011) Multifunctionality in dialog. Comp Speech Lang 25:222–245. doi:10.1016/j.csl. 2010.04.006

Bunt HC, Black W (eds) (2000) Abduction, belief and context in dialog: studies in computational pragmatics. John Benjamins Publishing Company, Amsterdam. doi:10.1075/nlp.1

Bunt HC, Alexandersson J, Choe J-W, Fang AC, HasidaK, PetukhovaV, Popescu-Belis A, Traum DR (2012a) ISO 24617-2: A semantically-based standard for dialog annotation. In: Proceedings of the 8th international conference on language resources and evaluation (LREC 2012), Istanbul, pp 430–437. http://www.lrec-conf.org/proceedings/lrec2012/pdf/180_Paper. pdf. Accessed 2 Mar 2016

Bunt HC, Kipp M, Petukhova V (2012b) Using DiAML and ANVIL for multimodal dialog annotation. In: Proceedings of the 8th international conference on language resources and evaluation (LREC 2012), Istanbul, pp 1301–1308. http://www.lrec-conf.org/proceedings/ lrec2012/pdf/1107_Paper.pdf. Accessed 2 Mar 2016

Carletta J, Isard A, Isard S, Kowtko J, Doherty-Sneddon G, Anderson A (1997) The reliability of a dialog structure coding scheme. Comput Linguist 23:13–31. http://dl.acm.org/citation.cfm?id= 972686. Accessed 20 Jan 2016

Clark HH (1996) Using language. Cambridge University Press, Cambridge. doi:10.1017/ cbo9780511620539

Clark HH, Brennan SE (1991) Grounding in communication. In: Resnick LB, Levine JM, Teasley SD (eds) Perspectives on socially shared cognition. American Psychological Association, Washington, pp 127–149. doi:10.1037/10096-006

Clark HH, Schaefer EF (1989) Contributing to discourse. Cogn Sci 13:259–294. doi:10.1207/ s15516709cog1302_7

Cooper R, Larsson S, Matheson C, Poesio M, Traum DR (1999) Coding instructional dialog for
    information states Trindi project deliverable D1.1. http://www.ling.gu.se/projekt/trindi//
    publications.html. Accessed 20 Jan 2016
Coupland N, Giles H, Wiemann J (eds) (1991) Miscommunication and problematic talk. Sage
    Publications, London
Eggins S, Slade D (2005) Analysing casual conversation. Equinox Publishing Ltd., Sheffield
Fernández R (2014) Dialog. In: Mitkov R (ed) The Oxford handbook of computational linguistics,
    2nd edn. Oxford University Press, Oxford. doi:10.1093/oxfordhb/9780199573691.013.25
Frampton M (2009) Reinforcement learning in spoken dialog systems: optimising repair strategies.
    VDM Verlag, Saarbrücken
Frampton M, Lemon O (2005) Reinforcement learning of dialog strategies using the user's last
    dialog act. In: Proceedings of 4th IJCAI workshop on knowledge and reasoning in practical
    dialog systems, Edinburgh. https://pureapps2.hw.ac.uk/portal/en/publications/reinforcement-
    learning-of-dialog-strategies-using-the-users-last-dialog-act(193e9575-2081-4338-b37a-
    d7a0c47e9dc9).html. Accessed 20 Jan 2016
Geis ML (2006) Speech acts and conversational interaction. Cambridge University Press,
    Cambridge
Griol D, Hurtado L, Segarra E, Sanchis E (2008) A statistical approach to spoken dialog systems
    design and evaluation. Speech Commun 50:666–682. doi:10.1016/j.specom.2008.04.001
Griol D, Callejas Z, López-Cózar R, Riccardi G (2014) A domain-independent statistical
    methodology for dialog management in spoken dialog systems. Comp Speech Lang 28:743–
    768. doi:10.1016/j.csl.2013.09.002
Grosz BJ, Sidner CL (1986) Attention, intentions, and the structure of discourse. Comput Linguist
    12(3):175–204. http://dl.acm.org/citation.cfm?id=12458. Accessed 20 Jan 2016
Gumperz J (1978) The conversational analysis of interethnic communication. In: Ross EL
    (ed) Interethnic communication. University of Georgia Press, Athens, pp 13–31
Hastie H, Poesio M, Isard S (2002) Automatically predicting dialog structure using prosodic
    features. Speech Commun 36(1–2):63–79. doi:10.1016/S0167-6393(01)00026-7
Hayashi M, Raymond G, Sidnell J (eds) (2013) Conversational repair and human understanding.
    Cambridge University Press, Cambridge
Heeman P, Allen JF (1994) Detecting and correcting speech repairs. In: Proceedings of the 32nd
    annual meeting of the Association of Computational Linguistics, Las Cruces, pp 295–302.
    doi:10.3115/981732.981773
Hirschberg J (2002) Communication and prosody: functional aspects of prosody. Speech Commun
    36(1–2):31–43. doi:10.1016/S0167-6393(01)00024-3
Hutchby I, Wooffitt R (2008) Conversation analysis. Polity Press, Oxford
Jokinen K, McTear M (2010) Spoken dialog systems. Synthesis lectures on human language
    technologies.   Morgan   and   Claypool   Publishers,   San   Rafael.   doi:10.2200/
    S00204ED1V01Y200910HLT005
Jurafsky D, Martin JH (2009) Speech and language processing: an introduction to natural language
    processing, computational linguistics, and speech recognition, 2nd edn. Prentice Hall, Upper
    Saddle River
Jurafsky D, Shriberg E, Biasca D (1997) Switchboard SWBD-DAMSL shallow-discourse-function
    annotation coders manual, Draft 13. University of Colorado, Boulder, CO. Institute of
    Cognitive Science Technical Report 97-02. https://web.stanford.edu/~jurafsky/ws97/manual.
    august1.html. Accessed 20 Jan 2016
Kipp M (2012) Multimedia annotation, querying and analysis in ANVIL. In: Maybury M
    (ed) Multimedia information extraction. IEEE Computer Society Press. doi:10.1002/
    9781118219546.ch21
Kowtko J, Isard S, Doherty GM (1993) Conversational games within dialog. Research paper
    HCRC/RP-31, Human Communication Research Centre, University of Edinburgh
Larsson S, Traum DR (2000) Information state and dialog management in the TRINDI dialog
    move engine toolkit. Nat Lang Eng 6(3–4):323–340. doi:10.1017/S1351324900002539
Levinson SC (1983) Pragmatics. Cambridge University Press, Cambridge

Lewis JR (2011) Practical speech user interface design. CRC Press, Boca Raton. doi:10.1201/b10461

Matheson C, Poesio M, Traum DR (2001) Modelling grounding and discourse obligations using update rules. In: Proceedings of the first annual meeting of the North American chapter of the ACL, Seattle, April 2001

McTear, M (2008) Handling miscommunication: why bother? In: Dybkjaer L, Minker W (eds) Recent trends in discourse and dialog. Springer, New York, pp 101–122. doi:10.1007/978-1-4020-6821-8_5

Narayanan S, Georgiou PG (2013) Behavioral signal processing: deriving human behavioral informatics from speech and language. Proc IEEE 101(5):1203–1233. doi:10.1109/JPROC.2012.2236291

Nass C, Brave S (2004) Wired for speech: how voice activates and advances the human-computer relationship. MIT Press, Cambridge

Nöth E, Batliner A, Warnke V, Haas J, Boros M, Buckow J, Huber R, Gallwitz F, Nutt M, Niemann H (2002) On the use of prosody in automatic dialog understanding. Speech Commun 36(1–2):45–62. doi:10.1016/S0167-6393(01)00025-5

Pentland A (2007) Social signal processing. Signal Process Mag 24(4):108–111. doi: 10.1109/MSP.2007.4286569

Rudnicky AJ, Wu X (1999) An agenda-based dialog management architecture for spoken language systems. In: Proceedings of IEEE automatic speech recognition and understanding workshop (ASRU99), Chichester, UK, pp 3–7. http://www.cs.cmu.edu/~xw/asru99-agenda.pdf. Accessed 20 Jan 2016

Sacks H (1984) On doing 'being ordinary'. In: Atkinson JM, Heritage JC (eds) Structures of social action: studies in conversation analysis. Cambridge University Press, Cambridge. doi:10.1017/CBO9780511665868.024

Sacks H, Schegloff EA, Jefferson G (1974) A simplest systematics for the organization of turn-taking for conversation. Language 50(4):696–735. doi:10.1353/lan.1974.0010

Schegloff EA (1968) Sequencing in conversational openings. Am Anthropol 70:1075–1095. doi:10.1525/aa.1968.70.6.02a00030

Schegloff EA (1982) Discourse as an interactional achievement: some uses of "uh huh" and other things that come between sentences. In: Tannen D (ed) Analysing discourse: text and talk. Georgetown University Roundtable on Languages and Linguistics 1981, Georgetown University Press, Washington, DC, pp 71–93

Schegloff EA, Sacks H (1973) Opening up closings. Semiotica 8(4):289–327. doi:10.1515/semi.1973.8.4.289

Schegloff EA, Jefferson G, Sacks H (1977) The preference for self-correction in the organisation of repair in conversation. Language 53:361–382. doi:10.2307/413107

Schröder M, Bevacqua E, Cowie R, Eyben F, Gunes H, Heylen D, ter Maat M, McKeown G, Pammi S, Pantic M, Pelachaud C, Schuller B, de Sevin E, Valstar M, Wöllmer M (2012) Building autonomous sensitive artificial listeners. IEEE Trans Affect Comput 3(2):165–183. doi:10.1109/T-AFFC.2011.34

Schuller B, Batliner A (2013) Computational paralinguistics: emotion, affect and personality in speech and language processing. Wiley, Chichester. doi:10.1002/9781118706664

Searle JR (1969) Speech acts. Cambridge University Press, Cambridge. doi:10.1017/CBO9781139173438

Searle JR (ed) (2013) Speech act theory and pragmatics. Springer, New York. doi:10.1007/978-94-009-8964-1

Sidnell J (2010) Conversation analysis: an introduction. Wiley-Blackwell, Chichester

Sidnell J, Stivers, T (eds) (2014) The handbook of conversation analysis. Wiley-Blackwell, Chichester. doi:10.1002/9781118325001

Sinclair JM, Coulthard M (1975) Towards an analysis of discourse. Oxford University Press, Oxford

Skantze G (2007) Error handling in spoken dialog systems—managing uncertainty, grounding and miscommunication. Ph.D. dissertation, KTH, Stockholm, Sweden

Skantze G, Hjalmarsson A (2013) Towards incremental speech generation in conversational systems. Comp Speech Lang 27(1):243–262. doi:10.1016/j.csl.2012.05.004

Stent A (2002) A conversation acts model for generating spoken dialog contributions. Comp Speech Lang 16:313–352. doi:10.1016/s0885-2308(02)00009-8

Tannen D (2001) You just don't understand: women and men in conversation. Ballentine Books, New York

Traum DR (1994) A computational theory of grounding in natural language conversation. Ph.D. dissertation, Department of Computer Science, University of Rochester, New York

Traum DR (2000) 20 questions for dialog act taxonomies. J Seman 17(1):7–30. doi:10.1093/jos/17.1.7

Traum DR, Hinkelmann EA (1992) Conversation acts in task-oriented spoken dialog. Comput Intell 8(3):575–599. doi:10.1111/j.1467-8640.1992.tb00380.x

Vinciarelli A, Pantic M, Bourlard H (2009) Social signal processing: survey of an emerging domain. Image Vis Compu 27(12):1743–1759. doi:10.1016/j.imavis.2008.11.007

Webber B, Egg M, Kordoni V (2012) Discourse structure and language technology. Nat Lang Eng 18(4):437–490. doi:10.1017/S1351324911000337

Wittgenstein L (1958) Philosophical investigations. Blackwell, Oxford

Chapter 4
Conversational Interfaces: Past and Present

Abstract Conversational interfaces have a long history, starting in the 1960s with text-based dialog systems for question answering and chatbots that simulated casual conversation. Speech-based dialog systems began to appear in the late 1980s and spoken dialog technology became a key area of research within the speech and language communities. At the same time commercially deployed spoken dialog systems, known in the industry as voice user interfaces (VUI), began to emerge. Embodied conversational agents (ECA) and social robots were also being developed. These systems combine facial expression, body stance, hand gestures, and speech in order to provide a more human-like and more engaging interaction. In this chapter we review developments in spoken dialog systems, VUI, embodied conversational agents, social robots, and chatbots, and outline findings and achievements from this work that will be important for the next generation of conversational interfaces.

4.1 Introduction

The idea of being able to talk to a machine has been around for a long time, but it was not until the late 1980s that researchers actually began creating speech-enabled interactive systems. Major advances in automatic speech recognition (ASR) technology meant that it was now possible to recognize a user's spoken input with a reasonable degree of accuracy. Up to this point, conversational interfaces had taken the form either of text-based dialog systems, where the main emphasis was on applying techniques from linguistics and artificial intelligence, or of chatbots that simulated conversation using simple pattern-matching techniques. Now speech is promising to become the prime choice for input when we speak to a conversational interface, whether it is a smartphone, a robot, a smartwatch, or some other types of devices.

In this chapter, we provide a brief overview of research and development in conversational interfaces and discuss those findings and achievements that we believe will be important for the next generation of conversational interfaces.

© Springer International Publishing Switzerland 2016
M. McTear et al., *The Conversational Interface*,
DOI 10.1007/978-3-319-32967-3_4

4.2   Conversational Interfaces: A Brief History

Since the 1980s, the idea of a conversational interface with which humans could interact for a variety of different purposes has been a focus of research and development in a number of different communities that have largely worked independently of one another. Spoken dialog systems (SDS) have been developed in academic and industrial research laboratories, while voice user interface (VUI) is the term used to refer to similar systems developed for commercial deployment. Embodied conversational agents (ECAs) are a type of spoken dialog system in which computer-generated animated characters combine speech with other modalities including facial expression, body stance, and hand gestures, while social robots, in a variety of different forms of embodiment, provide services, companionship, and entertainment. Finally, chatbots are yet another type of conversational system in which the emphasis has been on providing realistic simulations of casual conversation (or small talk) rather than helping humans to perform a particular task.

4.2.1   A Typical Interaction with a Spoken Dialog System

The following example illustrates the sorts of issues that were being addressed in spoken dialog systems research in the period from the early 1990s to up around 2000. The example is taken from an interaction with the Mercury system that was developed in the Spoken Language Systems Group of the MIT Computer Science and Artificial Intelligence Laboratory. Mercury provides telephone access to an online flights database. In this example, taken from a publication by Seneff and Polifroni (2000), the dialog begins with an open-ended question that invites the user to state their flight requirements and then continues with questions to elicit further information from the user (S = system, U = user):

S1: How may I help you?
U1: I want to travel to Washington Dulles on November fourth.
S2: From Boston to Washington Dulles on November 4. Can you provide the approximate departure time or airline preference?
U2: I need to arrive at around 9 a.m.
S3: Okay. Flights from Boston to Washington Dulles arriving at 9:00 a.m. on November 4.
One moment please.
The best matching flight is a US Airways flight arriving at 8:59 a.m. I also have a United flight arriving at 8:36 a.m.
U3: What time does the United flight leave?
S4: United flight 1819 departs at 7:00 a.m. from Boston. Shall I book this flight for you?

U4: Yes please.
S5: United flight 1819 has been added to your itinerary. You depart Boston at 7:00 a.m. and arrive in Washington Dulles at 8:36 a.m. Where would you like to go next?
U5: I would like to return to Boston.
S6: What date will you be returning on?
U6: The same day.
(dialog continues)

The task for the system is to elicit several items of information such as the departure and destination locations, desired departure date and time, airline preference, and whether a return flight is required. One way to do this would be to ask a separate question for each item of information, using *directed dialog* (also known as *system-initiative*). An alternative strategy is used in this example, in which an open-ended prompt (S1) invites the user to express the requirements more freely and U1 provides the destination and date of departure within a single utterance. Several questions are asked in directed dialog mode (S2, S4, S5, S6), but at U3 the user asks a question, thus taking the dialog initiative at this point. This type of dialog is referred to as *mixed-initiative*, as both the system and the user are able to ask questions. Mixed-initiative dialog provides a more natural interaction but requires more complex processing by the ASR and spoken language understanding (SLU) components.

Another point to note is that the system attempts to confirm its understanding of the user's utterances (S2, S3, S5). This is important as the system may have misrecognized or misunderstood the user's utterances (indeed, this strategy is also used for the same reasons by human agents in contact centers to verify that the user's requirements have been correctly understood). We will look in more detail at various aspects of dialog, such as dialog strategies, confirmations, and dialog control, in Chap. 10.

A key aim in spoken dialog research has been to make the interaction as natural as possible. An example is utterance U2 where the user responds, "I need to arrive around 9 a.m." in response to the question "Can you provide the approximate departure time or airline preference?" Taken literally this is not a direct response to the question and many systems would fail at this point because their recognition grammars would have been tuned to accept either a departure time or an airline code as a response. Or worse still, "9 a.m." might have been interpreted as the departure time. The restriction of recognition grammars to the words predicted in the user's response was a strategy for simplifying the ASR and SLU processes in the interests of improved accuracy. Although this strategy would most likely have been successful in a carefully designed directed dialog, it could have proved disastrous in a mixed-initiative dialog such as this one.

4.2.2 An Interaction that Goes Wrong

While the dialog in the preceding section ran smoothly with no apparent problems, it is frequently the case that dialogs break down for a variety of reasons. ASR errors have been a major concern for developers of spoken dialog systems and various methods have evolved to improve the performance of a system given that recognition accuracy cannot be guaranteed. The following example, from Bohus (2007: 23), illustrates some ASR errors that occurred in a spoken language dialog between a human user and an automated system (the system's interpretation of the user's input is in brackets and upper case):

S1: What city are you leaving from?
U1: Urbana Champaign [OKAY IN THAT SAME PAY].
S2: Sorry I'm not sure I understood what you said. What city are you leaving from?
U2: Urbana Champaign [FOR MINUTE SINCE HEY].
S3: I'm still having trouble understanding you. To go on, I need you to answer the following question. What city are you leaving from?
U3: Chicago [CHICAGO].
S4: Traveling from Chicago. Where would you like to go?
U4: Huntsville [SEOUL].
S5: Traveling to Seoul. What day did you need to travel?
U5: No no I'm traveling to Birmingham [THE TRAVELING TO BERLIN P M].
S6: Traveling in the afternoon. Okay what day would you be departing Chicago?
U6: Uh the tenth of august [AT THE TENTH OF AUGUST].
S7: A flight tomorrow. I have a Korean Airlines flight departing Chicago at 1:40 p.m., arrives Seoul at 5 p.m. the next day. Do you want to take that?
U7: My destination is Birmingham [FLIGHT DESTINATION MR WEEKEND].
S8: Traveling on Saturday, August 12th—I have a flight departing Chicago at 1:40 p.m. arrives Seoul at …

As can be seen from this example, the system (S) is unable to correctly recognize the user's (U) spoken input for the departure and arrival cities and also makes errors with the time and date. In the end, the system beliefs the user wants a flight from Chicago to Seoul whereas in fact what was requested was a flight from Urbana Champaign to Huntsville. While this example is extreme, it is typical of many real transactions with automated systems, particularly where there are problems affecting recognition accuracy such as unusual accents, interference on the channel, or background noise.

4.2.3 Spoken Dialog Systems

Research in spoken dialog systems (SDS) began around the late 1980s and has typically involved large interdisciplinary research groups in universities and industrial research laboratories. Before this, dialog systems in the 1960s and 1970s were text-based and were motivated by efforts to apply techniques from linguistics to applications involving dialog, for example, in systems such as BASEBALL (Green et al. 1963), SHRDLU (Winograd 1972), and GUS (Bobrow et al. 1977). By the early 1980s, researchers were beginning to realize that linguistic competence alone could not guarantee the "intelligence" of a conversational interface. Dialog systems should be able to deal with problematic aspects of communication, such as misconceptions and false assumptions on the part of the user. More generally, in order to be cooperative and helpful, they should also be able to recognize the intentions behind the user's utterances. Systems that were developed to meet these requirements drew on areas of artificial intelligence such as user modeling and planning (Allen 1995; Reilly 1987). For a more detailed account of early text-based dialog systems, see McTear (1987).

In the USA, one of the earliest spoken dialog projects was ATIS (Air Travel Information Service), funded by DARPA (Hemphill et al. 1990), while in Europe a major project funded by the European community at this time was SUNDIAL (McGlashan et al. 1992). In ATIS, the main focus was on developing and putting into practical use the input technologies of speech recognition and spoken language understanding that were required to make a flight reservation using spoken communication with a computer over the telephone. In contrast in SUNDIAL, there was a major focus on dialog management and on how to maintain a cooperative dialog with the user. ATIS also introduced the concept of collecting resources in the form of corpora—collections of dialogs from interactions with the system—that were available for future researchers through the Linguistic Data Consortium (LDC).[1] A further innovation was the establishment of the principle of regular evaluations and standard processes for evaluation—an important aspect of present-day spoken dialog systems research (see further Chap. 17).

Building on the results of ATIS and SUNDIAL, research in spoken dialog systems continued throughout the 1990s, supported by various European Community research programs, as well as by DARPA projects such as the Communicator Project, which involved a number of research laboratories and companies across the USA and also included several affiliated partner sites in Europe (Walker et al. 2001). There was also extensive research in dialog in other parts of the world, particularly in Japan, and more recently in Korea and China.

Spoken dialog technology continues to be an active research area, especially in university research laboratories. Some of the main findings from this research are discussed in Sect. 4.3.

---

[1]https://www.ldc.upenn.edu/. Accessed February 19, 2016.

4.2.4   Voice User Interfaces

Voice-user interfaces (VUI) also began around the 1990s with the realization that speech could be used to automate self-service tasks, such as call routing, directory assistance, information enquiries, and simple transactions. One of the first systems was AT&T's voice recognition call processing (VRCP) system (Wilpon et al. 1990) which greeted callers with the message "Please say collect, calling card, or third party." On recognizing the caller's input, the system transferred them to the requested service. The VRCP system became very successful commercially and has been used to answer billions of calls per year.

Up until the mid-1990s automated self-service was generally handled by inter-active voice response (IVR) systems in which the system used recorded spoken prompts (e.g., "press 1 for balance, press 2 to pay a bill"), and the user's response was in the form of key presses on the telephone keypad. The user's input mode is often referred to as DTMF (dual tone multifrequency) or touchtone. One of the most widely known applications is voice mail, but IVR technology has come to be used in almost any sizeable business to automate call routing and self-service (Brandt 2008).

HMIHY (How May I Help You), developed at AT&T (Gorin et al. 1997), was one of the first commercially deployed interactive speech systems to address the task of call routing on a larger scale. HMIHY's task is to identify the reason for the customer's call from unrestricted natural language input and then route the caller to the appropriate destination. Callers are greeted with an open-ended prompt ("How may I help you?") that encourages free-form spoken language input and then the system engages in dialog to confirm its understanding of the input and if necessary to request clarification, accept corrections, or collect additional information. By the end of 2001, HMIHY was handling more than 2 million calls per month and was showing significant improvements in customer satisfaction.

Although SDSs and VUIs use the same spoken language technologies for the development of interactive speech applications, there is a difference in emphasis between the two communities. Generally speaking, academic researchers focus on making new contributions to knowledge and on publishing their results in academic journals and conferences. VUI developers, on the other hand, are more concerned with addressing business needs, such as return on investment, as well as with human factors issues, such as usability and user satisfaction.

4.2.5   Embodied Conversational Agent, Companions, and Social Robots

ECAs are computer-generated animated characters that combine facial expression, body stance, hand gestures, and speech to provide a more human-like and more engaging interaction. ECAs are being employed increasingly in commercial applications, for example, to read news and answer questions about products on online shopping Web pages. Because ECAs have been perceived to be trustworthy, believable, and entertaining, they have been deployed in healthcare settings and also in role-plays, simulations, and immersive virtual environments. ECAs are discussed in greater detail in Chap. 15.

Artificial companions and social robots provide companionship and entertainment for humans. They can take the form of physical objects such as digital pets or robots, or they may exist virtually as software. Artificial companions and social robots can support activities of daily living and enable independent living at home for the elderly and for people with disabilities. They can also play an educational role for children, particularly for children with various types of disability. We discuss companions and social robots further in Chaps. 13 and 18.

4.2.6   Chatbots

Chatbots, also known as chatterbots, produce simulated conversations in which the human user inputs some text and the chatbot makes a response. One of the motivations for developers of chatbots is to try to fool the user into thinking that they are conversing with another human. To date most conversations with chatbots have been text-based, although some more recent chatbots make use of speech for input and output and in some cases also include avatars or talking heads to endow the chatbot with a more human-like personality. Generally, the interaction with chatbots takes the form of small talk as opposed to the task-oriented dialogs of SDSs and VUIs. Another difference is that the chatbot typically responds to the user's input rather than taking the initiative in the conversation, although in some cases the chatbot will ask questions to keep the conversation going. Conversations are generated using a stimulus–response approach in which the user's input is matched against a large set of stored patterns and a response is output (see further Chap. 7). This contrasts with the more complex mechanisms that are deployed in SDSs, VUIs, and ECAs.

Chatbots have their origins in the ELIZA system developed by Weizenbaum (1966). ELIZA simulates a Rogerian psychotherapist, often in a convincing way, and has inspired many generations of chatbot authors for whom a major motivation is to develop a system that can pass Turing's Imitation Game (Turing 1950). The Loebner Prize Competition, launched in 1991 by Dr. Hugh Loebner, has the aim of finding a chatbot that can pass the Imitation Game.[2]

Chatbots are being used increasingly in areas such as education, information retrieval, business, and e-commerce, for example, as automated online assistants to

---

[2]http://www.loebner.net/Prizef/loebner-prize.html. Accessed February 19, 2016.

complement or even replace human-provided service in a call center. IKEA has an automated online assistant called Anna, developed by Artificial Solutions.[3] Anna answers questions about IKEA products and opening hours, as well as questions about what is being served for lunch in the IKEA restaurant. Anna also displays emotions if she cannot find the requested information.

check out!

4.3   What Have We Learned so Far?

What have we learned from this work that is relevant to the next generation of conversational interfaces? In the following sections, we review some key findings and achievements grouped according to the following themes:

- Making systems more intelligent.
- Using incremental processing to model conversational phenomena.
- Languages and toolkits for developers.
- Large-scale experiments on system design using techniques from machine learning.

4.3.1   Making Systems More Intelligent

Spoken dialog systems research has until recently had its foundations in artificial intelligence, beginning with plan-based systems in the 1980s and continuing with models of conversational agency and rational interaction in the 1990s. Examples of this work are the TRAINS and TRIPS projects at the University of Rochester, in which intelligent planning and conversationally proficient artificial agents were developed that could engage in dialog with human participants to cooperatively reason about and solve a complex task (Allen 1995; Allen et al. 2001).

Information State Theory, also known as Information State Update (ISU), is a theoretically motivated attempt to characterize the dynamics of dialog, with its origins in work on dialog by Ginzburg (1996). An information state represents what is known at a given stage in a dialog. As the dialog progresses with the participants performing dialog moves, such as asking and answering questions and accumulating information, the information state is continually updated (Traum and Larsson 2003). A major thrust of the research was concerned with formalizing the components of the theory.

---

[3]https://www.chatbots.org/virtual_assistant/anna3/. Accessed February 19, 2016.

The contents of an information state may include information about the mental states of the participants (their beliefs, desires, intentions, obligations, and commitments), as well as information about the dialog (what utterances have been spoken, the dialog moves generated by these utterances, whether information is shared between the participants). Update rules consist of applicability conditions that define when a rule can be applied and effects that define the consequences of applying a rule.

The theory of information states was implemented as a dialog move engine (DME), which is a type of dialog management system. The toolkit TrindiKit was developed as an architecture for the DME, allowing dialog researchers to experiment with different kinds of information states and with rules for updating the states (Traum and Larsson 2003). TrindiKit was used in a number of prototype dialog systems, such as the Gothenburg Dialog System (GoDiS), developed at Gothenburg University by Cooper, Larsson, and others (Larsson et al. 1999). The TRINDI framework has been extended in a framework called DIPPER (Bos et al. 2003) and more recently a new toolkit called Maharani has been developed for ISU dialog management.[4]

Another line of research took the view that dialog can be characterized as an example of rational behavior involving the plans, goals, and intentions of the agents in the dialog. As an example, dialog acts such as clarification requests were explained in terms of a general principle of rational behavior in which agents have a commitment to being understood. The theoretical framework for rational agency was based on a set of axioms that formalized principles of rational action and cooperative communication (Cohen and Levesque 1990) and one example of the implementation of the theory was the ARTEMIS system, an agent technology developed at France Telecom-CNET as a generic framework for specifying and implementing intelligent dialog agents (Sadek and de Mori 1998). For recent approaches to plan, intent and activity recognition using statistical methods, see Sukthankar et al. 2014).

The use of artificial intelligence techniques in spoken dialog systems has diminished somewhat in recent years with the emergence of statistical and data-driven approaches. However, the models of conversational agency developed in this research are still of relevance for more advanced conversational user interfaces that go beyond simple information enquiries and transactions to complex tasks involving problem solving and reasoning. Allen et al. (2001) provide an informative example in which the agent has to engage in a chain of reasoning in order to correctly interpret otherwise ambiguous utterances using information from the current context. Recent theoretical work on computational theories of dialog includes Ginzburg (2015), Ginzburg and Fernández (2010) and Fernández (2014).

---

[4]http://www.clt.gu.se/research/maharani. Accessed February 19, 2016.

4.3.2 Using Incremental Processing to Model Conversational Phenomena

Traditional spoken dialog systems operate on a turn-by-turn basis in which the system waits until the user has completed an utterance—as detected by a silence threshold—and then begins the processes of interpreting the utterance, deciding on the next action to take, and planning and generating a response. The resulting latency between turns, although usually fairly brief, makes human–machine interaction appear stilted and unnatural, in contrast to dialog between humans where there is often no gap between the turns. Indeed, as described in Chap. 3, in human dialog there is often overlap at turn boundaries caused by the listener starting to speak before the current speaker has finished.

In order to be able to produce an appropriate responsive overlap, the participant who is currently listening has to monitor the ongoing utterance carefully, determine its meaning based on what has been heard so far, and plan an appropriate contribution—a process known as *incremental processing*. Evidence that humans engage in incremental processing can be found in numerous psycholinguistic studies of utterance interpretation and production (Levelt 1989; Clark 1996; Tanenhaus 2004). Incremental processing is deployed in Google Search to provide potential completions of a user's query while the user is still typing the query. This allows users to get results faster and also helps them with the formulation of their query.[5] Recent versions of Google Voice Search also make use of incremental speech processing by showing words as they are being recognized.

Recent work on incremental processing in dialog is motivated by the goal of developing technologies to model interactional phenomena in naturally occurring dialog with the aim of creating more efficient and more natural spoken dialog systems and of enhancing user satisfaction (Rieser and Schlangen 2011; Baumann 2013). One example of incremental processing in human–human conversations is the phenomenon of compound contributions—also known as collaborative completions or split utterances. Compound contributions, which occur when the listener cuts in on an ongoing utterance and offers a completion, have been found to constitute almost one-fifth of all contributions in a corpus study conducted by Howes et al. (2011). DeVault et al. (2011) implemented a virtual agent that could predict the remainder of an utterance based on an incremental interpretation of partial speech recognition results and by identifying the *point of maximal understanding*, i.e., the point where the system estimated that it was likely to have understood the intended meaning. At this point, the system was able to use the partial hypothesis to generate a completion. Other phenomena that have been studied include self-corrections (Buß and Schlangen 2011) and incremental speech generation (Skantze and Hjalmarsson 2013).

---

[5]http://www.google.com/instant/. Accessed February 19, 2016.

The main benefit of incremental processing is that dialog is more fluent and efficient. Skantze and Hjalmarsson (2013) compared incremental and non-incremental versions of their system and found that the incremental version was rated as more polite and efficient, as well as better at predicting when to speak. Aist et al. (2007) report similar findings. However, implementing incremental processing is technically challenging. In current conversational architectures, the information flow among the components is pipelined, which means that the output from one component is sent when it is complete as input to the next component in the pipeline. In a system using incremental processing, however, output is received incrementally and so the data that have to be processed are partial and potentially subject to revision in the light of subsequent output. Schlangen and Skantze (2011) present a general, abstract model of incremental dialog processing while Baumann (2013) provides a detailed discussion of architectural issues and models for incremental processing across the various components of a spoken dialog system.

Some open source software has been made available to developers wishing to implement incremental processing in their spoken dialog systems. INPROTK, which can be downloaded at Sourceforge,[6] was developed in the InPro project that ran initially at the University of Potsdam and subsequently at the University of Bielefeld. Another example is Jindigo, a framework for incremental dialog systems developed in Sweden at KTH.[7] Dylan (Dynamics of Language) is an implementation of dynamic syntax that uses a word-by-word incremental semantic grammar.[8]

4.3.3 Languages and Toolkits for Developers

Until around 2000 speech-based interactive voice response (IVR) systems were developed using Computer-Telephone Integration (CTI) technology that required developers to master intricate details of voice cards, application programming interfaces (APIs), circuit provisioning, and hardware configuration. In 1999, the VoiceXML Forum was founded by AT&T, IBM, Lucent, and Motorola with the aim of developing and promoting a standard World Wide Web Consortium (W3C) dialog markup language that would allow developers to create speech applications by building on the existing Web infrastructure, using standard Internet protocols and without the need for specialized APIs. As reported on the VoiceXML Forum, VoiceXML application development could be at least three times faster than development in traditional IVR environments.[9] Moreover, the use of a widely accepted standard such as VoiceXML would facilitate code portability and reuse. For reasons such as these, VoiceXML came to be widely adopted within the speech

---

[6]http://sourceforge.net/projects/inprotk/. Accessed February 19, 2016.

[7]http://www.speech.kth.se/prod/publications/files/3654.pdf. Accessed February 19, 2016.

[8]https://sourceforge.net/projects/dylan/. Accessed February 19, 2016.

[9]http://www.voicexml.org/about/frequently-asked-questions. Accessed February 19, 2016.

industry and there are now countless applications that have been developed using VoiceXML.

In addition to VoiceXML a number of other voice-related specifications have been produced by the W3C, including the following: Speech Recognition Grammar Specification (SRGS); Semantic Interpretation for Speech Recognition (SISR); Speech Synthesis Markup Language (SSML); Call Control eXtensible Markup Language (CCXML); Extensible MultiModal Annotation Markup Language (EMMA); Pronunciation Lexicon Specification (PLS); and State Chart XML: State Machine Notation for Control Abstraction (SCXML). These specifications are also used more generally for a range of applications in addition to their use to specify and implement conversational interfaces. URLs for the specifications can be found at the end of the References section.

Developing a conversational interface is a complex process involving the integration of several different technologies. The following platforms enable developers to sign up for a free account to create VoiceXML applications: Voxeo Evolution Portal[10] and Nuance Café[11] (formerly Bevocal Café). Chatbot applications can be created at Pandorabots[12] and ChatScript.[13] Chapter 7 features the Pandorabots platform, which supports the development of chatbots using Artificial Intelligence Markup Language (AIML), while Chap. 11 provides more detail on VoiceXML along with a tutorial showing how VoiceXML can be used to implement a spoken dialog application on the Voxeo platform.

4.3.4   Large-Scale Experiments on System Design Using Techniques from Machine Learning

Designers of commercially deployed conversational interfaces have to satisfy the frequently conflicting demands of business needs and of human factors requirements. Generally VUI design is based on best practice guidelines that experienced VUI designers have accumulated over the years (see, for example, Balentine and Morgan 2001; Balentine 2007; Cohen et al. 2004; Larson et al. 2005; Lewis 2011). An alternative approach that has emerged recently involves the use of data-driven methods to explore and evaluate different VUI design choices, making use of the large amounts of data generated by commercially deployed VUIs. Although this sort of data is usually not publicly available to researchers, a unique opportunity arose in the New York based company SpeechCycle where developers were given

---

[10]https://evolution.voxeo.com/. Accessed February 19, 2016.

[11]https://cafe.bevocal.com/. Accessed February 19, 2016.

[12]http://www.pandorabots.com/. Accessed February 19, 2016.

[13]http://chatscript.sourceforge.net/. Accessed February 19, 2016.

access to the logs of millions of calls and were able to conduct a series of research investigations into topics such as call flow optimization, prompt design, and statistical grammar tuning.

4.3.4.1 Call Flow Optimization

Usually the design of a VUI's call flow involves anticipating and handcrafting the various choices that the user can take at particular points in the dialog and the paths that the interaction may follow. With large systems a call flow design document may come to hundreds of pages of call flow graphs (see example in Paek and Pieraccini 2008: 719). This raises the issue of VUI completeness, i.e., whether it is possible to exhaustively explore all possible user inputs and dialog paths in order to determine the best strategies to implement in the call flow (Paek and Pieraccini 2008). While VUI completeness is likely to be extremely difficult to guarantee using traditional handcrafted techniques, the application of machine learning techniques to large amounts of logged data has the potential to automate the process. A typical approach is to design a call flow that includes potential alternative strategies, alternate randomly between the strategies and measure the effects of the different choices using metrics such as automation rate and average handling time.

As an example, Suendermann et al. (2010a) describe a call routing application that contained four questions in its business logic: service type (e.g., orders, billings), product (e.g., Internet, cable TV), actions (e.g., cancel, make payment), and modifiers (credit card, pay-per-view). Implementing the business logic one question at a time would have resulted in an average of 4 questions per interaction. However, using call flow optimization an average number of 2.87 questions was achieved, a reduction of almost 30 %.

The experiments were performed using a technique known as Contender (Suendermann et al. 2010a, 2011a, b). Contender is implemented as an input transition in a call flow with multiple output transitions representing alternative paths in the call flow. When a call arrives at an input transition, one of the choices is selected randomly. The evaluation of the optimal call flow path is based on a single scalar metric such as automation rate. This is similar to the reward function used in reinforcement learning, in which optimal dialog strategies are learnt from data (see further Chap. 10). The Contender technique has been described as a light version of reinforcement learning that is more applicable in commercial applications (Suendermann et al. 2010b).

The use of contenders enabled a large-scale evaluation of design strategies that would not be feasible using traditional handcrafted methods. For example, Suendermann and Pieraccini (2012) report that 233 Contenders were released into production systems and were used to process a total call volume of 39 million calls. Moreover, the practical value of call flow optimization using the contender technique was a 30 % reduction in the number of questions that the system had to ask to complete a transaction. Given the per-minute costs of interactive voice response hosting and a large monthly call volume, considerable savings could be made of up to five-to six-figure US dollars per month. Furthermore, a reduction in abandonment rates resulting from customer frustration and a quicker transaction completion rate resulted in improvements in user satisfaction ratings (Suendermann et al. 2010a).

4.3.4.2   Prompt Design

The prompts in a VUI are the outward manifestation of the system for the user that can convey a company's brand image through the persona (or personality) of the conversational interface. Prompts also instruct the user as to what they should say or do, so they should be designed to be clear and effective. The following is an example of a badly designed prompt (Hura 2008):

> I can help you with the following five options. You can interrupt me and speak your choice at any time. Please select one of the following: sign up for new service, add features to my service, move my existing service, problems with my satellite service, or ask a billing question.

In addition to the lengthy introduction, the menu options are long and complex as well as partially overlapping as four of the options use the word "service." These factors would make it difficult for the caller to locate and remember the option that they want. Best practice guidelines recommend a concise as opposed to a more verbose style for the system's initial prompt, thus saving time (and money for the provider by reducing the transaction time), as well as placing fewer demands on the caller's working memory (Balentine and Morgan 2001). Lewis (2011: 242–244) presents several examples of techniques for writing effective prompts.

Prompt design was investigated in the experimental work at SpeechCycle. The following is an example of alternative prompts at the problem capture phase of a troubleshooting application (Suendermann et al. 2011a):

> (a) Please tell me the problem you are having in a short sentence.
> (b) Are you calling because you lost some or all of your cable TV service? followed by (a) if the answer was "no."
> (c) (a) + or you can say "what are my choices" followed by a back-up menu if the answer was "choices."
> (d) (b) followed by (c) if the answer was "no."

Based on an analysis of over 600,000 calls, prompt (b) was found to be the winner. Determining this would have been more a matter of guesswork using traditional VUI design methods.

4.3.4.3 Design and Use of Recognition Grammars

In many VUIs, a simple handcrafted grammar is used in which the words and phrases that the system can recognize are specified in advance. Handcrafted grammars are simple to develop and easy to maintain and are appropriate as long as the user's input can be restricted to the items in the grammar. When more natural and more wide-ranging input is expected, other techniques such as statistical language models and statistical classifiers are more appropriate. However as these methods require a large amount of data to train, they are more difficult and more expensive to build, test, and maintain (see further, Chap. 8).

In order to investigate the effectiveness of statistical spoken language understanding grammars (SSLUs), Suendermann et al. (2009) employed a data-driven approach in which data were collected at each recognition point in a dialog and then used to generate and update SSLUs, using a corpus of more than two million utterances collected from a complex call routing and troubleshooting application in which call durations could exceed 20 min, requiring frequent use of recognition grammars to interpret the user's inputs. Generally, a call routing system begins with a general question such as "How may I help you?" as in the HMIHY system described earlier, the answer to the question would be interpreted using a semantic classifier, and the caller would be routed to the appropriate human agent. However, in a complex troubleshooting application the system takes on the role of the human agent and continues by asking a series of focused questions that are usually interpreted using specially handcrafted grammars. Suendermann et al. (2009) explored whether statistical grammars could be used more effectively at every recognition context in the interaction. Their method involved tuning the statistical grammars through a continuous process of utterance collection, transcription, annotation, language models and classifier training, most of which could be carried out automatically. After a three-month period of continuous improvement, they found that the performance of the statistical grammars outperformed the original rule-based grammars in all dialog contexts, including large vocabulary open-ended speech, more restricted directed dialogs, and even *yes/no* contexts.

4.4 Summary

In this chapter, we have looked at the origins of the conversational interface and have reviewed contributions from the research communities of spoken dialog systems, VUI, ECA, social robots, and chatbots. Although each of these communities shares the goal of facilitating human–machine interaction using spoken language, in some cases in combination with other modalities, they differ in their goals and methods. The main aim of academic researchers is to contribute to knowledge, while in industry the primary aim is to create applications that will bring commercial benefits. The motivation for some of the work on ECA and social robots has been to create social companions and virtual assistants, for example, in healthcare scenarios, while chatbots have been developed either to simulate conversation and fool judges, as in competitions such as the Loebner prize, to provide entertainment, or, more recently, to provide a service.

Despite these differences, several themes have emerged that will be significant for developers of future conversational interfaces. These include the following.

- Research in artificial intelligence has shown how conversational interfaces can be created that behave intelligently and cooperatively, and how context can be represented formally in terms of information state updates.
- Research on embodied virtual agents and social robots has aimed to endow conversational interfaces with human-like characteristics and the ability to display and recognize emotions.
- Research in incremental processing in dialog has investigated how the conversational interaction of systems can be made more human-like.
- Many useful tools and development environments have been produced that facilitate the development of conversational interfaces.
- Large-scale, data-driven experiments on various aspects of systems, such as dialog flow, prompt design, and the use of recognition grammars, have provided a methodology and tools that make development more efficient and more systematic.

A new dimension for conversational interfaces is that they will involve a whole new range of devices, possibly linked together in an Internet of Things. In the past, the interface to a spoken dialog system was usually a headset attached to a computer, while the typical interface for a voice user interface was a phone. With the emergence of smartphones and more recently of social robots, smartwatches, wearables, and other devices, new styles of interaction will emerge along with new challenges for implementation (see further Chap. 13).

Further Reading

There are several books on spoken and multimodal dialog systems, including the following: McTear (2004), López Cózar and Araki (2005), Jokinen (2009), and Jokinen and McTear (2010). Rieser and Lemon (2011), Lemon and Pietquin (2012), and Thomson (2013) cover statistical approaches. The chapters in Dahl (2004) present practical issues in dialog systems development. Chapter 17 of Allen (1995) provides an overview of AI-based approaches to conversational agency, while chapter 24 of Jurafsky and Martin (2009) reviews both AI-based as well as statistical approaches. Several chapters in Chen and Jokinen (2010) cover spoken dialog systems and ECA. See also Ginzburg and Fernandez (2010) on formal computational models of dialog. There are several chapters on interactions with robots, knowbots, and smartphones in Mariani et al. (2014). Nishida et al. (2014) investigates how to design conversational systems that can engage in conversational interaction with humans, building on research from cognitive and computational linguistics, communication science, and artificial intelligence.

Books on VUI include Cohen et al. (2004), Balentine (2007), Lewis (2011), Pieraccini (2012), and Suendermann (2011). See also various chapters in Kortum (2008).

Cassell et al. (2000) is a collection of readings on ECAs. Perez-Martin and Pascual-Nieto (2011) is a recent collection. The chapters in Trappl (2013) outline the prerequisites for a personalized virtual butler, including social and psychological as well as technological issues. Andre and Pelachaud (2010) provide a comprehensive review of research issues and achievements. See also Lester et al. (2004). Schulman and Bickmore (2009) describe the use of counseling dialogs using conversational agents.

Companions and social robots have been investigated in various European Union (EU)-funded projects, such as follows: COMPANIONS,[14] which focused specifically on interaction with social conversational agents; SEMAINE,[15] which involved ECA; LIREC,[16] which investigated how to build long-term relationships with artificial companions (agents and robots); and CHRIS,[17] which was concerned with the development of cooperative human robot interactive systems. Various social, psychological, ethical, and design issues related to long-term artificial companions are discussed in chapters by leading contributors from the field in Wilks (2010).

Research relevant to conversational interfaces is published in various speech and language journals, such as *Speech Communication, Computer Speech and Language, IEEE Transactions on Audio, Speech, and Language Processing, Computational Linguistics, Natural Language Engineering*. In 2009, a new journal *Dialog and Discourse* was launched with a focus mainly on theoretical aspects of dialog. Since 2000, there has been an annual SIGdial conference on topics related to dialog and discourse, while the Young Researchers Roundtable on Spoken Dialog Systems (YRRSDS) has held an annual workshop since 2005 to provide a forum for young researchers to discuss their research, current work, and future plans. The International Workshop on Spoken Dialog Systems (IWSDS) was first held in 2009 and takes place on an annual basis. There are also special tracks on dialog at conferences such as INTERSPEECH, ICASSP, ACL, and the Spoken Language Technology Workshop (SLT). The SemDial workshop series on the Semantics and Pragmatics of Dialog focuses on formal aspects of semantics and pragmatics in relation to dialog.

The main industry-based conference is SpeechTEK,[18] which is held annually (see also MobileVoice).[19] Speech Technology Magazine[20] is a free online magazine targeted at the speech industry.

---

[14]http://www.cs.ox.ac.uk/projects/companions/. Accessed February 2016.

[15]http://www.semaine-project.eu/. Accessed February 19, 2016.

[16]http://lirec.eu/project. Accessed February 19, 2016.

[17]http://www.chrisfp7.eu/index.html. Accessed February 19, 2016.

[18]http://www.speechtek.com. Accessed February 19, 2016.

[19]http://mobilevoiceconference.com/. Accessed February 19, 2016.

[20]http://www.speechtechmag.com/. Accessed February 19, 2016.

Exercise

Observing system demos provides insights into the capabilities of a system (as well as its limitations, although usually these will be avoided or downplayed on a demo). There are many online demos of spoken dialog systems. The following two systems provide flexible mixed-initiative interaction:

- MIT FlightBrowser Spoken Language System.[21]
- CMU Air Travel Reservation Dialog System.[22]

This video shows an interaction with the CLASSiC project prototype, a statistical spoken dialog system using reinforcement learning (see Chap. 10).[23]

This is an example of two embodied agents Ada and Grace that act as guides at the Museum of Science in Boston.[24]

Watch this amusing conversation between two chatbots.[25]

You can interact with a version of ELIZA here.[26]

References

Aist G, Allen JF, Campana E, Gallo CG, Stoness S, Swift M, Tanenhaus MK (2007) Incremental dialog system faster than and preferred to its nonincremental counterpart. In: Proceedings of the 29th annual conference of the cognitive science society. Cognitive Science Society, Austin, TX, 1–4 Aug 2007

Allen JF (1995) Natural language processing, 2nd edn. Benjamin Cummings Publishing Company Inc., Redwood, CA

Allen JF, Byron DK, Dzikovska M, Ferguson G, Galescu L, Stent A (2001) Towards conversational human-computer interaction. AI Mag 22(4):27–38

André E, Pelachaud C (2010) Interacting with embodied conversational agents. In: Chen F, Jokinen K (eds), Speech technology: theory and applications. Springer, New York, pp 122–149. doi:10.1007/978-0-387-73819-2_8

Balentine B (2007) It's better to be a good machine than a bad person. ICMI Press, Annapolis, Maryland

Balentine B, Morgan DP (2001) How to build a speech recognition application: a style guide for telephony dialogs, 2nd edn. EIG Press, San Ramon, CA

Baumann T (2013) Incremental spoken dialog processing: architecture and lower-level components. Ph.D. dissertation. University of Bielefeld, Germany

Bobrow DG, Kaplan RM, Kay M, Norman DA, Thompson H, Winograd T (1977) GUS: a frame-driven dialog system. Artif Intell 8:155–173. doi:10.1016/0004-3702(77)90018-2

Bohus D (2007). Error awareness and recovery in conversational spoken language interfaces. Ph.D. dissertation. Carnegie Mellon University, Pittsburgh, PA

---

[21]https://youtu.be/RRYj0SMhfH0. Accessed February 19, 2016.

[22]https://youtu.be/6zcByHMw4jk. Accessed February 19, 2016.

[23]https://youtu.be/lHfLr1MF7DI. Accessed February 19, 2016.

[24]https://youtu.be/rYF68t4O_Xw. Accessed February 19, 2016.

[25]https://youtu.be/vphmJEpLXU0. Accessed February 19, 2016.

[26]http://www.masswerk.at/elizabot/. Accessed February 19, 2016.

Bos J, Klein E, Lemon O, Oka T (2003) DIPPER: description and formalisation of an information-state update dialog system architecture. In: 4th SIGdial workshop on discourse and dialog, Sapporo, Japan, 5–6 July 2003. https://aclweb.org/anthology/W/W03/W03-2123.pdf

Brandt J (2008) Interactive voice response interfaces. In: Kortum P (ed) HCI beyond the GUI: design for haptic, speech, olfactory, and other non-traditional interfaces. Morgan Kaufmann, Burlington, MA:229-266. doi:10.1016/b978-0-12-374017-5.00007-9

Buß O, Schlangen D (2011) DIUM—an incremental dialog manager that can produce self-corrections. In: Proceedings of SemDial 2011. Los Angeles, CA, September 2011. https://pub.uni-bielefeld.de/publication/2300868. Accessed 20 Jan 2016

Cassell J, Sullivan J, Prevost S, Churchill E (eds) (2000) Embodied conversational agents. MIT Press, Cambridge, MA

Chen F, Jokinen K (eds) (2010) Speech technology: theory and applications. Springer, New York. doi:10.1007/978-0-387-73819-2

Clark HH (1996) Using language. Cambridge University Press, Cambridge. doi:10.1017/cbo9780511620539

Cohen MH, Giangola JP, Balogh J (2004) Voice user interface design. Addison Wesley, New York

Cohen P, Levesque H (1990) Rational interaction as the basis for communication. In: Cohen P, Morgan J, Pollack M (eds) Intentions in communication. MIT Press, Cambridge, MA:221–256. https://www.sri.com/work/publications/rational-interaction-basis-communication. Accessed 20 Jan 2016

Dahl DA (ed) (2004) Practical spoken dialog systems. Springer, New York. doi:10.1007/978-1-4020-2676-8

DeVault D, Sagae K, Traum DR (2011) Incremental interpretation and prediction of utterance meaning for interactive dialog. Dialog Discourse 2(1):143–170. doi:10.5087/dad.2011.107

Fernández R (2014). Dialog. In: Mitkov R (ed) Oxford handbook of computational linguistics, 2nd edn. Oxford University Press. Oxford. doi:10.1093/oxfordhb/9780199573691.013.25

Ginzburg J (1996) Interrogatives: questions, facts, and dialog. In: Lappin S (ed) Handbook of contemporary semantic theory. Blackwell, Oxford, pp 359–423

Ginzburg J (2015) The interactive stance. Oxford University Press, Oxford. doi:10.1093/acprof:oso/9780199697922.001.0001

Ginzburg J, Fernández R (2010) Computational models of dialog. In: Clark A, Fox C, Lappin S (eds) The handbook of computational linguistics and natural language processing. Wiley-Blackwell, Chichester, UK:429-481. doi:10.1002/9781444324044.ch16

Gorin AL, Riccardi G, Wright JH (1997) How may I help you? Speech Commun 23:113–127. doi:10.1016/s0167-6393(97)00040-x

Green BF, Wolf AW, Chomsky C, Laughery KR (1963) BASEBALL: an automatic question-answerer. In: Feigenbaum EA, Feldman J (eds) Computer and thought. McGraw-Hill, New York

Hempill CT, Godfrey JJ, Doddington GR (1990) The ATIS spoken language systems pilot corpus. In: Proceedings of the DARPA speech and natural language workshop, Hidden Valley, PA:96-101. doi:10.3115/116580.116613

Howes C, Purver M, Healey P, Mills G, Gregoromichelaki E (2011) On incrementality in dialog: evidence from compound contributions. Dialog Discourse 2(1):279–311. doi:10.5087/dad.2011.111

Hura S (2008) Voice user interfaces. In: Kortum P (ed) HCI beyond the GUI: design for haptic, speech, olfactory, and other non-traditional interfaces. Morgan Kaufmann, Burlington, MA:197-227. doi:10.1016/b978-0-12-374017-5.00006-7

Jokinen K (2009) Constructive dialog modelling: speech interaction and rational agents. Wiley, UK. doi:10.1002/9780470511275

Jokinen K, McTear M (2010) Spoken dialog systems. Synthesis lectures on human language technologies. Morgan and Claypool Publishers, San Rafael, CA. doi:10.2200/S00204ED1V01Y200910HLT005

Jurafsky D, Martin JH (2009) Speech and language processing: an introduction to natural language processing, computational linguistics, and speech recognition, 2nd edn. Prentice Hall, Upper Saddle River, NJ

Kortum P (ed) (2008) HCI beyond the GUI: design for haptic, speech, olfactory, and other non-traditional interfaces. Morgan Kaufmann, Burlington, MA

Larson JA (2005) Ten criteria for measuring effective voice user interfaces. Speech Technol Mag. November/December. http://www.speechtechmag.com/Articles/Editorial/Feature/Ten-Criteria-for-Measuring-Effective-Voice-User-Interfaces-29443.aspx. Accessed 20 Jan 2016

Larsson S, Bohlin P, Bos J, Traum DR (1999) TRINDIKIT 1.0 Manual. http://sourceforge.net/projects/trindikit/files/trindikit-doc/. Accessed 20 Jan 2016

Lemon O, Pietquin O (eds) (2012) Data-driven methods for adaptive spoken dialog systems: computational learning for conversational interfaces. Springer, New York. doi:10.1007/978-1-4614-4803-7

Lester J, Branting K, Mott B (2004) Conversational agents. In: Singh MP (ed) The practical handbook of internet computing. Chapman Hall, London. doi:10.1201/9780203507223.ch10

Levelt WJM (1989) Speaking. MIT Press, Cambridge, MA

Lewis JR (2011) Practical speech user interface design. CRC Press, Boca Raton. doi:10.1201/b10461

López Cózar R, Araki M (2005) Spoken, multilingual and multimodal dialog systems: development and assessment. Wiley, UK doi:10.1002/0470021578

Mariani J, Rosset S, Garnier-Rizet M, Devillers L (eds) (2014) Natural interaction with robots, knowbots and smartphones: putting spoken dialog systems into practice. Springer, New York doi:10.1007/978-1-4614-8280-2

McGlashan S. Fraser, N, Gilbert, N, Bilange E, Heisterkamp P, Youd N (1992) Dialogue management for telephone information systems. In: Proceedings of the third conference on applied language processing. Association for Computational Linguistics, Stroudsburg, PA:245-246. doi:10.3115/974499.974549

McTear M (1987) The articulate computer. Blackwell, Oxford

McTear M. (2004) Spoken dialogue technology: toward the conversational user interface. Springer, New York. doi:10.1007/978-0-85729-414-2

Nishida T, Nakazawa A, Ohmoto Y (eds) (2014) Conversational informatics: a data-intensive approach with emphasis on nonverbal communication. Springer, New York. doi:10.1007/978-4-431-55040-2

Paek T, Pieraccini R (2008) Automating spoken dialogue management design using machine learning: an industry perspective. Speech Commun 50:716–729. doi:10.1016/j.specom.2008.03.010

Perez-Martin D, Pascual-Nieto I (eds) (2011) Conversational agents and natural language interaction: techniques and effective practices. IGI Global, Hershey, PA doi:10.4018/978-1-60960-617-6

Pieraccini R (2012) The voice in the machine: building computers that understand speech. MIT Press, Cambridge, MA

Reilly RG (ed) (1987) Communication failure in dialog. North-Holland, Amsterdam

Rieser V, Lemon O (2011) Reinforcement learning for adaptive dialog systems: a data-driven methodology for dialog management and natural language generation. Springer, New York. doi:10.1007/978-3-642-24942-6

Rieser H, Schlangen D (2011) Introduction to the special issue on incremental processing in dialog. Dialog and Discourse 1:1–10. doi:10.5087/dad.2011.001

Sadek MD, De Mori R (1998) Dialog systems. In: De Mori R (ed) Spoken dialogs with computers. Academic Press, London, pp 523–561

Schlangen D, Skantze G (2011) A General, abstract model of incremental dialog processing. Dialog Discourse 2(1):83–111. doi:10.5087/dad.2011.105

Schulman D, Bickmore T (2009) Persuading users through counseling dialog with a conversational agent. In: Chatterjee S, Dev P (eds) Proceedings of the 4th international conference on persuasive technology, 350(25). ACM Press, New York. doi:10.1145/1541948.1541983

Seneff S, Polifroni J (2000) Dialog management in the mercury flight reservation system. In: Proceedings of ANLP-NAACL 2000, Stroudsburg, PA, USA, 11–16 May 2000. doi:10.3115/1117562.1117565

Skantze G, Hjalmarsson A (2013) Towards incremental speech generation in conversational systems. Comp Speech Lang 27(1):243–262. doi:10.1016/j.csl.2012.05.004

Suendermann D (2011) Advances in commercial deployment of spoken dialog systems. Springer, New York. doi:10.1007/978-1-4419-9610-7

Suendermann D, Pieraccini R (2012) One year of Contender: what have we learned about assessing and tuning industrial spoken dialog systems? In: Proceedings of the NAACL-HLT workshop on future directions and needs in the spoken dialog community: tools and data (SDCTD 2012), Montreal, Canada, 7 June 2012: 45–48. http://www.aclweb.org/anthology/W12-1818. Accessed 20 Jan 2016

Suendermann D, Evanini K, Liscombe J, Hunter P, Dayanidhi K, Pieraccini R (2009) From rule-based to statistical grammars: continuous improvement of large-scale spoken dialog systems. In: Proceedings of the IEEE international conference on acoustics, speech, and signal processing (ICASSP 2009), Taipei, Taiwan, 19–24 April 2009: 4713–4716. doi:10.1109/icassp.2009.4960683

Suendermann D, Liscombe J, Pieraccini R (2010a) Optimize the obvious: automatic call flow generation. In: Proceedings of the IEEE international conference on acoustics, speech, and signal processing (ICASSP 2010), Dallas, USA, 14-19 March 2010: 5370–5373. doi:10.1109/icassp.2010.5494936

Suendermann D, Liscombe J, Pieraccini R (2010b) Contender. In: Proceedings of the IEEE workshop on spoken language technology (SLT 2010), Berkeley, USA, 12–15 Dec 2010: 330–335. doi:10.1109/slt.2010.5700873

Suendermann D, Liscombe J, Bloom J, Li G, Pieraccini R (2011a) Large-scale experiments on data-driven design of commercial spoken dialog systems. In: Proceedings of the 12th annual conference of the international speech communication association (Interspeech 2011), Florence, Italy, 27–31 Aug 2011: 820–823. http://www.isca-speech.org/archive/interspeech_2011/i11_0813.html. Accessed 20 Jan 2016

Suendermann D, Liscombe J, Bloom J, Li G, Pieraccini R (2011b) Deploying Contender: early lessons in data, measurement, and testing of multiple call flow decisions. In: Proceedings of the IASTED international conference on human computer interaction (HCI 2011), Washington, USA, 16–18 May 2011: 747–038. doi:10.2316/P.2011.747-038

Sukthankar G, Goldman RP, Geib C, Pynadath DV, Bui HH (eds) (2014) Plan, activity, and intent recognition: theory and practice. Morgan Kaufmann, Burlington, MA

Tanenhaus MK (2004) On-line sentence processing: past, present and, future. The on-line study of sentence comprehension: ERPS, eye movements and beyond. In: Carreiras M, Clifton C Jr (eds) The on-line study of sentence comprehension. Psychology Press, New York: 371–392

Thomson B (2013) Statistical methods for spoken dialog management. Springer theses, Springer, New York. doi:10.1007/978-1-4471-4923-1

Trappl R (ed) (2013) Your virtual butler: the making-of. Springer, Berlin. doi:10.1007/978-3-642-37346-6

Traum DR, Larsson S (2003) The information state approach to dialog management. In: Smith R, Kuppevelt J (eds) Current and new directions in discourse and dialog. Kluwer Academic Publishers, Dordrecht: 325–353. doi:10.1007/978-94-010-0019-2_15

Turing AM (1950) Computing machinery and intelligence. Mind 59:433–460. doi:10.1093/mind/lix.236.433

Walker MA, Aberdeen J, Boland J, Bratt E, Garofolo J, Hirschman L, Le A, Lee S, Narayanan K, Papineni B, Pellom B, Polifroni J, Potamianos A, Prabhu P, Rudnicky A, Sanders G, Seneff S, Stallard D, Whittaker S (2001) DARPA communicator dialog travel planning systems: the June 2000 data collection. In: Proceedings of the 7th European conference on speech communication and technology (INTERSPEECH 2001), Aalborg, Denmark, 3–7 Sept 2001: 1371–1374. http://www.isca-speech.org/archive/eurospeech_2001/e01_1371.html

Weizenbaum J (1966) ELIZA—a computer program for the study of natural language communication between man and machine. Commun ACM 9(1):36–45. doi:10.1145/365153.365168

Wilpon JG, Rabiner LR, Lee CH, Goldman ER (1990) Automatic recognition of keywords in unconstrained speech using Hidden Markov models. IEEE T Speech Audi P 38(11):1870–1878. doi:10.1109/29.103088

Wilks Y (ed) (2010) Close engagements with artificial companions. Key social, psychological, ethical and design issues. John Benjamins Publishing Company, Amsterdam. doi:10.1075/nlp.8

Winograd T (1972) Understanding natural language. Academic Press, New York

W3C Specifications

CCXML http://www.w3.org/TR/ccxml/
EMMAhttp://www.w3.org/TR/2009/REC-emma-20090210/
Pronunciation Lexicon http://www.w3.org/TR/2008/REC-pronunciation-lexicon-20081014/
SISR http://www.w3.org/TR/semantic-interpretation/
SRGF http://www.w3.org/TR/speech-grammar/
SSML http://www.w3.org/TR/speech-synthesis/
State Chart XML http://www.w3.org/TR/2008/WD-scxml-20080516/
VoiceXML http://www.w3.org/TR/2007/REC-voicexml21-20070619/

Part II
Developing a Speech-Based Conversational Interface

Chapter 5
Speech Input and Output

Abstract When a user speaks to a conversational interface, the system has to be able to recognize what was said. The automatic speech recognition (ASR) component processes the acoustic signal that represents the spoken utterance and outputs a sequence of word hypotheses, thus transforming the speech into text. The other side of the coin is text-to-speech synthesis (TTS), in which written text is transformed into speech. There has been extensive research in both these areas, and striking improvements have been made over the past decade. In this chapter, we provide an overview of the processes of ASR and TTS.

5.1 Introduction

For a speech-based conversational interface to function effectively, it needs to be able to recognize what the user says and produce spoken responses. These two processes are known as ASR and TTS. A poor ASR system will result in the system failing to accurately recognize the user's input, while a poor TTS system produces output that in the worst case is unintelligible or unpleasant.

In this chapter, we provide an overview of the ASR and TTS processes, looking at the issues and challenges involved and showing how recent research developments have resulted in more accurate and more acceptable ASR and TTS systems.

5.2 Speech Recognition

ASR has been an active area of research for more than 50 years, but despite much promise, it is only recently that ASR has entered the mainstream with the emergence of voice-driven digital assistants on smartphones. One important factor in the adoption of speech as an input mode is that recognition accuracy has improved considerably in recent years. Now, spoken input has become attractive to a wider range of users compared with the relatively small number of professional enthusiasts who have used voice dictation to create legal and medical documents.

© Springer International Publishing Switzerland 2016
M. McTear et al., *The Conversational Interface*,
DOI 10.1007/978-3-319-32967-3_5

A primary goal of ASR research has been to create systems that can recognize spoken input from any speaker with a high degree of accuracy. This is known as speaker-independent large vocabulary continuous speech recognition (LVCSR). Progress toward LVCSR has been made over the past decades across a number of dimensions, including the following.

- **Vocabulary size**. The earliest ASR systems were able to recognize only a small number of words, such as variants on "yes or "no," or strings of digits. Gradually, the vocabularies have expanded, first to tens of thousands of words and now in current systems to vocabularies of millions of words.
- **Speaker independence**. Originally, users had to read hours of text to voice dictation systems to create a speaker-dependent model. Applications offering information and services to unknown, occasional callers—for example, to call centers—cannot be trained in advance, and so considerable effort has been directed toward making ASR speaker-independent. Problems still occur when speakers have atypical speech patterns, such as strong accents or speech disabilities.
- **Coarticulation**. Continuous speech is difficult to process because there is no clear marking of boundaries between words, and the pronunciation of an individual sound can be affected by coarticulation. For example, "good girl," when spoken fluently, is likely to sound like "goo girl" as the "d" in "good" may be assimilated to the "g" in "girl". Early ASR systems used isolated word recognition in which the speaker had to make a short pause between each word, but nowadays, ASR systems are able to handle continuous speech.
- **Read speech versus conversational speech**. Read speech is more regular and easier to process than spontaneous speech as the speaker does not have to plan what to say. The models for early voice dictation systems were based on read speech, but current ASR systems can handle spontaneously produced conversational speech, although some tasks, such as recognizing and transcribing speech between humans at multiparty meetings, still pose a challenge.
- **Robust speech recognition**. ASR tends to degrade under noisy conditions, where the noise may be transient, such as passing traffic, or constant, such as background music or speech. Techniques being researched to improve ASR robustness include signal separation, feature enhancement, model compensation, and model adaptation. The use of DNN -based acoustic models has also been found to improve robustness.
- **Microphone technology**. Generally, good ASR performance has required the use of a high-quality close-talking microphone. However, users are not likely to want to put on a headset to address an occasional command to a TV that supports voice input or to a device at the other side of the room. Also, when users talk through a smartphone, they might sometimes hold the device near their mouth and sometimes away from them in order to see the screen, so that the microphone is not always at the same distance. Recent advances in far-field speech recognition address this issue using microphone arrays and beam-forming technology that allows the user to address a device from any direction and at some distance from the microphones.

5.2.1 ASR as a Probabilistic Process

When a user speaks to a conversational interface, the ASR component processes the words in the input and outputs them as a string of text by matching the input against its models of the sounds and words of the language being spoken. This process is particularly difficult because spoken input is highly variable so that there can be no direct match between the model of a given phone (the smallest unit of sound) or word and its acoustic representation. The variation is caused by several factors such as:

- Inter-speaker variation due to physical characteristics such as the shape of the vocal tract, gender and age, speaking style, and accent.
- Intra-speaker variation due to factors such as emotional state, tiredness, and state of health.
- Environmental noise.
- Channel effects resulting from different microphones and the transmission channel.

What this means is that the pronunciation of a particular phone or word is not guaranteed to be the same each time, so that the recognition process will always be uncertain. For this reason, ASR is modeled as a stochastic process in which the system attempts to find the most likely phone, word, or sentence $\hat{W}$ from all possible phones, words, or sentences in the language L that has the highest probability of matching the user's input X. This can be formalized in term of conditional probability as follows:

$$\hat{W} = \underset{W \in L}{\operatorname{argmax}} \ P(W|X) \qquad (1)$$

Here, the function argmax returns the highest value for its argument $P(W|X)$. W is a sequence of symbols $W = w_1, w_2, w_3, ..., w_n$ in the acoustic model, and X is a sequence of segments of the input $X = x_1, x_2, x_3, ..., x_n$. This equation states that the estimate of the best symbol string involves a search over all symbol strings in the language to find the maximum value of $P(W|X)$, which is the probability of the string W given the acoustic input X.

As it is not possible to compute $P(W|X)$ directly, Bayes' rule can be applied to break it down as follows:

$$\hat{W} = \underset{W \in L}{\operatorname{argmax}} \ \frac{P(X|W)P(W)}{P(X)} \qquad (2)$$

Here, the denominator $P(X)$ can be ignored since the observation remains constant as we search over all sequences of symbols in the language. Thus, the equation becomes:

$$\hat{W} = \underset{W \in L}{\operatorname{argmax}} \ P(X|W)P(X) \qquad (3)$$

In this equation, $P(X|W)$ is the *observation likelihood,* while $P(X)$ is the *prior probability.* $P(X|W)$ represents the *acoustic model*, and $P(X)$ represents the *language model.* In the following sections, we describe the acoustic and language models and how they are combined to search for and output the most likely word sequence in a process known as decoding.

5.2.2 Acoustic Model

The acoustic model is the basis for estimating $P(X|W)$, the probability that a given sequence of words W will result in an acoustic output X. To develop the acoustic model, the system is trained on a corpus of phonetically transcribed and aligned data and machine-learning algorithms are used to estimate the model parameters.

The first stage of the ASR process involves capturing and analyzing the user's input, which is captured by a microphone in the form of an analogue acoustic signal and converted into a digital signal using an analogue-to-digital converter. The digital signal is enhanced to remove noise and channel distortions, and then, it undergoes processes of sampling, quantization, and feature extraction. Sampling involves measuring the amplitude of the signal at a given point in time. The number of samples to be taken is determined by the channel. For telephone speech, for example, 8000 measures are taken for each second of speech. As the amplitude measurements take the form of real values, quantization transforms these into integers to enable more efficient storage and processing. Feature extraction involves extracting those features that are relevant for ASR. Until recently, the most commonly used features were mel-frequency cepstral coefficients (MFCCs) that measure features such as frequency and amplitude.

Once the acoustic representation has been created, the input is divided into frames, usually 10 ms in length. Each 10-ms frame represents a vector of MFCC features, and the task of ASR is first to match the phones in its acoustic model to these frames, then combine the phones into words using a pronunciation dictionary, and finally estimate the most likely sequence of words given in a language model that specifies probabilities of word sequences. The first part of matching the input thus involves initially computing the likelihood of a feature vector x from an input frame given a phone q in the acoustic model, i.e., $p(x|q)$.

Hidden Markov models (HMMs) have been used in speech recognition since the late 1970s and in combination with Gaussian mixture models (GMMs) have been the dominant method for modeling variability in speech as well as its temporal sequential properties. DNNs have recently replaced GMM–HMM models and are now being used extensively in current ASR systems. In the following, we first briefly outline HMMs and GMMs and then describe recent developments using DNNs.

5.2.2.1 HMMs

An HMM is a probabilistic finite state automaton consisting of states and transitions between states. HMMs can be used to model various types of stochastic process

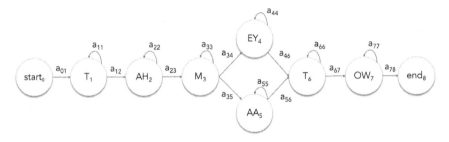

Fig. 5.1 An HMM with two pronunciations of the word "tomato"

involving sequences of events. In the case of ASR, they can model sequences such as sequences of phones that make up a word, or sequences of words that make up a sentence. Figure 5.1 shows a simple example of an HMM with two pronunciations of the word *tomato*.

In this model, each state represents a phone and a path through the HMM represents a temporal progression through the word "tomato." In LVCSR systems, phones are modeled at the level of subphones, usually called *triphones*, to account for variation in the acoustic properties of a phone over time. In recent work, parts of triphones are detected by detectors known as *senones*, allowing much more detailed contextual information to be captured. A tutorial on the basic concepts of speech recognition and on recent work on senones is provided in the Wiki for the CMU Sphinx recognizer.[1]

Note that each state has self-loops, e.g., a_{11} and a_{22}, so that the system can stay in a particular state for an arbitrary period of time. In this way, temporal variability in the speech signal is modeled.

HMMs are trained in order to estimate the two stochastic processes—the state transition probabilities and the parameters of the emitting probability densities. Traditionally, HMMs were trained using the forward–backward algorithm, also known as Baum–Welch algorithm, but more recently, discriminative learning algorithms have been used (He et al. 2008).

The input to the HMM is a sequence of observed feature vectors $o_1, o_2, \ldots, o_{n,}$ and the task is to find the best sequence of states that could have produced the input. Each state in the HMM has a probability distribution of possible outputs ($b_1(o_1)$, $b_2(o_2)$, and so on). As the system proceeds left-to-right from state s_i to state s_{i+1}, the phone in s_i is matched against an observed feature vector o_t and the output with the highest probability is returned. Because there is a distributional probability of outputs at each state and each output can occur in any state, it is possible to obtain the outputs from the model, but it is not possible to know which state an output came from. For this reason, the model is "hidden."

---

[1]http://cmusphinx.sourceforge.net/wiki/tutorialconcepts. Accessed February 20, 2016.

5.2.2.2 GMMs

Until recently, GMMs were used to compute the likelihood of an acoustic feature vector given an HMM state in terms of probability density functions. While it would be possible to estimate a single acoustic feature using a univariate Gaussian model, a multivariate Gaussian is required for MFCC vectors that typically consist of 39 features in LVCSR, and given that the distribution may be non-normal on account of variability in the spoken signal, a weighted mixture of multivariate distributions, as in a Gaussian mixed model, is required. GMMs have been found to be highly effective in various classification tasks in ASR and have been used in combination with HMMs to model the sequential properties of speech.

5.2.2.3 Deep Neural Networks

From around 2010, DNNs have replaced GMM–HMM models. DNNs are now used extensively in industrial and academic research as well as in most commercially deployed ASR systems. Various studies have shown that DNNs outperform GMM–HMM models in terms of increased recognition accuracy (Hinton et al. 2012; Seide et al. 2011).

A DNN differs from conventional neural networks in that it has multiple hidden units between the input and output layers, providing greater learning capacity as well as the ability to model complex patterns of data. These factors have been critical in enabling dramatic improvements in recognition accuracy due mainly to deeper and more precise acoustic modeling.

In acoustic modeling using GMMs, the acoustic input sequence is divided into frames, typically 10 ms in length, and the task is to map a unit in an HMM state, such as a phone, on to one or more frames. The role of the HMM model was to model temporal variability in the input, since the duration of a phone can vary considerably according to factors such as the speaker's speech rate.

However, there are several problems with the GMM–HMM model. Firstly, in order to simplify learning of the mapping between acoustic vector frames and HMM states, it is assumed that the current frame is conditionally independent from all other frames in the acoustic representation and that the current state is conditionally independent of all other states given the previous state. These conditional independence assumptions view speech as a sequence of discrete units and fail to capture the dynamic properties of speech in the temporal dimension—in particular, changes in the acoustic properties of phones across frames that reflect human speech production. Thus, in the more static approach required by the GMM–HMM model, the acoustic input is normalized as MFCCs and excludes paralinguistic information that has been shown to play a crucial role in human speech perception (Baker et al. 2009b). DNNs are able to model a richer set of features in the acoustic input. In particular, by taking as input several frames of acoustic feature vectors, up to about 150-ms duration compared with the 10-ms frames analyzed in the GMM–HMM approach, DNN models are able to capture spectro-temporal variations that give rise to different forms of a phone across time.

5.2.3 Language Model

The language model contains knowledge about permissible sequences of words and, when modeled stochastically, which words are more likely in a given sequence. There are two types of language model in common use:

- Handcrafted grammars, in which all the permissible word sequences are specified.
- N-gram language models that provide statistical information on word sequences.

Handcrafted grammars have been used mainly in commercially deployed VoiceXML-based applications where there is a finite number of ways in which the input to the application can be expressed so that it can be specified in advance. Grammars are useful for specifying well-defined sequences such as dates or times. They are also useful for developers who do not have access to expensive resources such as n-gram language models nor the data and tools to train their own models. However, grammar-based language models lack flexibility and break down if perfectly legal strings are spoken that were not anticipated at design time.

N-gram language models are used for large vocabulary speech recognition. N-grams are estimated from large corpora of speech data and are often specialized to a particular domain or a particular type of input. For example, the Google Speech API makes available language models for *free form* and *Web search*. The free-form language model is used to recognize speech as in the dictation of an e-mail or text message, whereas the Web search model is used to model more restricted forms of input such as shorter, search-like phrases, for example, "flights to London" or "weather in Madrid."

The basic idea behind n-grams is that a word can be predicted with a certain probability if we know the preceding $n - 1$ words. The following simple example will illustrate:

I want to book two seats on a flight to London

In this example, there are three occurrences of the acoustic sequence represented phonetically as /tu/, but in two cases, this is rendered as "to" and in one case as "two." How do we know which is which? N-gram information helps here as "to" is more likely following the word "want," while "two" is more likely following "book."

In the following section, we base our description on Chap. 4 of Jurafsky and Martin (2009). Estimating probabilities involves counting the occurrence of a word given the preceding words. For example, for the sentence "I want to book two seats on a flight to London," we might count how many times the word "to" is preceded in a corpus by the words "I want to book two seats on a flight" and estimate the probability. More formally, the probability of a word given a sequence of preceding words is as follows:

$$P(w_1^n) = P(w_1)P(w_2|w_1)P(w_3|P(w_1^2)...P(w_n|P(w_1^{n-1}) \tag{4}$$

However, estimating the probability of a word given a long sequence of preceding words is impractical as there would not be enough examples, even in a large body of data such as is available nowadays on the Web. For this reason, the task is approximated to a preceding sequence of one or a few words. Using one preceding word results in bigrams, which were used in ASR tasks for a number of years but have now been replaced by trigrams and higher-order n-grams as more data have become available. In the bigram case, the conditional probability of a word given the preceding word is $P(w_n|w_{n-1})$, and for a trigram, it is $P(w_n|w_{n-2}, w_{n-1})$. Substituting the bigram approximation into Eq. 4 and applying the joint rule of probability, the probability for the word sequence can be represented as follows:

$$P(w_1^n) \approx \prod_{k=1}^{n} P(w_k|w_{k-1}) \tag{5}$$

So, in order to estimate P(*I want to book two seats on a flight to London*), we calculate

$$P(I|<s>)P(\text{want}|I)P(\text{to}|\text{want})P(\text{book}|\text{to})P(\text{two}|\text{book})P(\text{seats}|\text{two})...$$

The probabilities are estimated using maximum likelihood estimation (MLE) in which the counts acquired from a training corpus are normalized. The amount of training data for n-gram models is typically many millions of words.

One of the problems that developers of n-gram language models face is sparse data so that a possible n-gram sequence may not have occurred in the data or with sufficiently low frequency to produce a poor estimate. Various smoothing techniques have been developed to overcome the sparse data problem (see, for example, the techniques described in Jurafsky and Martin 2009: 131–144).

5.2.4 Decoding

Given a sequence of acoustic likelihoods from the acoustic model, an HMM dictionary of word pronunciations, and a language model of n-grams, the next task is to compute the most likely word sequence. This task is known as *decoding*. Decoding involves searching through the HMM to find the most likely state sequence that could have generated the observation sequence. However, because the number of possible states is exponentially large, it is necessary to find a way to reduce the search. The Viterbi algorithm, which uses *dynamic programming*, has been widely applied for decoding in ASR (Viterbi 1967; Forney 2005[2]). Dynamic

---

[2]http://arxiv.org/abs/cs/0504020v2. Accessed February 20, 2016.

programming solves complex problems by breaking them down into simpler sub-problems. Viterbi stores intermediate results in the cells of a trellis, where each cell represents the probability of being in the current state after seeing the observations so far and having passed through the most probable state sequence to arrive at the current state, i.e., the probability of observing sequence $x_1 \ldots x_n$ and being in state s_j at time t. At time t, each state is updated by the best score from all states at time $t - 1$ and a backtracking pointer is recorded for the most probable incoming state. Tracing back through the backtracking pointers at the end of the search reveals the most probable state sequence.

5.3 Text-to-Speech Synthesis

TTS is the process of transforming written text to speech. Synthesized speech has improved in quality over the past few years and in many cases can sound almost human-like. The quality and intelligibility of synthesized speech play an important role in the acceptability of speech-based systems, as users often judge a conversational interface by the quality of the spoken output rather than its ability to recognize what they say.

TTS is used in many applications in which the message cannot be prerecorded and has to be synthesized on the fly. These include the following: screen reading for users with visual impairments, outputting messages at airports and railway stations, providing step-by-step navigation instructions for drivers, and speaking up to the minute news and weather reports. One of the best-known users of TTS is the astrophysicist Stephen Hawking who is unable to speak as a result of motor neuron disease.

In the following sections, we base our description on Chap. 8 of Jurafsky and Martin (2009). There are two stages in TTS: text analysis and waveform synthesis. In text analysis, the text to be spoken is transformed into a representation consisting of phonemes and prosodic information (referred to as a phonemic internal representation). Prior to the phonetic and prosodic analysis, the text is normalized, as described in the next section. In the second stage, the internal representation is converted to a waveform that can be output as spoken text. Figure 5.2 summarizes the main elements of the TTS process.

5.3.1 Text Analysis

The TTS component of a conversational interface receives a representation of the text to be spoken from the response generation component (see Fig. 2.5 in Chap. 2). This text may contain a number of features that need to be normalized before the text can be spoken. For example, if it is not clear where the sentence boundaries are, the text

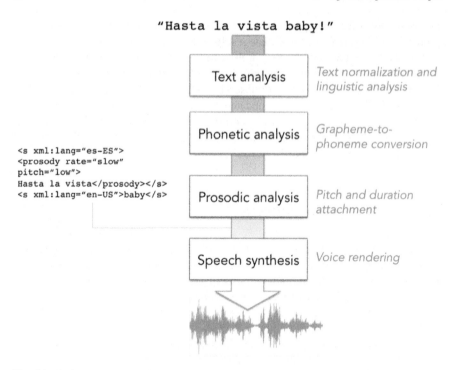

Fig. 5.2 Basic processes in a TTS system

will be spoken in a continuous stream without proper phrasing, making it sound unnatural and potentially more difficult to process. The text may also include abbreviations and acronyms that need to be transformed into a form that can be spoken. The same applies to other items such as dates, numbers, time, phone numbers, and currency values.

Determining sentence boundaries involves sentence tokenization. While it might appear that this should be a simple process with written text in which a full stop (period) represents the end of a sentence, in practice there can be ambiguity as full stops can also be used to indicate an abbreviation, as in "St.," or they can be part of a date, as in "12.9.97," or part of a sequence, as in "M.I.5". Although it is possible to handcraft a system to perform sentence tokenization, supervised machine-learning methods are generally used in which classifiers are trained on a set of features that distinguish between different markers of sentence boundaries.

Abbreviations, acronyms, and other nonstandard words need to be transformed so that they can be spoken correctly. Some examples are shown in Table 5.1.

As can be seen from these examples, the realization of nonstandard words depends on knowing their type. For example, "IKEA" should be pronounced as a word and not as a letter sequence. Often, the words are ambiguous: "2/3" may be a date, but it could also be a fraction ("two-thirds"), and depending on its position in the phrase, "st" could be "saint (St. John)" or "street (John St)," or both ("St. John

Table 5.1 Rendering abbreviations, acronyms, and other nonstandard words for TTS

Type	Example	Realization
Abbreviation	St	Street/saint
	Ms	Miz
	mph	Miles per hour
Letter sequence	HP	h p/Hewlett Packard
	dvd	d v d
	tv	t v/television
Acronym	IKEA, NATO, RAM, radar, laser	(pronounced as a word)
Date	2/3	The third of February (US) /the second of March (UK)
	March 8	March 8/the eighth of March
Number	1745	One seven four five/one thousand seven hundred and forty-five /seventeen hundred and forty-five
	1900	One nine oh oh/one nine double oh/nineteen hundred/one thousand nine hundred
Time	2:30	Two thirty/half past two
	7.45	Seven fourty-five/a quarter to eight /a quarter of eight
Phone numbers	653744	Six five three seven four four/sixty five thirty seven double four
Currency	£45.70	Forty-five seventy/forty-five pounds seventy pence

St = Saint John Street"). There are also dialect differences—for example, in how dates are spoken in the USA and UK (USA: "March eight" and UK: "the eighth of March"). Multifeature classifiers are used to classify the different types of nonstandard words so that they can be expanded into normal words for spoken output.

Homographs, i.e., words that have the same spelling but different pronunciations, are a problem for TTS. For example, *live* as a verb is pronounced to rhyme with "give" ("I live in London"), but as an adjective, it is pronounced to rhyme with "five" ("The Stones played a live concert in Hyde Park"). Usually, part of speech tagging can help to disambiguate homographs.

Once the text has been normalized, the next stages are phonetic and prosodic analysis. Phonetic analysis involves finding a pronunciation for each word in the text. For words that are known, the pronunciations can be looked up in a pronunciation dictionary that contains all the words of a language and their pronunciations. However, not all words may be represented in a pronunciation dictionary, so for the unknown words, the pronunciations are constructed using a process known as grapheme-to-phoneme conversion, in which rules that determine pronunciations based on spellings are applied to transform sequences of letters into sequences of phonemes. Another problem is that some words can be pronounced differently depending on their meaning. For example, in English, the word "bass" is pronounced /beis/ in a sentence such as "he plays bass guitar," but /baes/ in a sentence such as "I caught a bass when I was fishing". This can be solved using markup such as the Speech Synthesis Markup Language (SSML) (see Sect. 5.2.4).

Prosody includes phrasing, pitch, loudness, tempo, and rhythm. These are known as the *suprasegmental* features of speech as they extend over more than single segments of sounds. Prosody can indicate differences in the function of an utterance—for example, intonation can mark the difference between a question and a statement. Various prosodic features can also express emotional aspects such as anger or surprise. More generally, prosody is essential in order to make synthesized speech sound more natural. The main elements of prosody are as follows:

- Prosodic phrasing—how words and phrases group together as meaningful units. Often, punctuation helps to determine phrasing. For example, there should usually be a short pause following a comma. Syntactic information about the grouping of words into phrases is also useful. Given these sorts of information, classifiers can be trained to determine prosodic boundaries.
- Prosodic prominence—how some words in an utterance are made more salient through accentuation, emphasis, or longer duration. The information structure of an utterance determines which words should have prominence, as words that are accented are marked by the speaker as providing information that is assumed to be new to the hearer.
- Tune—for example, the use of a final rise in an utterance to indicate a question or a final fall to indicate a statement. Tune is also used to express attitudes, such as surprise.

Prosody is an active research area in TTS and sophisticated models have been developed, including ToBI (Tone and Break Indices) (Beckman et al. 2005) and Tilt (Taylor 2000).

5.3.2   Waveform Synthesis

The most popular technique for waveform synthesis in current TTS systems involves *concatenative synthesis*, in which segments of recorded speech are concatenated. Each recorded utterance is divided into units of various sizes, such as phones, diphones, syllables, words, phrases, and sentences, and the units are indexed based on acoustic features and stored in a very large database. When a text is to be synthesized, the best sequence of units that corresponds to the phonetic and prosodic specifications of the target representation of the text is selected from the database and combined to form spoken output. This process is known as unit selection. Unit selection has been found to produce more natural speech compared with other methods such as diphone synthesis, in which transitions between sounds are used as units. Sometimes, digital signal processing is applied to smooth concatenation points by making small pitch adjustments between neighboring units when there are pitch deviations.

In addition to concatenative approaches to TTS, other approaches include articulatory and biology-inspired methods. In Eunison,[3] a recent European Union (EU) funded project, a new voice simulator is being built based on physical first principles with a long-term aim of enabling, among other things, more natural speech synthesis. See also this video.[4]

5.3.3   Using Prerecorded Speech

Because synthesized speech is not as natural as human speech, prerecorded speech is often preferred over TTS in commercially deployed applications. This involves employing professional speakers, often referred to as voice talent, to record all the possible system outputs. While this is a feasible solution for a system with a limited range of output that can be predicted at design time, it cannot be used when the text is unpredictable—as in reading emails, news items, or text messages. In situations where most of the text is known in advance and some elements have to be inserted at runtime—as in flight announcements at airports—the output can sound jerky if there are pitch deviations between the recorded text and the inserted words. Lewis (2011:171–177) discusses the pros and cons of using prerecorded speech instead of or in combination with TTS.

5.3.4   Speech Synthesis Markup Language

SSML is a standard produced by the Voice Browser Group of the World Wide Web Consortium (W3C).[5] The aim of SSML is to provide a standard notation for the markup of text to be synthesized in order to override the default specifications of the TTS system. The markup can be applied to control various aspects of speech production. The following are some examples.

Structure analysis. Text can be divided into basic elements to provide more natural phrasing using <s> for sentence and <p> for paragraph, e.g.

```
<p>
<s>I want to fly to London on Friday </s>
<s>I need to be there by 3 p.m. </s>
<s>Do you have any business class seats? </s>
</p>
```

---

[3]http://www.eunison.eu/. Accessed February 20, 2016.
[4]https://www.youtube.com/watch?v=t4YzfGD0f6s&feature=youtu.be. Accessed February 20, 2016.
[5]http://www.w3.org/TR/speech-synthesis/. Accessed February 20, 2016.

Text normalization. Potentially ambiguous items are annotated so that they are spoken correctly, e.g.

```
<sub alias = "doctor" > Dr. </sub > Watson
130 Blenheim < sub alias = "drive" > Dr. </sub>
```

<say-as> This element specifies the type of text contained within the element and how to render it, e.g.

```
<say-as interpret-as = "vxml:date"> <value expr = "20151115"/> </say-as>
<say-as interpret-as = "vxml:currency"> <value expr = "USD20.54"/> </say-as>
```

<phoneme> The phoneme element provides a phonetic pronunciation for an item of text. For example, the following markup would be used to specify the US and UK pronunciations of the word "tomato," respectively:

```
You say <phoneme ph = 't ah0 m ey1 t ow0'> tomato </phoneme> , and I
say <phoneme ph = 't ah0 m aa1 t ow0'> tomato </phoneme>
```

<break/> This element specifies when to insert a pause in the text, e.g.

```
Welcome to hotel reservations.
<break time = "240 ms" />
How can I help you?
```

<prosody> This element specifies various prosodic features such as pitch, timing, pauses, speech rate, emphasis, and the relative timing of segments and pauses, e.g.

```
<prosody duration = "8000 ms"> This is an example with long duration
</prosody>
<prosody rate = "slow"> An example with slow rate of speech </prosody>
<prosody volume = "loud"> An example at a high volume </prosody>
<prosody pitch = "low"> This is an example of low pitch </prosody>
```

It should be noted that not all TTS engines implement all of the features listed in the SSML specification. For example, the Amazon Alexa Skills Kit supports a subset of the SSML tags.[6]

5.4 Summary

In this chapter, we have reviewed the technologies of ASR and TTS. Both technologies have developed considerably over the past few decades, and speech recognition in particular has begun to show impressive results with the adoption of DNNs. Speech recognition is never likely to be 100 % accurate, and in situations where the input is degraded, as in noisy environments, the word error rate is likely to increase. Developers of commercially deployed voice user interfaces often attempt to restrict the range of user inputs with carefully designed prompts, as discussed in Chap. 4. This strategy can also help to make users aware of the limitations of the system as a highly performing recognizer might generate expectations in the user that exceed what the system's dialog manager is able to deal with.

A recent development that affects the technologies of speech recognition and TTS is the role of personality and emotion in conversational interaction. In addition to recognizing the words spoken by the user, the system should also detect the emotions that are being conveyed in spoken utterances and make attributions about aspects of the user's personality. On the output side, the system should express its emotions and personality. Indeed, a whole tradition of creating *persona* to convey a company's brand, or the use of avatars, talking heads, and social robots, has been building on these technologies. We will examine the roles of emotion and personality in more detail in Chaps. 14 and 15.

Having looked at the technologies involved in speech recognition and TTS, we turn in the next chapter to some practical applications of these technologies in which we will introduce the HTML5 Web Speech API (Web SAPI) that can be used to add speech input and output to Web pages and the Android Speech APIs that can be used for speech input and output on mobile devices.

Further Reading

Speech Recognition

Professor Fred Jelinek and his colleagues at IBM Research pioneered the statistical methods that are the basis of current speech recognition systems (Jelinek 1998). See also Rabiner and Juang (1998). Jurafsky and Martin (2009) is the standard text on speech processing and natural language processing, covering both topics in detail. Huang et al. (2001) also provide a detailed account of both aspects

---

[6]https://developer.amazon.com/public/solutions/alexa/alexa-skills-kit/docs/speech-synthesis-markup-language-ssml-reference. Accessed February 20, 2016.

of spoken language processing. Deng and Yu (2013) and Yu and Weng (2015) describe the deep learning approach to ASR using DNNs. Pieraccini (2012) provides an accessible overview of speech recognition technology and its history. For a comprehensive account of mathematical models for speech technology, see Levinson (2005). Esposito et al. (2016) present recent advances in nonlinear speech processing aimed at modeling the social nature of speech including personality traits and emotions (see further Chap. 14).

There are also several book chapters reviewing speech recognition technology, including Furui (2010), Huang and Deng (2010), and Renals and Hain (2010). Baker et al. (2009a, b) review the current state of speech recognition and outline a number of challenges for the future. Deng and Li (2013) provide a comprehensive overview of machine-learning paradigms for speech recognition.

Text-to-Speech Synthesis

Recent books on TTS include Holmes and Holmes (2001), Dutoit (2001), and Taylor (2009). Chapter 16 of Huang et al. (2001) and Chap. 8 of Jurafsky and Martin (2009) provide detailed coverage of speech synthesis along with chapters on phonetics and digital speech processing. See also Aaron et al. (2005), Suendermann et al. (2010), Pieraccini (2012), and Black's comprehensive online tutorial (Black 2000).

Conferences and Journals

Speech recognition and text-to-speech synthesis are main topics at the annual ICASSP and INTERSPEECH conferences and at the IEEE ASR and Understanding (ASRU) Workshop, as well as at the commercially oriented SpeechTEK conference.

Key journals are as follows: IEEE Transactions on Audio, Speech, and Language Processing (since 2014 renamed as IEEE/ACM Transactions on Audio, Speech, and Language Processing); Speech Communication; Computer Speech and Language.

Exercises

1. Speech recognition. For those readers wishing to acquire a detailed knowledge of ASR, there is an online course at MIT by Dr. James Glass and Professor Victor Zue, leading researchers in spoken language technologies.[7] The lecture notes can be downloaded along with a list of exercises.[8]
2. Text-to-speech synthesis. The Smithsonian Speech Synthesis History Project collected a history of speech synthesis with voices from 1922 to the 1980s. Audio clips are available at Dennis Klatt's History of Speech Synthesis page.[9]

---

[7]http://ocw.mit.edu/courses/electrical-engineering-and-computer-science/6-345-automatic-speech-recognition-spring-2003/index.htm. Accessed February 20, 2016.

[8]http://ocw.mit.edu/courses/electrical-engineering-and-computer-science/6-345-automatic-speech-recognition-spring-2003/assignments/. Accessed February 20, 2016.

[9]http://www.cs.indiana.edu/rhythmsp/ASA/Contents.html. Accessed February 20, 2016.

Listening to these clips will provide an interesting perspective on how TTS has progressed over the years.

Compare the samples from the Smithsonian Speech Synthesis History Project with some current TTS systems that provide online demos. Some examples are as follows: IVONA Text-to-Speech,[10] NaturalReader,[11] Cepstral,[12] DioSpeech,[13] Festival,[14] MaryTTS.[15]

References

Aaron A, Eide E, Pitrelli JF (2005) Conversational computers. Sci Am June: 64–69. doi:10.1038/scientificamerican0605-64

Baker J, Deng L, Glass J, Khudanpur S, Lee C-H, Morgan N, O'Shaughnessy D (2009a) Developments and directions in speech recognition and understanding, Part 1. Sig Process Mag IEEE 26(3):75–80. doi:10.1109/msp.2009.932166

Baker J, Deng L, Khudanpur S, Lee C-H, Glass J, Morgan N, O'Shaughnessy D (2009b) Updated MINDS report on speech recognition and understanding, Part 2 signal processing magazine. IEEE 26(4):78–85. doi:10.1109/msp.2009.932707

Beckman ME, Hirschberg J, Shattuck-Hufnagel S (2005) The original ToBI system and the evolution of the ToBI framework. In: Jun S-A (ed) Prosodic typology—the phonology of intonation and phrasing, Chapter 2. Oxford University Press, Oxford, pp 9–54. doi:10.1093/acprof:oso/9780199249633.003.0002

Black A. (2000) Speech synthesis in Festival: a practical course on making computers talk. http://festvox.org/festtut/notes/festtut_toc.html. Accessed 20 Jan 2016

Deng L, Li X (2013) Machine learning paradigms for speech recognition: an overview. IEEE T Speech Audi P 21 (5) May 2013:1061–1089. doi:10.1109/tasl.2013.2244083

Deng L, Yu D (2013) Deep learning: methods and applications. Found Trends Signal Process 7(3–4):197–386. doi:10.1561/2000000039

Dutoit T (2001) An introduction to text-to-speech synthesis. Springer, New York. doi:10.1007/978-94-011-5730-8

Esposito A, Faundez-Zanuy M, Cordasco G, Drugman T, Solé-Casals J, Morabito FC (eds) (2016) Recent advances in nonlinear speech processing. Springer, New York

Forney GD Jr (2005) The Viterbi algorithm: a personal history. http://arxiv.org/abs/cs/0504020v2. Accessed 20 February 2016

Furui S (2010) History and development of speech recognition. In: Chen F, Jokinen K (eds) Speech technology: theory and applications. Springer, New York:1–18. doi:10.1007/978-0-387-73819-2_1

He X, Deng L, Chou W (2008) Discriminative learning in sequential pattern recognition. IEEE Signal Process Mag 25(5):14–36. doi:10.1109/msp.2008.926652

---

[10]https://www.ivona.com/. Accessed February 20, 2016.

[11]http://www.naturalreaders.com/index.html. Accessed February 20, 2016.

[12]http://www.cepstral.com/en/demos. Accessed February 20, 2016.

[13]http://speech.diotek.com/en/. Accessed February 20, 2016.

[14]http://www.cstr.ed.ac.uk/projects/festival/onlinedemo.html. Accessed February 20, 2016.

[15]http://mary.dfki.de/. Accessed February 20, 2016.

Hinton G, Deng L, Yu D, Dahl GE, Mohamed A-R, Jaitly N, Senior A, Vanhoucke V, Nguyen P, Sainath TN, Kingsbury B (2012) Deep neural networks for acoustic modeling in speech recognition: the shared views of four research groups. IEEE Signal Process Mag 82:82–97. doi:10.1109/msp.2012.2205597

Holmes J, Holmes W (2001) Speech synthesis and recognition. CRC Press, Boca Raton

Huang X, Acero A, Hon H-W (2001) Spoken language processing: a guide to theory, algorithm, and system development. Prentice Hall, Upper Saddle River, NJ

Huang X, Deng L (2010) An overview of modern speech recognition. In: Indurkhya N, Damerau FJ (eds) Handbook of natural language processing. CRC Press, Boca Raton, pp 339–366. http://research.microsoft.com/pubs/118769/Book-Chap-HuangDeng2010.pdf. Accessed 20 Jan 2016

Jelinek F (1998) Statistical methods for speech recognition. MIT Press, Massachusetts

Jurafsky D, Martin JH (2009) Speech and language processing: an introduction to natural language processing, computational linguistics, and speech recognition, 2nd edn. Prentice Hall, Upper Saddle River, NJ

Levinson SE (2005) Mathematical models for speech technology. Wiley, Chichester, UK

Lewis JR (2011) Practical speech user interface design. CRC Press, Boca Raton. doi:10.1201/b10461

Pieraccini R (2012) The voice in the machine: building computers that understand speech. MIT Press, Cambridge, MA

Rabiner L, Juang B-H (1998) Fundamentals of speech recognition. Prentice Hall, Upper Saddle River

Renals S, Hain T (2010) Speech recognition. In: Clark A, Fox C, Lappin S (eds) The handbook of computational linguistics and natural language processing. Wiley-Blackwell, Chichester, UK, pp 299–322. doi:10.1002/9781444324044.ch12

Seide F, Li G, Yu D (2011) Conversational speech transcription using context-dependent deep neural networks. In: Proceedings of the 12th annual conference of the international speech communication association (INTERSPEECH 2011). Florence, Italy, 27–31 Aug 2011, pp 437–440

Suendermann D, Höge H, Black A (2010) Challenges in speech synthesis. In: Chen F, Jokinen K (eds) Speech technology: theory and applications. Springer, New York, pp 19–32. doi:10.1007/978-0-387-73819-2_2

Taylor P (2000) Analysis and synthesis using the tilt model. J Acoust Soc Am 107(3):1697–1714. doi:10.1121/1.428453

Taylor P (2009) Text-to-speech synthesis. Cambridge University Press, Cambridge. doi:10.1017/cbo9780511816338

Viterbi AJ (1967) Error bounds for convolutional codes and an asymptotically optimum decoding algorithm. IEEE T Inform Theory 13(2):260–269. doi:10.1109/TIT.1967.1054010

Yu D, Deng L (2015) Automatic speech recognition: a deep learning approach. Springer, New York. doi:10.1007/978-1-4471-5779-3

Web Pages

Comparison of Android TTS engines http://www.geoffsimons.com/2012/06/7-best-android-text-to-speech-engines.html

Computer Speech and Language http://www.journals.elsevier.com/computer-speech-and-language/

EURASIP journal on Audio, Speech, and Music Processing http://www.asmp.eurasipjournals.com/

History of text-to-speech systems www.cs.indiana.edu/rhythmsp/ASA/Contents.html

IEEE/ACM Transactions on Audio, Speech, and Language Processing http://www.signalprocessingsociety.org/publications/periodicals/taslp/

International Journal of Speech Technology http://link.springer.com/journal/10772

Resources for TTS http://technav.ieee.org/tag/2739/speech-synthesis

Speech Communication http://www.journals.elsevier.com/speech-communication/

SSML: a language for the specification of synthetic speech- http://www.w3.org/TR/2004/REC-speech-synthesis-20040907/

Chapter 6
Implementing Speech Input and Output

Abstract There are a number of different open-source tools that allow developers to add speech input and output to their apps. In this chapter, we describe two different technologies that can be used for conversational systems, one for systems running on the Web and the other for systems running on mobile devices. For the Web, we will focus on the HTML5 Web Speech API (Web SAPI), while for mobile devices we will describe the Android Speech APIs.

6.1 Introduction

There are a number of different open-source tools as well as commercially available products that allow developers to add speech input and output to their apps. In this chapter, we describe two different technologies that can be used for conversational systems, one that runs on the Web and the other on mobile devices. For the Web, we will focus on the HTML5 Web Speech API (Web SAPI), while for mobile devices we will describe the Android Speech APIs.

There are several reasons why we have chosen these technologies:

- The HTML5 Web Speech API makes it easy to add speech recognition and text-to-speech synthesis (TTS) to Web pages, allowing fine control and flexibility in versions of Chrome from version 25 onward.[1, 2]
- There is a proliferation of Android devices so that it is appropriate to develop applications for this platform.[3]
- The Android Speech API is open source and thus is more easily available for developers and enthusiasts to create apps, compared with some other operating

---

[1]https://developers.google.com/web/updates/2013/01/Voice-Driven-Web-Apps-Introduction-to-the-Web-Speech-API?hl=en. Accessed February 21, 2016.

[2]https://developers.google.com/web/updates/2014/01/Web-apps-that-talk-Introduction-to-the-Speech-Synthesis-API?hl=en. Accessed February 21, 2016.

[3]https://en.wikipedia.org/wiki/Android_(operating_system). Accessed February 21, 2016.

© Springer International Publishing Switzerland 2016
M. McTear et al., *The Conversational Interface*,
DOI 10.1007/978-3-319-32967-3_6

systems. Anyone with an Android device can develop their own apps and upload them to their device for their own personal use and enjoyment.

There are several alternatives to the Google Speech APIs that might be suitable for those who wish to experiment further. We provide a list of these in Tables 6.1 and 6.2.

Table 6.1 Open-source tools for text-to-speech synthesis

Name	Web site
AT&T speech API	http://developer.att.com/apis/speech
Cereproc (academic license)	https://www.cereproc.com/en/products/academic
DioTek	http://speech.diotek.com/en/
Festival	http://www.cstr.ed.ac.uk/projects/festival/
FLite (CMU)	http://www.festvox.org/flite/
FreeTTS	http://freetts.sourceforge.net/docs/index.php
Google TTS	http://developer.android.com/reference/android/speech/tts/TextToSpeech.html
iSpeech TTS	http://www.ispeech.org/
MaryTTS, DFKI, Germany	http://mary.dfki.de/
Microsoft	http://www.bing.com/dev/speech

Table 6.2 Tools for speech recognition

Name	Technology	Web site
Android ASR	API	http://developer.android.com/reference/android/speech/SpeechRecognizer.html
AT&T speech	API	http://developer.att.com/apis/speech
CMU PocketSphinx		http://www.speech.cs.cmu.edu/pocketsphinx/
CMU Sphinx	HMM	http://cmusphinx.sourceforge.net/
CMU statistical language modeling toolkit	Language modeling	http://www.speech.cs.cmu.edu/SLM_info.html
HTK, Cambridge	HMM	http://htk.eng.cam.ac.uk/
iSpeech		http://www.ispeech.org/
Julius (Japanese)	HMM	http://julius.sourceforge.jp/en_index.php
Kaldi	Deep neural net	http://kaldi.sourceforge.net/about.html
RWTH ASR, Aachen, Germany	HMM	http://www-i6.informatik.rwth-aachen.de/rwth-asr/
Microsoft SDK		http://www.bing.com/dev/speech

The code corresponding to the examples in this chapter is in GitHub, in the ConversationalInterface repository,[4] in the folder called *chapter6*.

6.2 Web Speech API

The Web Speech API is a subset of HTML5 that aims to enable Web developers to provide speech input and output on Web browsers. The API is not tied to a specific automatic speech recognition (ASR) or TTS engine and supports both server-based and client-based ASR and TTS. The API specification[5] was published in 2012 by the Speech API W3C Community Group[6] and adopted by important companies in the sector. The API is not a standard, nor is it included in any standardization track, but it is supported by some Web browsers such as Google Chrome (speech input and output) and Safari (speech output).

The API provides all the necessary elements for TTS and ASR. TTS has been implemented so that different prosody parameters can be tuned, and if there is no speech synthesizer installed, the text is shown visually on the Web. ASR captures audio from the microphone and streams it to a server where the speech recognizer is located. The API is designed to enable brief (one-shot) speech input as well as continuous speech input. ASR results are provided to the Web along with other relevant information, such as n-best hypotheses and confidence values.

The code for this section is in the GitHub ConversationalInterface repository, in the folder */chapter6/WebSAPI/*. You can open the example of the Web Speech API with Google Chrome to see how it works. Bear in mind that ASR only works when the code is hosted on a server; thus, in order to test the examples with speech input you will need to upload the files to a server. To make it simpler for you, we have uploaded the files to our own server.[7] You can open them directly with Chrome and see how they work.

6.2.1 Text-to-Speech Synthesis

In this section, we will explain the basics of TTS capabilities of the Web SAPI, for which a complete specification is available.[8]

---

[4]http://zoraidacallejas.github.io/ConversationalInterface/. Accessed March 2, 2016.

[5]https://dvcs.w3.org/hg/speech-api/raw-file/9a0075d25326/speechapi.html. Accessed February 21, 2016.

[6]http://www.w3.org/community/speech-api/. Accessed February 21, 2016.

[7]https://lsi.ugr.es/zoraida/conversandroid/. Accessed March 2, 2016.

[8]https://dvcs.w3.org/hg/speech-api/raw-file/tip/speechapi.html#tts-section. Accessed February 21, 2016.

TTS can be performed by using the `SpeechSynthesisUtterance` class and the `SpeechSynthesis` interface that together provide a series of attributes, methods, and events that allow full control over the TTS engine. We will discuss some of the most relevant of these.

- Attributes:

  - *text*: the text to be synthesized.
  - *lang*: the language for the synthesis.
  - *voiceURI*: the location of the synthesis engine (if not specified, the default speech synthesizer is used).
  - *volume*: volume in a range [0, 1] where 1 is the default level.
  - *rate*: speech range in a range [0.1, 10], where 1 is the default rate, 2 is twice as fast as the default, 0.5 is half as fast as the default, and so on.
  - *pitch*: Pitch in a range [0, 2] where 1 is the default level.

- Methods:

  - *speak*: adds the text to the queue for its synthesis.
  - *cancel*: empties the queue. If any phrase is being synthesized, it is interrupted.
  - *pause*: pauses the phrase being synthesized.
  - *resume*: restarts the synthesis of a paused phrase.
  - *getVoices*: shows the voices available for speech synthesis.

- Events:

  - *start*: fired when the synthesis starts.
  - *end*: fired when the phrase has been synthesized successfully.
  - *error*: fired when an error occurs during the synthesis.
  - *pause*: fired when the synthesis has been paused.
  - *resume*: fired when the synthesis has been resumed.

Here is an example with the minimum structure for a script using the Web SAPI. We have created a new instance of a `SpeechSynthesisUtterance`, initializing it with a text; then, we initiate the synthesis with the `start` method. In this case, the text is synthesized when the page is loaded (Code 6.1).

```
<script>
   var utt = new SpeechSynthesisUtterance('Hello world!');
   window.speechSynthesis.speak(utt);
</script>
```

Code 6.1 Fragment from `tts_min.html`

As shown above, there are many other attributes that can be specified in addition to the text to be synthesized. Code 6.2 shows how they can be configured.

As it is possible to control the synthesis from the HTML code, it is fairly easy to decide the moment when an utterance is started, paused, or resumed using the visual interface. Code 6.3 shows how to start synthesizing a text that has been introduced by the user in the Web. Since it is necessary to pay attention to message encoding when using languages with special characters (e.g., Spanish), we have included a small function to address this issue.

The use of events provides an additional level of control, as it allows developers to manage the different situations that may appear. In Code 6.4, we can observe how we can couple each of the events (e.g., msg.onStart) with a certain functionality (e.g., show the message "the synthesis has started" in a TextArea).

```
<script>
    var utt = new SpeechSynthesisUtterance();
    utt.volume = 1;
    utt.rate = 1.5;
    utt.pitch = 1;
    utt.lang = 'en-EN';
    utt.text = 'This is a prosody test';
    speechSynthesis.speak(utt);
</script>
```

Code 6.2 Fragment from tts_att.html

```
<body>
(...) <button id="btn" onclick="start_talking()">Say it</button>
</body>
<script>
    var utt = new SpeechSynthesisUtterance();
    function start_talking() {
        utt.text =
encode_utf8(document.getElementById('message').value);
        speechSynthesis.speak(utt);
    }
    function encode_utf8(s) {
        return unescape(encodeURIComponent(s));
    }
</script>
```

Code 6.3 Fragment from tts_btn.html

```
<script>
   var msg = new SpeechSynthesisUtterance();
   msg.onstart = function(){

        document.getElementById('feedback').innerHTML =

        'The synthesis has started'; }

   (...)
</script>
```

Code 6.4 Fragment from `tts_events.html`

6.2.2 Speech Recognition

Probably you are not aware that your speech browser has ASR capabilities. This is the case for Google Chrome as can be seen at the Web Speech API Demonstration page.[9] Be careful to enable the Web browser to make use of your microphone when you are prompted to do so.

As with the TTS examples, we will now show some of the basic capabilities of ASR in the Web SAPI.[10] We will use mainly the `SpeechRecognition` interface that provides the following attributes, methods, and events:

- Attributes:

  - *grammar*: a SpeechGrammar object with the recognition grammar.
  - *lang*: the language for recognition.
  - *continuous*: should be false for a single "one-shot" recognition result and true for zero or more results (e.g., dictation).
  - *maxAlternatives*: maximum number of recognition alternatives (value of N for the n-best list). The default value is 1.
  - *serviceURI*: localization of the recognition engine (if not specified, the default one is used).

- Methods:

  - *start*: starts listening.
  - *stop*: stops listening and tries to obtain a recognition result from what has been heard up to that moment.
  - *abort*: stops listening and recognizing and does not provide any result.

- Events:

  - *audiostart:* fired when the audio starts being captured.

---

[9]https://www.google.com/intl/en/chrome/demos/speech.html. Accessed February 21, 2016.
[10]https://dvcs.w3.org/hg/speech-api/raw-file/9a0075d25326/speechapi.html#speechreco-section. Accessed February 21, 2016.

- *soundstart*: fired when sound has been detected.
- *speechstart*: fired when the speech that will be used for recognition has started.
- *speechend*: fired when the speech that will be used for recognition has ended.
- *soundend*: fired when sound is no longer detected.
- *audioend*: fired when the audio capture is finished.
- *result*: fired when the speech recognizer returns a result.
- *nomatch*: fired when the speech recognizer returns a result that does not match the grammar.
- *error*: fired when an error occurs during speech recognition. Speech recognition errors can be further studied by using the Speech RecognitionError interface that provides 8 error codes corresponding to the most common error events.[11]
- *start*: fired when the recognition service has begun to listen to the audio with the intention of recognizing the input.
- *end*: fired when the service has been disconnected.

The list of recognition results is managed in the Web SAPI by a `SpeechRecognitionList` object, which is a collection of one or more `SpeechRecognitionResult` objects depending on whether it was a non-continuous or continuous recognition. The `SpeechRecognitionList` has a `length` attribute (size of the list), and the item getter, which returns the `SpeechRecognitionResult` in a certain position in the list. The `SpeechRecognitionResult` has two attributes: `length` (number of recognition hypotheses) and `final` (to check whether it is a final or interim recognition result), and the item getter that allows access to each of the elements in the best list.

For example, in a one-shot speech recognition scenario, the user says "hello how are you?" and there are the following three best recognition results:

hello how are to?
hello Howard you?
hey Jo are you?

In this example, as it is a one-shot interaction, the system does not generate interim results so the `SpeechRecognitionList` will have `length=1`, and if we access the single `SpeechRecognitionResult` it contains, it will have `length=3` (as the *n*-best list contains 3 results), and `final=true` (as it is the final system guess, not an interim result that may be corrected during continued interaction). In a scenario of continuous speech recognition, the recognizer may correct interim recognition results, as shown in the Chrome demo.[12] Thus, there may be `SpeechRecognitionLists` with lengths greater than 1, and the `SpeechRecognitionResults` may not be final.

---

[11] https://dvcs.w3.org/hg/speech-api/raw-file/tip/speechapi.html#speechreco-error.   Accessed February 21, 2016.

[12] https://www.google.com/intl/en/chrome/demos/speech.html. Accessed February 21, 2016.

These objects can be accessed via a `SpeechRecognitionEvent` using its `results` attribute. In addition, the `SpeechRecognitionAlternative` object presents a simple view of the results with two attributes: `transcript` (a concatenation of the consecutive `SpeechRecognitionResults`) and confidence (a value in the range [0, 1], where a higher number indicates that the system is more confident in the recognition hypothesis).

We will now show how to use these objects in simple programs. In order to run your applications, you must:

- Upload them to a server. It can be a local server (e.g., using Apache).[13]
- Use Chrome and give permission to use the microphone when prompted.
- It is also desirable to set the Javascript console to debug errors.

In Code 6.6, when the `onresult` event is thrown, it generates a `SpeechRecognitionEvent` called `event`, and we query the results attached to the event (`SpeechRecognitionList`), which is a list of size `length`. Each of the elements of that list (`SpeechRecognitionResult`) has a list of N elements corresponding to the n-best recognition results (in this case $N <= 5$). Thus, the result attached to the event is an array of size `length`x5. In the `for` loop, we choose the 1-best element (position 0) for each recognition result and concatenate it to form the final `transcript` shown in the interface.

Thus, each element of `event.results` is a recognized word along with its alternatives. In this way, `event.results[0]` is the n-best list corresponding to the first word recognized and `event.results[0][0]` is the most probable hypothesis for that first word (1-best) (Code 6.5).

In Code 6.6 we present an example in which we show the n-best alternatives, confidence values, and intermediate recognition results.

This can be easily translated into a selection list as shown in Code 6.7. As can be observed, we only have to include the text corresponding to the transcription of each recognition result (`result[i].transcript`) into the options of a selection HTML form item (`select.options`).

Finally, Code 6.8 presents an example with continuous speech recognition that shows the difference between final and interim recognition results.

6.3   The Android Speech APIs

To add voice to our Android apps, we will work with two packages from the Android Speech API: `android.speech` and `android.speech.tts`.

`Android.speech.tts` contains, among other things, the following classes and interfaces:

---

[13]Remember that we have uploaded to http://lsi.ugr.es/zoraida/conversandroid for you, so that you do not have to set a server to try the examples provided in the book.

```
<body>
  (...)
  <button id="start_button"

    onclick="startListening(event)">Press and talk</button>
  <br/><br/>You have said...
  <br/><label id="feedback" name="feedback">...</label>
</body>

<script>
  var recognition = new webkitSpeechRecognition();

  recognition.onresult = function(event) {
    final_transcript = ' ';
    for (var i = 0; i < event.results.length; ++i) {
      final_transcript += event.results[i][0].transcript;
    }
    feedback.innerHTML = final_transcript;
  };

  function startListening(event) {
    recognition.lang = 'en-EN';
    recognition.continuous = false;
    recognition.maxAlternatives = 5;
    recognition.start();
  }
</script>
```

Code 6.5 Fragment from `asr_simple.html`

```
recognition.onresult = function(event) {
  final_transcript = '';
  for (var i=0; i < event.results.length; ++i) {
    for(var j=0; j<recognition.maxAlternatives; ++j){
      final_transcript +=

        'results['+i+']['+j+'] - Transcript=\''

        + event.results[i][j].transcript+ '\' - Confidence='

        + event.results[i][j].confidence + '<br/>';
    }
  }
  feedback.innerHTML = final_transcript;
};
```

Code 6.6 Fragment from `asr_details.html`

- `TextToSpeech` (which provides access to TTS).
- `UtteranceProgressListener` (which provides access to the progress of an utterance through the TTS queue).

```
<body>
  (...)

  <button id="start_button" onclick=

    "startListening(event)">Press and talk</button>
    <br/><br/>Choose the result you prefer:

  <select id="select"></select>
</body>

(...)
<script>
  var recognition = new webkitSpeechRecognition();

  recognition.onresult = function(event) {
    if (event.results.length > 0) {
      var result = event.results[0];
      for (var i = 0; i < result.length; ++i) {
        var text = result[i].transcript;
        select.options[i] = new Option(text, text);
      }
    }
  }

  (...)
</script>
```

Code 6.7 Fragment from `asr_optionlist.html`

- `TextToSpeech.Engine` and `TextToSpeech.EngineInfo` which, respectively, provide constants and parameters for controlling the TTS, and information about the TTS engines installed.

  `Android.speech` contains the interfaces and classes for ASR, including:

- `RecognizerIntent` (to support ASR by starting an Intent).
- `SpeechRecognizer` (the class that provides access to the ASR service).
- `RecognitionListener` (an interface to receive notifications from the ASR).

In the following sections, we describe how to use these resources to add TTS and ASR capabilities to Android apps. Remember that you can find the code in GitHub, in the ConversationalInterface repository, in the folder called *chapter6*. It can be opened and executed in Android Studio. Bear in mind that ASR is not available on virtual devices, so you will need to use an Android device to execute the files. Also Android Studio supports Git as a version control system, so you may find it handy to directly fetch the code from our GitHub repository into your Android Studio.

```
recognition.onresult = function(event) {
  var final = "";
  var interim = "";
  for (var i = 0; i < event.results.length; ++i) {
    if (event.results[i].final) {
      final += event.results[i][0].transcript;
    } else {
      interim += event.results[i][0].transcript;
    }
  }
  feedback_final.innerHTML += final;
  feedback_interim.innerHTML += interim;
};
```

Code 6.8 Fragment from `asr_continuous.html`

6.3.1 Text-to-Speech Synthesis

The components of the Google TTS API (package `android.speech.tts`) are documented at this Web page.[14] Interfaces and classes are listed, and further details can be obtained by clicking on these.

TTS has been supported in Android since API level 4 and all devices that are above this level come with a TTS engine. However, some devices have limited storage and capabilities and may lack some TTS resources or may not have certain languages available, which is why it is good practice to start TTS by checking the availability of the TTS engine.

As can be observed in the code of the `SimpleTTS` app, the first thing we do once the activity is initialized is to invoke the `initTTS` method to check whether there is a TTS engine installed in the device. In order to do that, we create an `Intent` and start it by indicating that we expect to receive a result. An `Intent` is an abstract representation of an action corresponding to an activity or service that can be initiated from other activities.[15] In our case, we will use the `ACTION_CHECK_TTS_DATA` Intent that is already provided by the Android API in the `TextToSpeech` class. Then, we start it with the `startActivityForResult` method, which receives as arguments the Intent and an integer request code that univocally identifies it. We have declared a constant `TTS_DATA_CHECK` as the request code to identify the check TTS installation Intent. Once the check is finished, the `onActivityResult` method is transparently invoked (Code 6.9).

The `onActivityResult` method is invoked by all Intents initiated with `startActivityForResult`. Thus, the first thing we must do is to check the Intent for which we are currently receiving a result. To do this, we can check that the

---

[14]http://developer.android.com/reference/android/speech/tts/package-summary.html.   Accessed February 22, 2106.

[15]If you are not familiar with the term, you can learn more here: http://developer.android.com/guide/components/intents-filters.html. Accessed February 22, 2016.

```
private int TTS_DATA_CHECK = 12; // Request code

(...)

private void initTTS() {
(...)
//Check if the engine is installed, when the check is

   finished, the onActivityResult method is automatically

   invoked
   Intent checkIntent =

       new Intent(TextToSpeech.Engine.ACTION_CHECK_TTS_DATA);
       startActivityForResult(checkIntent, TTS_DATA_CHECK);
}
```

Code 6.9 Create and start and Intent that looks for a TTS engine in the device (fragment from SimpleTTS.java)

request code received corresponds to the request code used to initiate the Intent (in our case TTS_DATA_CHECK). Then, we manage the result of the Intent. If there is a TTS engine installed in the device (the result code is CHECK_VOICE_DATA_PASS), we create a TextToSpeech object and save it in the variable mytts. Now, we can use mytts everywhere else in the code when we need to use the TTS engine. When the TTS engine is initialized, we show a Toast[16] and set the synthesis language to US English. If there is no TTS engine installed, then we try to launch its installation from Google Play (Code 6.10).

Now that we can control the TTS Engine through mytts, we can synthesize texts as we wish. The SimpleTTS app (see Fig. 6.1) shows a simple behavior: The user can type a message in a text field and press a button[17] to hear it synthesized back.

This behavior is controlled in a clickListener method attached to the button where the speak method is invoked. The speak method used to accept three parameters, but that version has been recently deprecated and now it accepts four:

1. The text to be synthesized;
2. The queuing strategy; that is, as it is possible that the TTS engine is busy synthesizing other texts, we must indicate whether we add the new message to the queue or we remove the current entries in the queue to process it;

---

[16]A Toast is a small text pop up shown on the device to provide simple feedbacks and notifications. Learn more in: http://developer.android.com/guide/topics/ui/notifiers/toasts.html. Accessed February 22, 2016.

[17]The input_text and speak_button elements are defined in the simpletts.xml layout file.

```
protected void onActivityResult(int requestCode, int result-
Code, Intent data) {


// Check that the result received is from TTS_DATA_CHECK
if (requestCode == TTS_DATA_CHECK) {
   // There is a TTS Engine available in the device
   if (resultCode==TextToSpeech.Engine.CHECK_VOICE_DATA_PASS){
      // Create a TextToSpeech instance
      tts = new TextToSpeech(this, new OnInitListener() {
         public void onInit(int status) {
            if (status == TextToSpeech.SUCCESS) {
               Toast.makeText(SimpleTTS.this, "TTS initialized",

                  Toast.LENGTH_LONG).show();
               if (tts.isLanguageAvailable(Locale.US) >= 0)
                  tts.setLanguage(Locale.US);
            }

            (...)

      }});
   } else {
      // The TTS is not available, we will try to install it:
      Intent installIntent = new Intent();
      installIntent.setAction

         (TextToSpeech.Engine.ACTION_INSTALL_TTS_DATA);
      (...)

      startActivity(installIntent);
   }
}}
```

Code 6.10 Manage the result for the Intent that looks for a TTS engine in the device (fragment from SimpleTTS.java)

3. Specific synthesis parameters to tune the engine;
4. A String identifier for the synthesis request. To avoid errors, we have to check the SDK version to use the new or old version.

However, starting TTS with an Intent in this way does not make full use of the potential provided by the API. In order to do that, we can employ a more sophisticated approach. The RichTTS app shows how to do it. On the one hand, we have added a listView to select the synthesis language and a slider to select the volume level (Fig. 6.2).

These elements could also have been added to SimpleTTS; that is, they can also be used with the Intent approach, as we have just added new code to our previous button click listener (Code 6.11) to set the language and volume (in bold in Code 6.12). As can be observed, now we are sending TTS parameters when

Fig. 6.1 Interface of the
`SimpleTTS` app

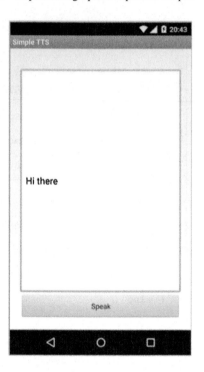

invoking the `speak` method. Since Android version Lollipop, these parameters are included in a `Bundle`, in earlier versions they are provided within a `HashMap`.

On the other hand, we now make use of the `UtteranceProgressListener` class and the `OnInitListener` interface instead of using an `Intent`.

`UtteranceProgressListener`[18] is an abstract class with methods that are automatically invoked:

- When an utterance starts being synthesized (onStart).
- When an utterance has successfully completed processing (onDone).
- When it has been stopped in progress or flushed from the synthesis queue (onStop).
- When an error occurs during processing (onError).

Most of these methods are abstract, and thus, their code must be provided in the subclasses. This makes sense, as there is no general management for situations such

---

[18]http://developer.android.com/reference/android/speech/tts/UtteranceProgressListener.html. Accessed February 22, 2016.

Fig. 6.2 Interface of the
RichTTS app

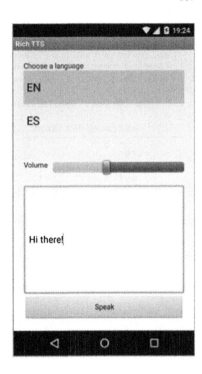

as these and so it is left to the programmer to indicate what to do when these events
happen. The challenge here is that our RichTTS class already is a subclass of
Activity (extends Activity), and as Java does not support multiple inheritance
(i.e., in Java a class can only have one superclass), it is not possible for RichTTS to
be an Activity and an UtteranceProgressListener simultaneously. To
get around this problem, we set an UtteranceProgressListener only for our
particular TextToSpeech instance (mytts) and indicate how each event will be
managed. In our case, we will just show toasts[19] (Code 6.13).

We must be careful when implementing the methods, as in different Android
versions different approaches are used. In versions earlier than API level 15 (Ice
Cream Sandwich), the deprecated OnUtteranceCompletedListener was
used. From API level 15 on, UtteranceProgressListener replaced it.
Recently, in API level 21 (Lollipop), the onError method changed, and now
instead of having only one parameter (the utterance id), it has two (utterance id and

---

[19]In order to show toasts from a listener, make sure that you run them on the UI thread.

```
speakButton.setOnClickListener(new OnClickListener() {
@Override
public void onClick(View v) {
  String text = inputText.getText().toString();
  if (text != null && text.length() > 0) {

    if (Build.VERSION.SDK_INT>=21)

      mytts.speak(text, TextToSpeech.QUEUE_ADD, null,"msg");

    else //Deprecated

      mytts.speak(text, TextToSpeech.QUEUE_ADD, null);
  }
}});
```

Code 6.11 Initiate synthesis when the button is pressed (fragment from `SimpleTTS.java`)

```
public void onClick(View v) {
  String text = inputText.getText().toString();
  if (text != null && text.length() > 0) {
    //Set the language selected from the listView

    if(language!=null)
      mytts.setLanguage(new Locale(language));

    //Read volume from seekbar. The seekbar allows choosing

      integer values from 0 to 10, we must translate them to

      a float from 0 to 1
    int vol=

      ((SeekBar)findViewById(R.id.volumeBar)).getProgress();
    float volumeLevel = (float) vol/10;


    if(Build.VERSION.SDK_INT >=21) {
      Bundle tts_params = new Bundle();
      tts_params.putFloat(TextToSpeech.Engine.KEY_PARAM_VOLUME,

        volumeLevel);
      mytts.speak(text, TextToSpeech.QUEUE_ADD,

        tts_params,"msg");

    } else {
      HashMap<String,String> tts_params =

        new HashMap<String, String>();
      tts_params.put(TextToSpeech.Engine.KEY_PARAM_VOLUME,

        Float.toString(volumeLevel));
      mytts.speak(text, TextToSpeech.QUEUE_ADD, tts_params);
    }
  }
}
```

Code 6.12 Including language and volume when initiating speech synthesis (fragment from `RichTTS.java`)

```
mytts.setOnUtteranceProgressListener(


  new UtteranceProgressListener() {


    @Override
    public void onDone(String utteranceId) {
      runOnUiThread(new Runnable() {
        public void run() {
          Toast.makeText(RichTTS.this,"Mission accomplished",

             Toast.LENGTH_LONG).show();
        }
      });
    }

    @Override
    public void onError(String utteranceId, int errorCode) {
      runOnUiThread(new Runnable() {
        public void run() {
          Toast.makeText(RichTTS.this,"Oops! there was an

             error :S", Toast.LENGTH_LONG).show();
        }
      });
    }

    @Override
    public void onStart(String utteranceId) {
      runOnUiThread(new Runnable() {
        public void run() {
          Toast.makeText(RichTTS.this, "TTS started!",

             Toast.LENGTH_LONG).show();
        }
      });
    }

    (...)
});
```

Code 6.13 Managing TTS events with UtteranceProgressListener (fragment from the setTTS method of `RichTTS.java`)

error code). All these changes can be handled by checking the SDK version programmatically (see the `setTTS` method in `RichTTS.java`).

The second mechanism that we are using to make a richer use of TTS is to implement the `OnInitListener` interface that offers another abstract method (`onInit`) that is called to signal the completion of the TTS engine initialization. Again, we have implemented it to show a `toast` (Code 6.14).

```
public class RichTTS extends Activity implements
android.speech.tts.TextToSpeech.OnInitListener{

   (...)

  public void onInit(int status) {
    if (status != TextToSpeech.ERROR)
      Toast.makeText(RichTTS.this, "TTS initialized",

        Toast.LENGTH_LONG).show();
    else
      Log.e(LOGTAG, "Error initializing the TTS");
  }

}
```

Code 6.14 Implementation of the OnInitListener interface RichTTS.java

6.3.2 Speech Recognition

Similar to TTS, there are different ways in which ASR can be introduced in Android apps:

1. Using the RecognizerIntent class, by creating an Intent for ASR.
2. Using the SpeechRecognizer class to start the ASR service and implementing the RecognitionListener interface that provides methods that implement the management of the different events that may occur.

Both can use the same ASR engine, and thus, the ASR performance is equivalent when using either strategy. The main difference is that using an Intent (option 1) is easier to program and requires fewer lines of code. However, it provides a less rich event control compared with the use of the SpeechRecognizer class (option 2). Also if the app should be constantly listening, it is necessary to use option 2, as the Intent stops listening when the user remains silent for a certain period of time.

SimpleASR presents an example using option 1. The user can select the ASR parameters (language model and maximum number of results) and then press a button. When the button is pressed, the app starts listening and provides a list of recognition results along with their recognition confidence scores (Fig. 6.3).

In the onClickListener of the button, we first check whether ASR is available on the device by making sure it is not a virtual device and that the device is able to resolve the RecognizerIntent. If everything is ok, we set the recognition parameters and invoke the listen method in which we actually start the Intent (Code 6.15).

Fig. 6.3 Interface of the SimpleASR app

In the listen method, we can see how we create a RecognizerIntent to attempt to recognize speech for which many extras can be specified (see the specification of the RecognizerIntent class for more details[20]). In our app, we have just included the required EXTRA_LANGUAGE_MODEL and the EXTRA_MAX_RESULTS parameters. With respect to the language models, there are two options:

- LANGUAGE_MODEL_FREE_FORM for free-form speech recognition with a more dictation-like language model.
- LANGUAGE_MODEL_WEB_SEARCH that uses a language model based on Web search that is more keyword based and less determined by language modeling rules.

The EXTRA_MAX_RESULTS parameter is used to limit the maximum number of recognition hypotheses returned by the speech recognizer.

As the implementation of the API is likely to stream audio to remote servers where the ASR is actually performed, it is necessary to start the Intent with startActivityForResult and not just with startActivity. The result is processed in the onActivityResult method (Code 6.16).

---

[20]http://developer.android.com/reference/android/speech/RecognizerIntent.html.

```
public void onClick(View v) {

//Speech recognition does not currently work on simulated de-
vices,it the user is attempting to run the app in a simulated
device they will get a Toast

if("generic".equals(Build.BRAND.toLowerCase())){
  Toast toast = Toast.makeText(getApplicationContext(),

  "ASR is not supported on virtual devices",

  Toast.LENGTH_SHORT);
  toast.show();
  Log.d(LOGTAG, "ASR attempt on virtual device");

}else{

  List<ResolveInfo> intActivities =

    packM.queryIntentActivities(

    new Intent(RecognizerIntent.ACTION_RECOGNIZE_SPEECH), 0);

  //If speech recognition is supported, then set parameters

    and start listening

  if (intActivities.size() != 0) {

    setRecognitionParams();

    listen();


  //If speech recognition is not supported
  } else {
      Toast toast = Toast.makeText(getApplicationContext(),

        "ASR not supported", Toast.LENGTH_SHORT);
      toast.show();
      Log.d(LOGTAG, "ASR not supported");
  }

}}
```

Code 6.15 Checking that speech recognition is supported (fragment from SimpleASR.java)

In the onActivityResult method, if the ASR is successful (RESULT_OK), two main pieces of data are retrieved:

- The *n*-best list of recognition hypotheses, which is a collection of Strings representing the alternative phrases recognized by the system.

```
private void listen()   {
  Intent intent = new

      Intent(RecognizerIntent.ACTION_RECOGNIZE_SPEECH);
  // Specify language model
  intent.putExtra(RecognizerIntent.EXTRA_LANGUAGE_MODEL,

      languageModel);
   // Specify max number of recognition results

  intent.putExtra(RecognizerIntent.EXTRA_MAX_RESULTS,

      numberRecoResults);

   // Start listening
  startActivityForResult(intent, ASR_CODE);

}
```

Code 6.16 Setting the ASR parameters (fragment from SimpleASR.java)

- A vector of floats representing the confidence scores corresponding to each of the n recognition hypotheses. This information has only been available since API level 14 (Ice Cream Sandwich), so we must make sure that we do not try to request it in devices that do not support it.

In our app, these two pieces of information are combined in the output to show a list view in the graphical interface with all the recognition hypotheses and their confidence scores (Code 6.17).

The RichASR app presents an implementation that uses option 2 and thus offers a more fine-grained coverage of recognition events. In our case, we use a simple method to handle these events by displaying messages in a TextView on the interface, as shown in Fig. 6.4.

In the RichASR app, we will use a SpeechRecognition object called myASR that is initialized when the activity starts, after checking that the device supports speech recognition (Code 6.18).

The listen method, which is invoked when pressing the button, is similar to the previous example, but this time we also indicate the recognition language. However, in this case, we do not use startActivityForResult, but use the Intent as a parameter to the startListening method of a SpeechRecognizer object that we have called myASR (Code 6.19).

As shown in Code 6.18, it is possible to set a RecognitionListener for the SpeechRecognizer object. This involves implementing the RecognitionListener interface (RichASR implements RecognitionListener), which defines numerous methods that are invoked when different ASR events occur. For example:

```
protected void onActivityResult(int requestCode, int result-
Code, Intent data) {


if (requestCode == ASR_CODE)   {
 if (resultCode == RESULT_OK)   {
   if(data!=null) {
         //Retrieves the N-best list and the confidences from

            the ASR result

      ArrayList<String> nBestList =

         data.getStringArrayListExtra

         (RecognizerIntent.EXTRA_RESULTS);


      float[] nBestConfidences = null;

      if (Build.VERSION.SDK_INT >= 14)

        nBestConfidences = data.getFloatArrayExtra

           (RecognizerIntent.EXTRA_CONFIDENCE_SCORES);


      ArrayList<String> nBestView = new ArrayList<String>();

      for(int i=0; i<nBestList.size(); i++){
         if(nBestConfidences!=null){
            if(nBestConfidences[i]>=0)
              nBestView.add(nBestList.get(i) + " (conf: " +

                 String.format("%.2f", nBestConfidences[i])

                 + ")");
            else
              nBestView.add(nBestList.get(i) +

                 " (no confidence value available)");
         } else
           nBestView.add(nBestList.get(i) +

                 " (no confidence value available)");
      }
         //Includes the collection in the ListView of the GUI
      setListView(nBestView);
(...)

}}}}
```

Code 6.17 Retrieving speech recognition results (fragment from SimpleASR.java)

Fig. 6.4 Interface of the
RichASR app

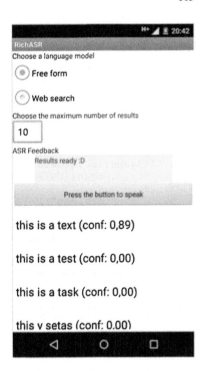

- When the ASR engine is ready to listen (onReadyForSpeech).
- When it encounters an error (onError).
- When it detects speech (onBeginningOfSpeech).
- When speech stops (onEndOfSpeech).
- When it receives buffered and partial results (onBufferReceived, onPartialResults).
- When it finishes recognizing (onResults).

We have implemented a very simple behavior for these methods in RichASR, which involves showing a message in a TextView in the interface (called feedbackTxt), except that with the onResults method we show the ListView with the recognition results and their confidences.

Each time that ASR is used in an app, either with Intent or with SpeechRecognition (option 1 or 2), the following permissions must be specified in the Manifest file:

- Recording audio, thus giving access to the device microphone.
- Including access to the Internet to stream audio for speech processing.

```
public void initASR() {

    // find out whether speech recognition is supported
    List<ResolveInfo> intActivities =

        this.getPackageManager().queryIntentActivities(
        new Intent(RecognizerIntent.ACTION_RECOGNIZE_SPEECH), 0);


    //Speech recognition does not work on simulated devices

    if("generic".equals(Build.BRAND.toLowerCase())){
            Log.e(LOGTAG, "ASR is not supported on virtual

            devices");
    } else {
      if (intActivities.size() != 0) {
        myASR = SpeechRecognizer.createSpeechRecognizer

          (getApplicationContext());
        myASR.setRecognitionListener(this);
      }
    }
}
```

Code 6.18 Instantiating a SpeechRecognizer in `RichASR.java`

- Checking the Internet connection to check whether the device is connected to Internet before attempting to start ASR (Code 6.20).

Previously when the permissions were declared in the Manifest and users wanted to install the app, they would be warned about the permissions it required and these would be granted if they continued with the installation. However, since Android Marshmallow (API level 23), users can grant permissions to apps while the apps are running. According to Google[21] this gives more control to users over the app as they may grant some permissions (e.g., accessing the contact list) and revoke others (e.g., tracking the device's location), and they can do this at any time and not only during installation.

Thus, as permissions may be revoked at any time, we cannot take for granted that our apps have the permissions that are required to perform certain tasks. In particular, when using ASR, we cannot assume that our app has permission to record audio in the user's device. That is why we have created the `checkASRPermission` method, which is invoked every time we attempt to listen. As can be observed in Code 6.21, we check whether the permission is granted, and if not, we explicitly request it. This request is an asynchronous process for which the callback is the `onRequestPermissionsResult` method that informs the user whether the permission was granted or denied.

---

[21] http://developer.android.com/intl/es/training/permissions/requesting.html.

```
public void listen(final Locale language, final

  String languageModel, final int maxResults)
{

(...)

  Intent intent =

    new Intent(RecognizerIntent.ACTION_RECOGNIZE_SPEECH);

    // Specify the calling package to identify the application
    intent.putExtra(RecognizerIntent.EXTRA_CALLING_PACKAGE,

      getPackageName());

    // Specify language model
    intent.putExtra(RecognizerIntent.EXTRA_LANGUAGE_MODEL,

      languageModel);

    // Specify how many results to receive

    intent.putExtra(RecognizerIntent.EXTRA_MAX_RESULTS,

      maxResults);

    // Specify recognition language
    intent.putExtra(RecognizerIntent.EXTRA_LANGUAGE,

      language);

  myASR.startListening(intent);

(...)
}
```

Code 6.19 Fragment of the listen method in `RichASR.java`

```
<uses-permission android:name=

  "android.permission.INTERNET" />
<uses-permission android:name=

  "android.permission.RECORD_AUDIO"/>
<uses-permission android:name=

  "android.permission.ACCESS_NETWORK_STATE" />
```

Code 6.20 Fragment of the AndroidManifest.xml file in `SimpleASR` and `RichASR`

6.3.3   Using Speech for Input and Output

Now, we are ready to combine rich ASR and TTS management into an app that has speech input and output. We already have the necessary resources, and the

```
public void checkASRPermission() {
  if (ContextCompat.checkSelfPermission(

     getApplicationContext(),

     Manifest.permission.RECORD_AUDIO) !=

              PackageManager.PERMISSION_GRANTED) {
     // If  an explanation is required, show it
     if (ActivityCompat.shouldShowRequestPermissionRationale(

     this, Manifest.permission.RECORD_AUDIO))

       Toast.makeText(getApplicationContext(), "SimpleASR

         must access the microphone in order to perform

         speech recognition", Toast.LENGTH_SHORT).show();
     // Request the permission.

     // Callback in "onRequestPermissionResult"
     ActivityCompat.requestPermissions(this,

       new String[]{Manifest.permission.RECORD_AUDIO},
       MY_PERMISSIONS_REQUEST_RECORD_AUDIO);

  }
}
```

Code 6.21 Fragment to check the permission required to recognize speech

challenge is now to create a mechanism to make our code reusable so that we can use the same speech management code for every app that we want to build. We show how this can be done in the TalkBack app. The app contains two java files: VoiceActivity and MainActivity. VoiceActivity contains the code to process ASR and TTS with a rich event control, and MainActivity provides an example showing how it can be used within a particular app.

VoiceActivity is an abstract class created so that it can be extended by the main activity of apps that include speech input and output. The initialization, listen, and speak methods are already implemented, and developers using this class only need to provide code for the abstract methods that indicate how to respond to the different ASR and TTS events, a management issue that changes from app to app.

VoiceActivity arranges together the code that uses SpeechRecognizer and TextToSpeech in order to manage ASR and TTS. As the names of the ASR and TTS events are similar (e.g., onError), we have redirected some methods to others, for example, in pieces of code such as Code 6.22.

As can be observed now that we are paying attention to the complete picture, the reason for these redirections is to "rename" the methods so that they can be easily understood outside the VoiceActivity class. Thus, public void onError

```
//TTS

myTTS.setOnUtteranceProgressListener(new
UtteranceProgressListener()
{

  @Override
  public void onError(String utteranceId) {
    onTTSError(utteranceId);
  }

  (...)

}

//ASR

(...)

@Override
public void onError(int errorCode) {
  processAsrError(errorCode);
}
```

Code 6.22 Fragment of the VoiceActivity class (TalkBack app)

(String utteranceId) and public void onError(int errorCode)
are very similar, but public void onTTSError(String utteranceId)
and public void processASRError(int errorCode) are very easily
distinguishable.

As can be observed, the methods to manage the TTS events (onTTSDone,
onTTSError and onTTSStart) have not been implemented in the
VoiceActivity class and thus are defined as abstract. For ASR, we are just
interested in some of these situations, managed by the following abstract methods:

- processAsrResults processes recognition results.
- processAsrReadyForSpeech to process the situation when the ASR is
  ready to listen.
- processError which is used not only to cope within the onError event, but
  also when the ASR result was NOMATCH; that is, the engine heard speech, but it
  could not be linked to any recognizable phrase.

The idea is that the actual code to manage these events is provided as near to the
interface as possible, in the class that implements VoiceActivity. This way
VoiceActivity is very versatile and can be used as a basic structure to include
speech input and output capabilities in any app.

In the onRequestPermissionsResult (Code 6.21) that we have implemented in VoiceActivity, we now invoke the abstract methods
showRecordPermissionExplanation and onRecordAudio
PermissionDenied, so that each app can, respectively, implement its own

management of the situation in which it is necessary to explain why the app must have a certain permission and when it is not possible to perform speech recognition because the user does not grant access to the microphone.

We have also encoded a more detailed control of the locale. When using TTS, it is important to indicate the language that is being employed, as even a single word may be pronounced differently in different languages (e.g., "Paris" is pronounced differently in French and English), or even in the same language, (e.g., "tomato" has different pronunciations in British and American English). In the VoiceActivity class, we have included different methods to manage different situations:

- If the user indicates both language and country code (e.g., en-US), then we try to use them both.
- If only the language code is indicated (e.g., en), then the default country code, which is determined by checking the device settings, is used.
- If the language is not specified, then the language and country are determined according to the device settings.

As it may happen that a locale is specified but is not available in the device, we check that the language is available using the isLanguageAvailable method of the TextToSpeech class (Code 6.23).

Another detail that we have not yet discussed is the inclusion of an utterance id in the speak methods. This has been included to have a means of controlling

```
public void setLocale(String languageCode,

                       String countryCode) throws Exception{
  if(languageCode==null) {
    setLocale();
      throw new Exception("Language code was not provided,

      using default locale");
  } else {
    if(countryCode==null)
      setLocale(languageCode);
    else {
      Locale lang = new Locale(languageCode, countryCode);
      if (myTTS.isLanguageAvailable(lang) ==

          TextToSpeech.LANG_COUNTRY_VAR_AVAILABLE )
        myTTS.setLanguage(lang);

      else {
        setLocale();
        throw new Exception("Language or country

          code not supported, using default locale");
      }
    }
  }
}
```

Code 6.23 Setting Locale for TTS (fragment of the VoiceActivity class in TalkBack app)

```
@Override
public void onTTSDone(String uttId) {
    if(uttId.equals(ID_PROMPT_QUERY.toString())) {
        runOnUiThread(new Runnable() {
            public void run() {
                startListening();
            }
        });
    }
}
```

Code 6.24 Using the prompt IDs to start listening after a question is posed (fragment of the VoiceActivity class in TalkBack app)

cases such as which utterance is ready to be synthesized; which utterance has already been synthesized; or whether an error was encountered while synthesizing.

MainActivity presents an example of how to use the VoiceActivity class. It is a very simple application that asks the user to say something, and then, it synthesizes back the recognized String, but it shows clearly how ASR and TTS can be synchronized using the event control.

MainActivity is defined as a subclass of VoiceActivity and thus must implement all the abstract methods; that is, it must say how to manage the TTS and ASR events. In this case, it is simple: When the user presses the button (onClick), we synthesize a message asking the user to say something; when the synthesis is finished (onTTSDone), we start listening; and when we obtain recognition results (processASRResults), we synthesize back what was recognized by the system. Also, we process other events (e.g., when the system is ready to start listening —processAsrReadyForSpeech) to change the appearance of the button so that the user knows when the system is listening.

We have defined two ids for the utterances: ID_PROMPT_QUERY and ID_PROMPT_INFO to distinguish prompts functioning as questions from prompts functioning as statements or information providing. Using these identifiers, when a prompt is synthesized and the onTTSDone method is executed, if the prompt was a question, we immediately start listening, which is not the case when the prompt was a statement (Code 6.24).

6.4   Summary

In this chapter, we have shown how speech input and output can be processed in Android devices. We have covered different approaches from simple Intents to more complex processing using specific classes and interfaces related to speech recognition and synthesis that cover the most frequent events. Finally, we have discussed an application that considers all the different factors and have created a class called VoiceActivity that can be used for any app that uses speech input and output. This code will be used in other chapters of the book as the speech front-end to many different services.

Further Reading

We have focused in this chapter on the facilities provided by Google, but there are many tools that can be used for speech recognition and synthesis within Android. Some well-known tools for text-to-speech synthesis and speech recognition are listed in Tables 6.1 and 6.2.

Wolf Paulus has created an Android app called Horsemen of Speech Recognition that integrates several ASR engines into a single app.[22] You can try different inputs with each ASR engine and examine the different results returned.

Web Speech API Tutorials

Tutorials with topics already covered in this chapter:

- Voice-Driven Web Apps: Introduction to the Web Speech API.[23]
- How to use the Web Speech API.[24]
- Using the Web Speech API to Create Voice-Driven HTML5 Games.[25]
- Working with the Web Speech API.[26]

  More advanced tutorials:

- Auto-translate. This tutorial uses the Web Speech API to input speech from the microphone (speech to text) and speech synthesis (text to speech) to play back your translated speech using the Google Translate API.[27]
- Using Voice to Drive the Web: Introduction to the Web Speech API.[28]

Exercises
Web Speech API

1. Prosody: Implement a Web form with a text area to input a text, different elements to choose prosody features, and a button to start synthesizing the text using the prosody selected. Hint.[29]

---

[22]https://play.google.com/store/apps/details?id=com.techcasita.android.reco. Accessed February 22, 2016.

[23]http://updates.html5rocks.com/2013/01/Voice-Driven-Web-Apps-Introduction-to-the-Web-Speech-API. Accessed February 22, 2016.

[24]http://stiltsoft.com/blog/2013/05/google-chrome-how-to-use-the-web-speech-api/. Accessed February 22, 2016.

[25]http://html5hub.com/using-the-web-speech-api/#i.1vh8jnvry3ex4w. Accessed February 22, 2016.

[26]http://grahamhinchly.wordpress.com/2013/11/14/working-with-the-web-speech-api/. Accessed February 22, 2016.

[27]http://www.moreawesomeweb.com/demos/speech_translate.html. Accessed February 22, 2016.

[28]http://www.adobe.com/devnet/html5/articles/voice-to-drive-the-web-introduction-to-speech-api.html. Accessed February 22, 2016.

[29]http://www.broken-links.com/tests/webspeech/synthesis.php. Accessed February 22, 2016.

2. Voices: Implement a form that shows the voices available in your browser and allows you to select one of them for speech synthesis. Hints:

  – Demo of Web Speech API Speech Synthesis interface (text to speech).[30]
  – Getting started with the Speech Synthesis API.[31]
  – Web Apps that talk—Introduction to the Speech Synthesis API.[32]

3. Recognition configuration: Implement a Web form in which the speech recognition parameters can be configured from the interface.

Android

1. Engines: Extend the RichTTS app with a ListView in which the user can choose the TTS engine employed. Hint.[33]
2. TTS to file: Extend the RichTTS app to save the synthesized message into a file. Hint.[34]
3. Web search: Extend the TalkBack app to perform a Web search with the recognition result obtained. Hint.[35]

---

[30]http://html5-examples.craic.com/google_chrome_text_to_speech.html. Accessed February 22, 2016.

[31]http://blog.teamtreehouse.com/getting-started-speech-synthesis-api. Accessed February 22, 2016.

[32]https://developers.google.com/web/updates/2014/01/Web-apps-that-talk-Introduction-to-the-Speech-Synthesis-API. Accessed February 22, 2016.

[33]http://developer.android.com/reference/android/speech/tts/TextToSpeech.html#getEngines(). Accessed February 22, 2016.

[34]http://developer.android.com/reference/android/speech/tts/TextToSpeech.html#synthesizeToFile (java.lang.CharSequence,android.os.Bundle,java.io.File,java.lang.String). Accessed February 22, 2016.

[35]http://developer.android.com/reference/android/content/Intent.html#ACTION_WEB_SEARCH.

Chapter 7
Creating a Conversational Interface Using Chatbot Technology

Abstract Conversational interfaces can be built using a variety of technologies. This chapter shows how to create a conversational interface using chatbot technology in which pattern matching is used to interpret the user's input and templates are used to provide the system's output. Numerous conversational interfaces have been built in this way, initially to develop systems that could engage in conversation in a human-like way but also more recently to create automated online assistants to complement or even replace human-provided services in call centers. In this chapter, some working examples of conversational interfaces using the Pandorabots platform are presented, along with a tutorial on AIML, a markup language for specifying conversational interactions.

7.1 Introduction

In Chap. 6, we showed how to add speech input and output to a mobile app using the Google Speech APIs. However, speech input and output are only one part of the tasks that we might require from a conversational interface. Our query might be about the weather in London or for directions to the nearest Starbucks. We will want our query to be interpreted by the conversational interface as a request to answer a question or to carry out some action. We will also want the app to respond with something related to what we asked for, such as a spoken answer to our question about the weather or a display of information such as a map with the requested directions.

Consider once again the components of a spoken language-based conversational interface that we described in Chap. 2 (Fig. 7.1).

As we can see, once the user's input has been recognized by the speech recognition component, it has to be interpreted in order to determine its meaning. In some approaches, this might involve a thorough analysis of the input using techniques from spoken language understanding (SLU)—for example, a grammar to represent the permissible inputs and a parser to apply the grammar to the input and to extract a semantic representation. Then, the dialog manager has to decide what actions to

© Springer International Publishing Switzerland 2016
M. McTear et al., *The Conversational Interface*,
DOI 10.1007/978-3-319-32967-3_7

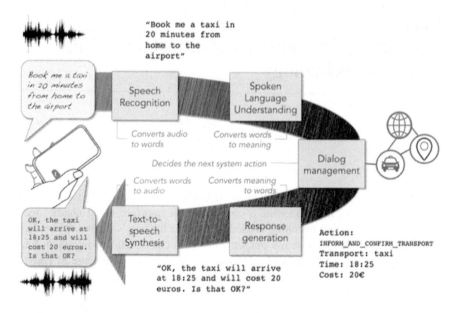

Fig. 7.1   The components of a spoken language-based conversational interface

DM   take, using as a basis this semantic representation and other relevant information such as the current context. This will result in some output being generated—words to be spoken and possibly output in other modalities, such as images, lists, and maps. For these tasks, additional components are required, in particular: SLU (see Chaps. 8 and 9), dialog management (DM) (see Chaps. 10 and 11), and response generation (RG) (see Chap. 12).

In this chapter, we will present a simpler approach that has been widely applied in chatbot technology, as illustrated in Fig. 7.2. In this approach, the input is matched against a large store of possible inputs (or patterns) and an associated response is outputted. The chatbot approach was first used in the ELIZA system (Weizenbaum 1966) and has continued until the present day in the form of apps that provide an illusion of conversation with a human as well as in areas such as education, information retrieval, business, and e-commerce, for example, as automated online assistants to complement or even replace human-provided services in a call center (see further Chap. 4).

More recently, chatbot technology has been extended to support the development and deployment of virtual personal assistants by incorporating methods for interpreting commands to the device or queries to Web services—for example, to search for information on the Internet, access information on the device, such as contacts and calendars, perform a task on the device such as launching an app, setting an alarm, or placing a call.

We will use the Pandorabots platform, a popular Web service that enables developers to create and host chatbots, to show how a conversational interface can be created with chatbot technology. We first introduce Pandorabots and then

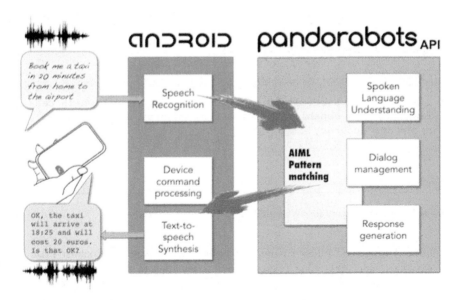

Fig. 7.2 Using Pandorabots for spoken language understanding, dialog management, and response generation

provide a brief overview of AIML (Artificial Intelligence Markup Language), which is used to specify conversations for a chatbot on the Pandorabots platform. Following this, we will show how to embed a Pandorabots chatbot in an Android app and how to provide speech input and output. We will then show how the chatbot can be extended to handle commands to the device and queries to Web services. In the final section, we will show how this approach can be further extended.

The code corresponding to the examples in this chapter is in GitHub, in the folder *chapter7* of the ConversationalInterface[1] repository.

7.2 Introducing the Pandorabots Platform

Pandorabots is a bot-hosting service launched in 2002 that enables chatbot developers (referred to in Pandorabots as botmasters) to develop, test, and deploy chatbots (or more simply bots) without requiring a background in programming.[2] AIML was developed by Dr. Richard Wallace as a language for specifying conversations with chatbots and was used by Wallace to develop the chatbot ALICE

---

[1]http://zoraidacallejas.github.io/ConversationalInterface/. Accessed March 2, 2016.
[2]http://www.pandorabots.com/. Accessed February 20, 2016.

```
<category>

<pattern> WHAT ARE YOU </pattern>

<template>

I am the latest result in artificial intelligence, which can
reproduce the capabilities of the human brain with greater
speed and accuracy.

</template>

</category>
```

Code 7.1 AIML category for "What are you?"

(Artificial Linguistic Internet Computer Entity) which won the Loebner Prize in 2000, 2001, and 2004. The Loebner prize is awarded to the chatbot that in an annual competition is considered by judges to be the most human-like. Other award winning bots developed using AIML include Mitsuki, Tutor, Izar, Zoe, and Professor. Currently more than 221,000 chatbots in many languages are hosted on the platform. The platform has recently been revamped so that in addition to the original chatbot-hosting server there are now facilities on a Developers Portal to support the deployment of chatbots on the Web and on mobile devices.

Many chatbots that are currently available on mobile devices were created using Pandorabots and AIML. These include the following: Voice Actions by Pannous (also known as Jeannie), Skyvi, Iris, English Tutor, BackTalk, Otter, and Pandorabot's own CallMom app. CallMom can perform the same sorts of tasks as other chatbots but also includes a learning feature so that it can learn personal preferences and contacts and can be taught to correct speech recognition errors. More information about Pandorabots and chatbots in general can be found at the ALICE A.I. Foundation site.[3] See also the Chatbots.org website.[4]

As mentioned in the previous section, in order to simulate conversation, chatbot technology makes use of pattern matching in which the user's input is matched against a large set of stored patterns and a response is output that is associated with the matched pattern. The technique was first used in the ELIZA system and has been deployed in subsequent chatbots ever since. Authoring a chatbot on Pandorabots involves creating a large number of AIML categories that at their most basic level consist of a pattern against which the user's input is matched and an associated template that specifies the chatbot's response. Code 7.1 is a simple example of an AIML category.

---

[3]http://www.alicebot.org/. Accessed February 20, 2016.
[4]https://www.chatbots.org/. Accessed February 20, 2016.

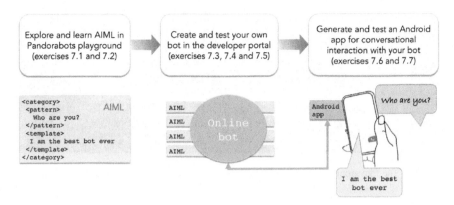

Fig. 7.3 The steps required to generate a bot

   In this example, if the user's input is matched against the text in the pattern, the system executes the contents of the template. Matching in AIML is done by the Graphmaster, which stores all the patterns in a graph. The graph consists of a collection of nodes called Nodemappers that map the branches from each node. The root of the Graphmaster is a Nodemapper with a large number of branches, one for each of the first words of the patterns. In the case of ALICE, there were around 40,000 patterns with about 2000 different first words (Wallace 2003). Matching involves traversing the graph on a word-by-word basis from the root to a terminal node. Interestingly, this process bears some similarity to the process of incremental processing by humans where an input sentence is analyzed on a word-by-word basis as opposed to waiting for the utterance to be completed (Crocker et al. 1999; see also Chap. 4). More detail about pattern matching in AIML can be found in Wallace (2003) and also at the AliceBot Web page.[5] A formal definition of the Graphmaster and matching is provided in Wallace (2009).
   Once a pattern has been matched, the contents of the associated template are output. In the example above, the contents are in the form of text, so the output is a message consisting of that text, but the template can also contain executable code and various tags that can be used to compute more complex and more flexible responses, as will be explained in Sect. 7.6.
   In the remainder of this chapter, we will provide a series of exercises showing the steps required to generate a bot (see Fig. 7.3).

Exercise 7.1: Creating a bot in Pandorabots Playground Pandorabots provides a Web-based platform called Playground for testing bots developed in AIML. In this section, we describe how to create an account in Playground and develop your

[5]http://www.alicebot.org/documentation/matching.html. Accessed February 20, 2016.

Fig. 7.4 The Playground home page (reproduced with permission from Pandorabots, Inc.)

first bot using a built-in library of AIML files. The AIML files for the bot that we will build in the next sections are in GitHub ConversationalInterface repository in the folder *chapter7/AIML*.

1. To access Playground, go to the Pandorabots Web page[6] and click on the Playground tab, or you can also go directly to the Playground page.[7] Here you can sign up for a 10-day free trial account and log in using your Facebook, Google, Twitter, or Yahoo account. You will find a QuickStart tutorial, a more detailed tutorial, and a tab for FAQs, as shown in Fig. 7.4.
2. Once you have created an account, you can start to create bots, interact with them and develop them further. The following are brief instructions. For more detailed instructions, see the Playground Web site.
3. Click on the My Bots tab to see a list of your bots. (If you have just signed up for an account, the list will be empty.)
4. Click on the Create Bot tab to create a bot. Give it a name, for example, "talkbot."
5. You will have a set of default AIML files.
6. To get started, you can make use of the chatbot base called Rosie, which is a collection of AIML files. Rosie provides conversational interaction. The files in the Rosie chatbot base allow you to get started quickly without having to write any AIML code. This page[8] includes a number of other useful resources. Go to the page and upload the files found under the lib directory.
7. Select your bot. This will bring up the editor screen for your bot (Fig. 7.5).
8. Click on the Train tab. Now you can test your bot by asking questions.

---

[6]http://www.pandorabots.com/. Accessed February 20, 2016.

[7]https://playground.pandorabots.com/en/. Accessed February 20, 2016.

[8]https://github.com/pandorabots/rosie. Accessed February 20, 2016.

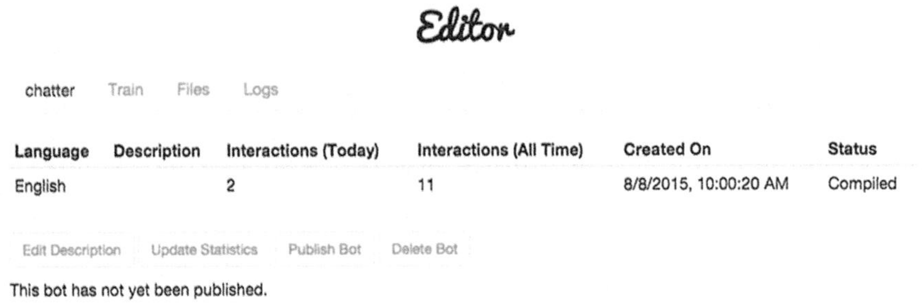

Editor

chatter   Train   Files   Logs

Language	Description	Interactions (Today)	Interactions (All Time)	Created On	Status
English		2	11	8/8/2015, 10:00:20 AM	Compiled

Edit Description   Update Statistics   Publish Bot   Delete Bot

This bot has not yet been published.

Fig. 7.5 The Playground editor screen (reproduced with permission from Pandorabots, Inc.)

7.3   Developing Your Own Bot Using AIML

As you may have noticed, the chatbot bases provided by Pandorabots cover many of the aspects of conversational speech. However, if you want your bot to be able to respond to questions in a specific domain, you will need to create new AIML code. In this section, we provide a brief tutorial on AIML that we will use as a basis for creating a chatbot that provides answers to frequently asked questions in the domain of type 2 diabetes. A more comprehensive tutorial can be found on the Pandorabots Playground page.

The following are some of the questions that a user might ask:

> What is type 2 diabetes?
> What are the main symptoms of type 2 diabetes?
> What is the main cause of type 2 diabetes?
> How do you treat type 2 diabetes?
> Can type 2 diabetes be cured?
> Does exercise help?
> Tell me about blood sugar.
> What are the long term complications?

To be able to answer questions such as these, we need to create categories that will specify the range of inputs and the answers that we wish to associate with them.

7.3.1   Creating Categories

A first step is to collect questions and their answers and to create categories consisting of patterns and templates, for example (Code 7.2):

```
<category>

<pattern>what is Type 2 diabetes</pattern>

<template>type 2 diabetes is a metabolic disorder that is
characterized by high blood glucose in the context of insulin
resistance and relative insulin deficiency</template>

</category>
```

Code 7.2 AIML category for "What is type 2 diabetes?"

```
<category>

<pattern>what are the causes </pattern>

<template>

   <srai>what is the main cause of type 2 diabetes</srai>

</template>

</category>
```

Code 7.3 AIML category for "What is type 2 diabetes?" including an <srai> tag

However, it will quickly become clear that there are many ways of asking the same question, for example:

What is the main cause of type 2 diabetes?
What causes type 2 diabetes?
What are the causes of type 2 diabetes?
What is the main cause?
What are the causes?
How do you get type 2 diabetes?

Rather than list all of these synonymous questions and their answers (which in this case would all be the same), AIML makes use of a technique in which synonymous questions are mapped on to one canonical pattern using the <srai> tag, as in (Code 7.3).

In this example, the bot's response is retrieved recursively by finding the response in the category that contains the pattern of the canonical pattern what is the main cause of type 2 diabetes. However, given the possible combinatorial explosion of synonymous questions, specifying all the required symbolic reductions remains a formidable task.

7.3.2 Wildcards

The use of wildcards also enables similar patterns to be mapped on to one canonical form. For example, the wildcard * matches one or more words in the input, so that the following patterns can be treated as variants at different levels of expressiveness:

> What are the main symptoms of type 2 diabetes?
> What are the * symptoms of type 2 diabetes?
> What are the * symptoms *?

Another wildcard ^ can match zero or more words, so that in the following code, the input "what are the main symptoms" would be matched along with other inputs including words after the word "symptoms" (Code 7.4).

The words captured by a wildcard can be repeated back in the template by using the `<star/>` tag. For example, in response to a question about symptoms, the human might say, "I feel tired," for which the pattern and template in Code 7.5 would cover any symptom mentioned.

This will allow an interaction such as:

> Human: I feel tired.
> Bot: When do you feel tired?

There can be more than one wildcard in a pattern, and they can be repeated back in the template by using `<star index = "x">`, where x corresponds to the position of the wildcard in the input (Code 7.6).

This will allow an interaction such as:

> Human: I feel tired and I have a headache.
> Bot: So you feel tired and you have a headache.

While wildcards are a useful way of avoiding redundancy in patterns, as in any other system using wildcards there is the potential for false positives, where an input may be matched against an unintended canonical pattern, resulting in

```
<pattern>what are the main symptoms ^ </pattern>
```

Code 7.4 AIML code using the ^ wildcard

```
<category>

<pattern>I feel * </pattern>

<template>When do you feel <star/></template>

</category>
```

Code 7.5 AIML code using the * wildcard and the `<star/>` tag

```
<category>

  <pattern>I feel * and I * </pattern>

  <template>So you feel <star/> and you

    <star index = "2" /> </template>

</category>
```

Code 7.6 AIML code using the `<star/>` tag at several positions

unexpected system responses. Looking again at Code 7.5, if the human inputs a time expression, this will be matched and repeated in the bot's response, resulting in an incongruous output:

> Human: I feel tired when I get up in the morning.
> Bot: When do you feel tired when you get up in the morning?

Careful testing of patterns would help to avoid this sort of problem, but it may be difficult to test exhaustively for every possible input and how it would be matched against patterns containing wildcards.

7.3.3   Variables

Variables can be used to set and retrieve information from an interaction that is stored as a predicate, for example (Code 7.7). The value of the `medication` predicate can be retrieved later using `<get>`, as in (Code 7.8).

```
<category>

  <pattern>my main medication is * </pattern>

  <template>OK, I have noted that your main medication is

    <set name = "medication"<star/></set> </template>

</category>
```

Code 7.7 AIML code that defines a predicate called "medication"

```
<template>you said that your main medication is <get name =
"medication"/></template>
```

Code 7.8 AIML code that uses the value of a predicate

7.3.4 Sets and Maps

Repetition can be avoided in patterns where a word forms part of a set—for example, all color words can be declared in a set of colors so that separate patterns do not have to be written for inputs such as:

> I like red cars.
> I like blue cars.
> I like green cars.

To create a set, click on the `plus` button of the Files tab of the Editor and select the set type. For example, create a file called `color` and add some colors. The set can then be used in a category containing a pattern as shown in Code 7.9.

This technique can be used to address the problem of lexical alternation, for example, dealing with morphological variants of a word, as in the forms "find," "finding," and "found."

```
<category>

<pattern>I like the color <set>color</set></pattern>

<template>That's interesting, so do I </template>

</category>
```

Code 7.9 An AIML set

Table 7.1 Sets in AIML

State set	State2capital set
Alabama	Alabama:Montgomery
Arizona	Arizona:Phoenix
California	California:Sacramento
...	...

```
<category>

  <pattern>What is the capital of <set>state</set></pattern>

  <template><map name="state2capital"><star/></map></template>

</category>
```

Code 7.10 An AIML map

Maps are used to specify associations between sets. For example, a set could be a list of US states and a map could be a function that associates an element in the set of states with an element in a set of state capitals, as in Table 7.1.

Using maps, it is then possible to use a single category as shown in Code 7.10 to ask and answer questions about any US state, for example:

> What is the capital of Alabama?
> What is the capital of Arizona?
> What is the capital of California?

Pandorabots has some built-in sets and maps for collections such as natural numbers, singular and plural nouns. For further details on sets and maps, see Wallace (2014b).

Useful! → **7.3.5 Context**

There are several mechanisms in Pandorabots for dealing with aspects of the conversational context. The simplest of these is the tag <that> which enables the system to remember the last sentence it has said. Consider the following example:

> Human: I have problems with my blood sugar.
> Bot: is blood sugar your main problem?
> Human: yes.
> Bot: what medication are you taking for your blood sugar?

```
<category>

 <pattern>^blood sugar^</pattern>

 <template>is blood sugar your main problem </template>

</category>

<category>

 <pattern>yes<pattern>

 <that>is blood sugar your main problem</that>

 <template> what medication are you taking for your blood
 sugar </template>

</category>
```

Code 7.11 Dealing with context using the AIML `<that>` tag

This interaction can be specified as in Code 7.11. Here, the second category is bound to a context where it will only be matched if the last sentence spoken by the bot was "Is blood sugar your main problem?".

A more wide-ranging mechanism involves setting a topic. For example, the topic could be set to `blood sugar` and only those patterns within the scope of the topic will be matched. This is illustrated in Code 7.12.

AIML has a number of other features including the ability to learn from information provided by the human interlocutor. For more details, consult the Pandorabots tutorial[9] or the AIML2.0 specification (Wallace 2014a).

One of the main advantages of the pattern-matching approach as implemented in AIML and similar languages is its simplicity. AIML makes use of an efficient matching algorithm that conducts a graph search through the prestored patterns on a word-by-word basis so that responses can be provided in real time. Also, given that the application is typically the simulation of conversation, it is not always critical if the user's input cannot be matched against a pattern. In this case, the ultimate default category is invoked to help keep the conversation going. For example, the chatbot can say something like "tell me more about yourself," which may cause the user to say something that will be matched by a pattern. Indeed, the art of chatbot programming is to provide sufficient responses in the ultimate default category that can be generated randomly and that can help the system recover from failures to match the user's input. This feature is probably not so useful in applications like the diabetes bot that simulate the answering of frequently asked questions in a particular domain.

---

[9]https://playground.pandorabots.com/en/tutorial/. Accessed February 20, 2016.

```
<topic name = "blood sugar">

<category>

  <pattern>^blood sugar^</pattern>

  <template> blood sugar is a common problem in diabe-
tes</template>

</category>

<category>

  <pattern>how can I lower it<pattern>

  <template> there are various medications that will help you
lower your blood sugar </template>

</category>

<category>

  <pattern>what would you recommend<pattern>

  <template> you should consult your doctor as the medication
to be prescribed will depend on other factors  </template>

</category>
```

Code 7.12 Setting a topic in AIML

Exercise 7.2: Creating the diabetes bot

1. Click on the Create Bot tab to create a bot. You can call it "Diabetes."
2. You will have a set of default AIML files. Add a new file called "questions."
   You can upload a sample file from the ones you will find in the
   ConversationalInterface repository /chapter7/AIML/.
3. Select your bot. This will bring up the editor screen for your bot, as shown in
   Fig. 7.6.
4. Click on the Train tab. Now you can test your bot by asking questions.

You will soon find that your bot is unable to answer all of your questions, in which case you will need to add more categories. You can edit and further develop the bot by following the more detailed instructions in the Playground tutorials. If you wish to make your bot available to other members within the Clubhouse—a community of other botmasters—to do this you need to click on the tab Publish Bot.

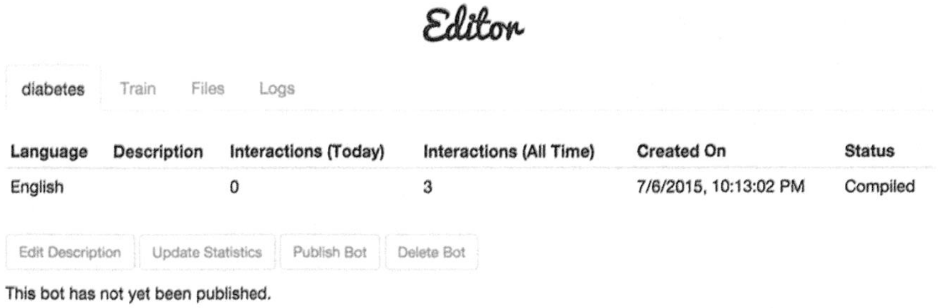

Editor

diabetes	Train	Files	Logs

Language	Description	Interactions (Today)	Interactions (All Time)	Created On	Status
English		0	3	7/6/2015, 10:13:02 PM	Compiled

Edit Description | Update Statistics | Publish Bot | Delete Bot

This bot has not yet been published.

Fig. 7.6 The Playground editor screen for the diabetes bot (reproduced with permission from Pandorabots, Inc.)

7.4   Creating a Link to Pandorabots from Your Android App

Once you have sufficiently developed your bots in Playground, you are ready to deploy them. In this section, we show how to embed a Pandorabots bot into an Android app. There are several reasons why you might wish to embed your bot in an app:

- The Playground only allows you to interact with your bot using the Web interface provided.
- The Playground allows you to make your bot available to other registered bot masters by publishing it in the Clubhouse, but you cannot make it publicly accessible on a Web site or by deploying it as a mobile application. For this, you need to create an account on the Developer Portal.
- By embedding the bot in an Android app, your app can be deployed on an Android mobile phone or tablet and can be made available to others to use, just like any other Android app.
- You can provide a speech-based front end to your app using the Google Speech APIs.
- You can add additional functions, such as making commands to access device functions—for example, to launch an app or to check the time. You can also link to other Web services, such as search and maps.

7.4.1   Creating a Bot in the Developer Portal

Exercise 7.3: Signing up for an account on the Developer Portal   The Developer Portal provides all the necessary tools and SDKs for deploying bots anywhere. To sign up to the Developer Portal:

```
$ npm install -g pb-cli
```

Code 7.13 Installing the CLI in node.js

```
pb upload myfile.aiml
```

Code 7.14 Using the pb command to upload a file

1. Click on the Dev Portal tab on the Pandorabots main page or go directly to the Developer Portal.[10]
2. Sign up for an account. You have to register for a plan. Accounts are free for a 10-day trial period after which different plans are available depending on needs. You should try to iron out any problems with your AIML code in Playground before moving your bot over to the Developer Portal. If you find that you need to make more API calls than allowed on your plan, you can upgrade your plan, as required.
3. Once your account and plan have been approved, you can retrieve your user_key and application_id, which are required in order to make API calls.

Exercise 7.4: Using the APIs to create a bot, upload and compile files, and talk to a bot There are two ways to use the APIs:

1. Using the Pandorabots CLI (command line interface).
2. Using an HTTP client-like cURL.

Both are supported, though the easiest way is to use the Pandorabots CLI.[11]

The CLI can be installed by going to the Developer Portal home page and scrolling down to the section entitled Getting Started, which is below the pricing information. Detailed instructions are also provided here.[12]

The CLI is written in Javascript, so it is first necessary to setup node.js before installing the CLI. An installer for node.js for both OS X and Windows users is available here.[13]

Node.js includes npm, a Javascript package manager that you can use to install the CLI using the command (Code 7.13).

This will install the CLI and make the pb and pandorabots commands available for use in the command line, for example (Code 7.14).

The CLI needs to be configured using a JSON file called chatbot.json in order to allow these commands. The chatbot.json configuration file stores

---

[10]https://developer.pandorabots.com/. Accessed February 20, 2016.

[11]https://github.com/pandorabots/pb-cli. Accessed February 20, 2016.

[12]http://blog.pandorabots.com/introducing-the-pandorabots-cli/. Accessed February 20, 2016.

[13]http://nodejs.org/download/. Accessed February 20, 2016.

basic information about your application: the app_id, the user_key, and the botname. When a CLI command is run, all the required information is added from chatbot.json. To create this file, you can use the init command and the CLI will prompt for all the required information. First create a directory for chatbot.json, then run init, as shown in Code 7.15.

This will now allow various commands to be run, such as shown in Code 7.16. A complete list of the commands is available here.[14]

```
$ mkdir mydirectory

$ cd mydirectory

$ pb init

(when information has been added

$ pb create
```

Code 7.15 Configuring the CLI

```
$ pb create
$ pb upload
$ pb compile
$ pb talk
```

Code 7.16 Examples of pb commands

```
$ pb upload example.aiml
```

Code 7.17 The pb command for uploading a file

```
$ curl -v -X PUT
'https://aiaas.pandorabots.com/bot/APP_ID/BOTNAM
E/file/example.aiml?user_key=USER_KEY'   --data-
binary @home/mybot/example.aiml
```

Code 7.18 The curl command for uploading a file

---

[14]https://github.com/pandorabots/pb-cli. Accessed February 20, 2016.

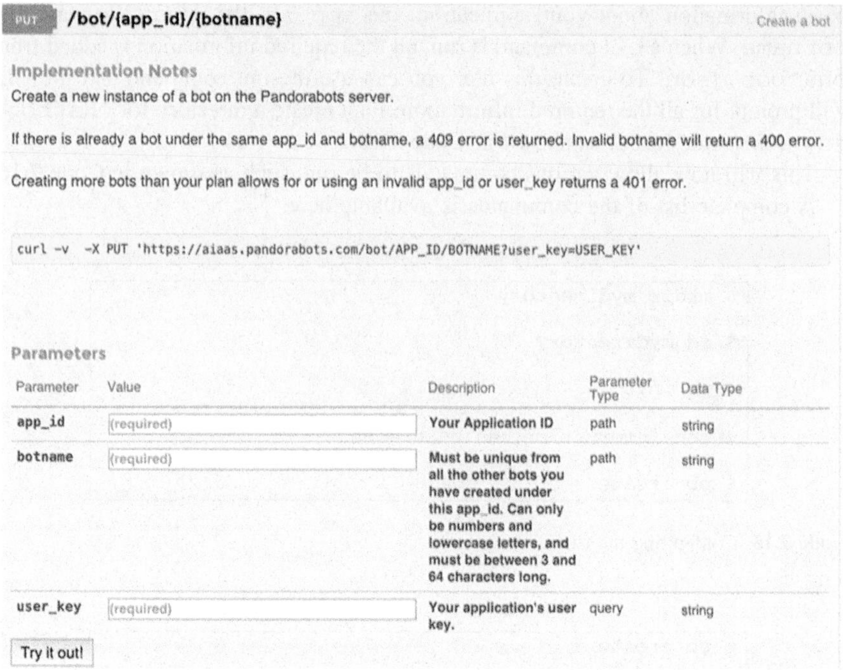

PUT   **/bot/{app_id}/{botname}**   Create a bot

Implementation Notes

Create a new instance of a bot on the Pandorabots server.

If there is already a bot under the same app_id and botname, a 409 error is returned. Invalid botname will return a 400 error.

Creating more bots than your plan allows for or using an invalid app_id or user_key returns a 401 error.

```
curl -v  -X PUT 'https://aiaas.pandorabots.com/bot/APP_ID/BOTNAME?user_key=USER_KEY'
```

Parameters

Parameter	Value	Description	Parameter Type	Data Type
app_id	(required)	Your Application ID	path	string
botname	(required)	Must be unique from all the other bots you have created under this app_id. Can only be numbers and lowercase letters, and must be between 3 and 64 characters long.	path	string
user_key	(required)	Your application's user key.	query	string

Try it out!

Fig. 7.7 Creating a bot in the Developer Portal (reproduced with permission from Pandorabots, Inc.)

Using cURL to talk to the API requires more complex commands. For example, to upload a file using the CLI, you would type the command shown in Code 7.17.

However, using curl you would have to insert the values for user_key, app_id and botname and type something like the command shown in Code 7.18.

Some API resources that provide assistance with API calls are available here.[15] For example, the resource in Fig. 7.7 shows an alternative way to create a bot.

These resources are also useful for observing the feedback provided to the API call and the text of the API calls, as this may be useful when linking to an app programmatically.

Exercise 7.5: Creating and testing a bot on the Pandorabots Developer Portal

1. Create a bot using either the CLI or the cURL command.
2. Upload one or more AIML files to your bot.
3. Compile the bot (Note: each time a file is modified or uploaded the bot must be compiled in order for the changes to be available in a conversation.
4. Talk to the bot using the input patterns in your AIML files.

---

[15]https://developer.pandorabots.com/docs. Accessed February 20, 2016.

7.4.2 Linking an Android App to a Bot

In this section, we describe how to link from an Android app to a Pandorabots bot that has been created on the Developer Portal. This is shown in the `TalkBot` app. The app works as follows: the user can press a button to say something to the bot, the recognized utterance is passed on to Pandorabots and using its corresponding AIML code it generates a response that is retrieved by the Android app. If the response is simple (e.g., a text), it is synthesized back to the user, whereas if the response includes mobile functions (see Sect. 7.5), they are executed and the results are synthesized to the user (e.g., checking the battery level and informing about it).

We have arranged the classes in this app in different packages (folders) as shown in Fig. 7.8.

- In *Pandora,* we have included the classes to connect to Pandorabots (`PandoraConnection`), process the results received (`PandoraResultProcessor`, `FindLocation`), and manage possible errors (`PandoraException`, `PandoraErrorCode`).
- In *VoiceInterface,* we include the `VoiceActivity` class to process the speech interface. This is exactly the same class as was used in the `TalkBack` app (Chap. 6).
- In the root, we have the `MainActivity` class. This class specifies the main behavior of our app.

The `MainActivity` class is very similar to the one presented for the `TalkBack` app in Chap. 6. In `TalkBack`, in order to demonstrate speech recognition and TTS we simply took the best result from speech recognition and spoke it out using TTS (Code 7.19).

Now in `TalkBot`, we want to send the recognized result of the user's input to the Pandorabots service and get a response. The following code accomplishes this and then calls a method to process the response (Code 7.20).

Additionally in the `catch` section, there is a call to the method `processBotErrors`, which deals with a number of possible errors such as invalid keys and ids for connecting to Pandorabots, no match for the input, and Internet connection errors.

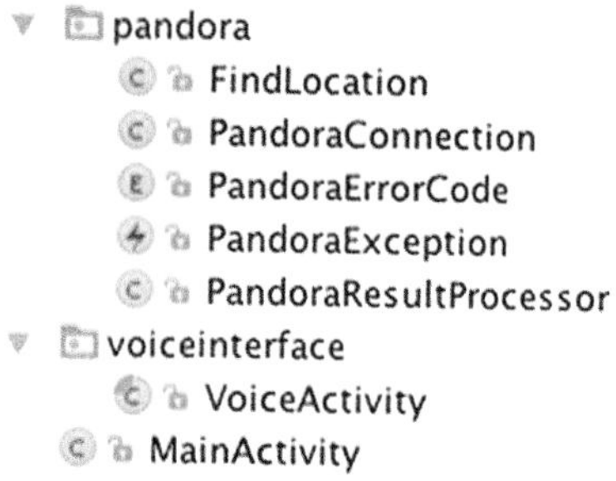

Fig. 7.8 Packages and classes for the TalkBot app

▼ 🗁 pandora
    Ⓒ 🗎 FindLocation
    Ⓒ 🗎 PandoraConnection
    Ⓔ 🗎 PandoraErrorCode
    ⚡ 🗎 PandoraException
    Ⓒ 🗎 PandoraResultProcessor
▼ 🗁 voiceinterface
    Ⓒ 🗎 VoiceActivity
  Ⓒ 🗎 MainActivity

```
String bestResult = nBestList.get(0);

(...)
speak(bestResult, "EN", ID_PROMPT_INFO);
```

Code 7.19 Fragment of the `processAsrResults` method in the `MainActivity` class (TalkBack app, Chap. 6)

```
String userQuery = nBestList.get(0);

(...)
try {
    String response = pandoraConnection.talk(userQuery);
    processBotResults(response);
} catch (PandoraException e) {
    processBotErrors(e.getErrorCode());
}
```

Code 7.20 Fragment of the `processAsrResults` method in the `MainActivity` class (TalkBot app)

When a response is returned from Pandorabots, it is handled in the `processBotResults` method (Code 7.21).

Here, there are two cases to consider:

1. The result is in the form of text to be output as a spoken response. In this case, the method `removeTags` is called to remove any HTML tags. The resulting string is spoken using TTS.
2. The result contains an `<oob>` tag, indicating that the response requires further processing to determine what sort of mobile function is being requested. In this case, the result is sent to the method `processOobOutput` in the class `OOBProcessor`. This will be discussed further in Sect. 7.5.

Connecting with Pandorabots

Some parameters are declared in `MainActivity` that are required to make a connection to your bot on Pandorabots. You must insert your own values (Code 7.22).

These values are used in the `PandoraConnection` class to establish the connection with Pandorabots. We have included them in `MainActivity` to make the `PandoraConnection` class independent of the actual bot used. By doing this, you can use `PandoraConnection` every time you want to use a bot in your Android apps without changing a single line of the code and just adjusting these parameters in the initial activity that uses the code (`MainActivity` in our case).

```
public void processBotResults(String result){


   // Speak out simple text from Pandorabots after removing

     any HTML content
   if(!result.contains("<oob>")){
     result = removeTags(result);
     (...)

     speak(result,"EN",ID_PROMPT_INFO);
     (...)
   }
   // Send responses with <oob> for further processing
   else{

     PandoraResultProcessor oob=

        new PandoraResultProcessor(this, ID_PROMPT_INFO);
     (...)

     oob.processOobOutput(result);
     (...)
   }
}
```

Code 7.21 Fragment of the processBotResults method in the MainActivity class (TalkBot app)

```
private String host = "aiaas.pandorabots.com";
private String userKey = "Your user key here";
private String appId = "Your app id here";
private String botName = "The name of your bot here";
PandoraConnection pandoraConnection = new PandoraConnec-
tion(host, appId, userKey, botName);
```

Code 7.22 Initializing the Pandorabots connection parameters in the MainActivity class (TalkBot app)

PandoraConnection is a simplified version of a class created by Richard Wallace as a Java API to the Pandorabots service.[16] The class has been edited and simplified to adapt it to the requirements of our Android app. The class does the following:

1. The connection parameters specified in MainActivity are initialized.
2. The user's input string is sent to the chatbot on the Pandorabot's service and the bot's response is returned as a JSON object from which the responses to be returned to MainActivity are extracted. In order to do that, the apache http client libraries are employed, which requires including them as dependencies in the build.gradle file (Code 7.23).

---

[16]https://github.com/pandorabots/pb-java. Accessed February 20, 2016.

```
Content content =

    Request.Post(uri).execute().returnContent();
String response = content.asString();
JSONObject jObj = new JSONObject(response);
JSONArray jArray = jObj.getJSONArray("responses");
for (int i = 0; i < jArray.length(); i++) {
    responses += jArray.getString(i).trim();
}
```

Code 7.23 Fragment of the talk method in the PandoraConnection class (TalkBot app)

Fig. 7.9 Interface of the
TalkBot app

Exercise 7.6: Speech enabling conversational interaction in the TalkBot app
Run the app and interact with it using inputs similar to those that you used to
interact with the bot in Playground. Remember that for the code to work you must
include your own connection parameters. Figure 7.9 shows the interface for the
app.

7.5 Introducing Mobile Functions

Chatbot technology has been extended to support the development of virtual personal assistants that can carry out commands to the device and answer queries to Web services. A recent addition to AIML supports these mobile functions. The following is an example (Code 7.24).

Here, the <oob> tags separate content within the template that is not part of the response to be spoken to the user: In this case, including a tag indicating that the task involves search and that the content of the search is the item retrieved from the wildcard *. In other words, if the user says "Show me a Web site about speech recognition" the bot outputs the words "let's try a Google Search" and encloses the command to search for speech recognition within <oob> tags. Figure 7.10 shows the result of this interaction:

In the next section, we explain how to realize this interaction in our Android app.

7.5.1 Processing the <oob> Tags

In TalkBot, we capture the user's input using Google speech recognition and send the text to Pandorabots. Once the text is matched against a pattern in the AIML file, the content of the template is sent back to the MainActivity class. If the content is only text, it is spoken out using the Google TTS.

However, if the template contains an <oob> tag, we need to check for this and take some action. In this case, we call a new class OOBProcessor in MainActivity that contains a number of methods for processing the content of the <oob> tag and executing the required commands.

First, we need to separate the items in the template into <oob> content and text to be spoken to the user. This processing is done in the PandoraResultProcessor class. The processOobOutput method processes the JSON object returned from Pandorabots and extracts the content of the label that, as shown in Code 7.25.

```
<category>

<pattern>SHOW ME A WEBSITE ABOUT * </pattern>

<template> Let's try a google search

   <oob><search><star/></search></oob>

</template>

</category>
```

Code 7.24 Including mobile functions in AIML with <oob> tags

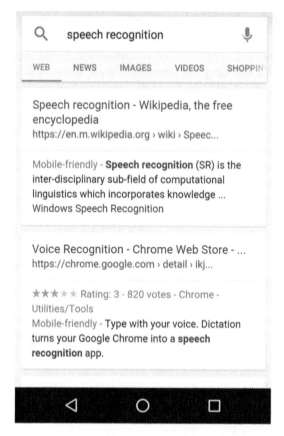

Fig. 7.10 Response to the query "show me a Web site about speech recognition." Accessed at 22:38 on February 17, 2016. Google and the Google logo are registered trademarks of Google Inc., used with permission

```
that:
<oob><launch>calendar</launch></oob>launching
calendar
```

Code 7.25 JSON object returned from query to Pandorabots

The text within the <oob> tags is assigned to a variable `oobContent` and the remaining text is assigned to `textToSpeak` (Code 7.26).

Next, we analyze `oobContent` to determine the type of command—for example, "search," "battery," and "maps"—that is contained in the tag embedded within the <oob> tag, using the `processOobContent` method. In this case, we extract `<launch>calendar</launch>`, which enables us to determine that the command is to launch an app and that the app is the calendar. With this

```
oobContent: <launch>calendar</launch>

textToSpeak: launching calendar
```

Code 7.26 Assigning the text with the <oob> tags

```
launchApp(app,textToSpeak)
```

Code 7.27 Calling the `launchApp` method

information, we can call the `launchApp` method with the required parameters (Code 7.27).

Other commands, such as "search" and "maps" are handled in a similar way. However, each command has to be implemented appropriately, depending on how it is handled in Android. For example, to launch an app we need to check whether that app actually exists on the device, as the user could speak the name of an app and have the name recognized, but this does not guarantee that there is an app of that name on that particular device. The `launchApp` method does the following:

1. Gets a list of app names and package names on the device.
2. Checks if the app requested is on the device.
3. If not, reports to the user.
4. If yes, gets the package name of the requested app.
5. Launches the app.

7.5.2 Battery Level

The `batteryLevel` method checks the level of the battery on the user's device given an input such as "what is my battery level?" The following code launches an Android intent to check the battery level and convert the raw battery level to a percentage number that can be spoken out (Code 7.28).

7.5.3 Search Queries

Two categories are tagged with an <oob> tag that includes <search> (Code 7.29).

Given inputs such as "tell me about speech recognition," the text of the search query is extracted and passed to the Android `ACTION_WEB_Search` intent to be executed and the text in the template is spoken using TTS (Code 7.30).

```
private void batteryLevel() throws Exception {
   (...)
   Intent batteryIntent = ctx.registerReceiver(null,

      new IntentFilter(Intent.ACTION_BATTERY_CHANGED));
   int rawlevel = batteryIntent.getIntExtra("level", -1);
   double scale = batteryIntent.getIntExtra("scale", -1);
   double level = -1;
   int pct;

   if (rawlevel >= 0 && scale > 0) {
     level = rawlevel / scale;
     pct = (int) (level * 100);   //Conversion from the raw

                              battery level to a percentage
     ((VoiceActivity) ctx).speak("Your battery level is " +

     String.valueOf(pct) + "per cent", "EN", msgId);
   }
   (...)
}
```

Code **7.28** The `batteryLevel` method of the `PandoraResultProcessor` class (`TalkBot` app)

```
<category><pattern>Tell me about *</pattern>

<template>Searching for <star/>
<oob><search><star/></search></oob></template>

</category>

<category><pattern>What is *</pattern>

<template>Searching for information about <star/>
<oob><search><star/></search></oob></template>

</category>
```

Code **7.29** AIML code including a `<search>`

```
Intent intent = new Intent(Intent.ACTION_WEB_SEARCH);
intent.putExtra(SearchManager.QUERY, query);
(...)
ctx.startActivity(intent);
```

Code **7.30** Fragment of the `search` method of the `PandoraResultProcessor` class (`TalkBot` app)

7.5.4   Location and Direction Queries

In this section, we show how to make queries about locations and directions, for example:

Where is New York?
Show me directions from New York to Boston.

We distinguish between absolute queries, such as these, in which all the parameters of the query are mentioned explicitly, and relative queries, in which the user's current location is assumed implicitly, as in:

Find the nearest Starbucks.
Show me directions to Boston.

The motivation for making this distinction is to avoid potential problems with the use of contextual information. For example, given the following sequence:

Show a map of Boston.
Find the nearest Starbucks.

It is possible that Boston could be assigned as the current context so that the next question is interpreted as

Find the nearest Starbucks in Boston.

instead of

Find the nearest Starbucks relative to my current location.

Without direct access to the context mechanisms being applied, it is not possible to resolve this. Indeed, in interaction between humans it is not always clear whether a subsequent question relates to a previous one or whether it is part of a new topic.

In order to deal with this in AIML, we identify those inputs that contain absolute queries and those that contain relative queries by inserting an additional tag

<myloc> into relative queries and adding additional code to find the user's current location. The following is a high-level overview of these processes.

Identify location queries

1. Check if oobContent contains <map>.
2. Parse to extract the values for the map query (mapText, textToSpeak).
3. Call the mapSearch method.

mapSearch

1. Speak the content of textToSpeak.
2. Replace spaces in mapText with "+".
3. Use the ACTION_VIEW action and specify the location information in the intent data with the Geo URI scheme.
4. Start activity (Code 7.31).

Relative queries are marked up in AIML shown in Code 7.32.
A relative query results in the following actions:

1. The tag <myloc> causes the FindLocation class to be instantiated in order to find the latitude and longitude of the user's current location.
2. The ACTION_VIEW action is called with the values for latitude (lat) and longitude (lng) for the user's current location found.

Directions are handled in a similar way. For example, if the input is a relative directions query, as in "directions to Boston," the <myloc> tag causes

```
Intent geoIntent = new Intent(
        android.content.Intent.ACTION_VIEW,
        Uri.parse("geo:"+ lat + "," + lng + "?q=" + mapText));
ctx.startActivity(geoIntent);
```

Code 7.31 Fragment of the mapSearch method of the PandoraResultProcessor class (TalkBot app)

```
<category>

<pattern>FIND THE NEAREST STARBUCKS </pattern>

<template>I'm looking on the map
<oob><map><myloc>Starbucks</myloc>
</map></oob>

</template>

</category>
```

Code 7.32 AIML relative query

```
FindLocation findLocation = new FindLocation(ctx);
double lat = findLocation.getLatitude();
double lng = findLocation.getLongitude();
Uri uri = Uri.parse("http://maps.google.com/maps?saddr=" +
lat + "," + lng + "&daddr=" + to);
Intent intent = new Intent(Intent.ACTION_VIEW, uri);
```

Code 7.33 Fragment of the getDirections method in the PandoraResultProcessor class (TalkBot app)

FindLocation to be called to find the latitude and longitude values of the user's current location, assuming the current location as the origin of the directions query (Code 7.33).

Exercise 7.7: Testing the app

1. Run the sample code with a range of inputs that includes oob processing. You can find the oob.aiml file in the *chapter7/AIML* folder in our GitHub ConversationalInterface repository.
2. Note any queries that do not work.
3. Try to determine the problem, for example:

    – Is it a speech recognition error?
    – Is it due to missing categories in AIML?
    – Are additional or modified Java methods required?

7.6   Extending the App

We can extend the app in a number of ways. For example, we could add more AIML categories to allow a wider range of inputs. Another extension would be to add more commands to the device and queries to Web services. The CallMom app[17] illustrates how this can be done.

In our application, we made use of Google Search and Google Maps to provide responses to queries. CallMom also consults a number of external knowledge sources, including Wolfram Alpha, DbPedia, Trueknowledge.com, Answers.com, Weather Service, various shopping sites, and other Pandorabots.

Most chatbot markup languages nowadays have methods for representing information. For example, in AIML 2.1 there is a facility to create ontologies that enable the chatbot to make use of and reason with knowledge, while ChatScript has a facility for invoking WordNet ontologies. Another approach extends the reference implementation of AIML to enable the extraction of domain knowledge from semantic Web ontologies using a scripting language called OwlLang and to store new knowledge obtained from the conversations in the ontologies (Lundqvist et al. 2013).

---

[17]http://callmom.pandorabots.com/static/callmombasic/features.html. Accessed February 20, 2016.

7.7   Alternatives to AIML

AIML is a widely used markup language for specifying chatbots. However, there are some alternatives, the most notable of which is ChatScript. ChatScript was developed in 2010 by Wilcox (2011a, b) and is used mainly to provide natural language understanding capabilities for characters in games, but has also been used for the chatbot Rose that won the Loebner Prize competition in 2014. While AIML's pattern matching is word-based, in ChatScript it is meaning-based, supporting sets of words called concepts to represent synonyms, as shown in Code 7.34.

This allows rules to be written that respond to all sorts of meat (Code 7.35).

Here, the input pattern (in parentheses) contains the concept "meat" that can be matched by any of the words in the concept ~meat. The chatbot's response is the text following the input pattern.

ChatScript is available as open source[18], and there is also a tutorial on how to build a conversational bot using ChatScript.[19]

Other alternatives to AIML are Api.ai[20] and Wit.ai.[21] Chapter 9 shows how to use the Api.ai platform to extract a semantic analysis from the user's input.

7.8   Some Ways in Which AIML Can Be Further Developed

In this section, we review some ways in which AIML has been extended as well as some suggestions for further developments.

7.8.1   Learning a Chatbot Specification from Data

Creating a chatbot in a language such as AIML typically involves hand coding a large number of categories, a process that can take several years if starting from scratch. Developers creating a chatbot on the Pandorabots Web site can make use of libraries of AIML categories to get started. For commercial developers on a special license, there is also a tool called Pattern Suggester that is part of Program AB, the most recent reference implementation of AIML 2.0. Pattern Suggester helps to automate the process of creating new patterns through a type of unsupervised

---

[18]http://sourceforge.net/projects/chatscript/. Accessed February 20, 2016.

[19]http://inspiredtoeducate.net/inspiredtoeducate/learn-to-build-your-own-conversational-robot-using-chatscript/. Accessed February 20, 2016.

[20]https://api.ai/. Accessed February 20, 2016.

[21]https://wit.ai/. Accessed February 20, 2016.

```
concept: ~meat ( bacon ham beef meat flesh veal
lamb chicken pork steak cow pig )
```

Code 7.34 Declaring a concept in ChatScript

```
s: ( I love ~meat ) Do you really? I am a vegan
```

Code 7.35 Using a concept in ChatScript

learning for patterns.[22] In one experiment, by searching through 500,000 inputs in logs from the CallMom app, the Pattern Suggester was able to find new patterns and create graphs at a rate of 6 categories per minute (Wallace 2014c).

A similar approach is the use of machine-learning techniques to read text from a corpus and convert it to the required AIML format. Abu Shawar and Atwell (2005) trained a bot using text from the Dialog Diversity Corpus, the spoken part of the British National Corpus (BNC), and online FAQ (Frequently Asked Questions) Web sites. They were able to generate more than one million categories extracted from the BNC. FAQs are a good corpus source as they have a clear turn-taking structure that can be easily adapted to the AIML pattern-template format. Several FAQ chatbots were generated, including one using the FAQ of the School of Computing at the University of Leeds, and a Python tutor trained on the public domain Python programming language FAQ Web site. De Gasperis et al. (2013) describe an algorithm in which texts in a corpus are used in a bottom-up procedure that chooses portions of text to be used as answers along with a keyword analysis of each piece of selected text to build questions. Each text representing an answer is then associated with possible questions and their formal variants (or paraphrases). Wu et al. (2008) describe an approach involving automatic chatbot knowledge acquisition from online forums using rough sets and ensemble learning.

AIML 2.0 contains learning features that enable the system to be taught new information and other chatbots such as Cleverbot,[23] Jabberwacky,[24] and Kyle[25] are also able to learn. Cleverbot employs a data mining approach in which it memorizes everything that is said to it and then searches through its saved conversations to find a response to new input, while Jabberwacky models the way humans learn language, facts, contexts, and rules. Kyle models the way humans learn language, knowledge, and context, making use of the principles of positive and negative feedback.

---

[22]https://code.google.com/p/program-ab/. Accessed February 20, 2016.

[23]http://www.cleverbot.com/. Accessed February 20, 2016.

[24]http://www.jabberwacky.com/j2about. Accessed February 20, 2016.

[25]http://www.leeds-city-guide.com/kyle. Accessed February 20, 2016.

7.8.2   Making Use of Techniques from Natural Language Processing

One potential criticism of chatbot technology is that it does not make use of theoretically driven approaches and dialog technology but instead uses a simple pattern-matching approach within a stimulus-response model. It could be argued that incorporating additional technologies into AIML would make the authoring process more difficult. It would be useful to conduct empirical studies to ascertain the effectiveness of the additional technologies for the authoring process as well as for pattern matching and RG. As it is, pattern matching in AIML is fast and efficient, even when searching a large number of patterns. Moreover, from a practical viewpoint it could be argued that in reality most language use in inter-action with a chatbot does not need to address the ambiguous and complex sentences that are the concern of theoretical linguists and that a stimulus–response model has the merits of simplicity and practical utility (see discussion of this issue by Wallace.[26]

Nevertheless, there have been some useful suggestions as to how AIML could be enhanced using techniques from natural language processing, most notably in a paper by Klüwer (2011). One problem concerns the authoring of patterns. In order to be able to handle surface variation in input, i.e., alternative syntactic structures and alternative lexical items, an AIML botmaster has to manually create a large number of alternative patterns. Klüwer describes some natural language processing technologies that could be used to optimize pattern authoring. For example, surface variation in patterns could be addressed by using dependency structures (see Chap. 8) rather than surface strings so that all variations on a sentence with the same dependency structure would be associated with a single pattern. To handle sentences that have the same meaning but different surface forms—for example, the active and passive forms of a sentence—a semantic analysis of the different forms of the sentence would abstract from their surface forms and allow the different forms to be associated with a single pattern. These techniques could also help to address the problem of false positives when the user's input is matched erroneously with patterns including wildcards that are used to cover variations in surface structure.

Natural language processing technology could also be used to generate alternative output to allow for greater flexibility. With current chatbots, the output is generally static, having either been defined manually as a system response or assembled from templates in which some variables are given values at runtime. In AIML, it is possible to code a set of alternative responses that are generated randomly and there is also a <condition> tag that allows particular actions in a template to be specified conditionally. ChatScript makes use of a C-style scripting language that can be used along with direct output text to produce more flexible

---

[26]http://www.pandorabots.com/pandora/pics/wallaceaimltutorial.html. Accessed February 20, 2016.

responses. For an approach to the automatic generation of output from abstract representations, see Berg et al. (2013).

7.9   Summary

In this chapter, we have shown how to create a chatbot that can engage in conversational interaction and also perform functions on a mobile device such as launching apps and accessing Web services. We have used the Pandorabots platform to run the chatbot and have specified the conversational interaction using AIML. Using this approach, we have not needed to implement components for SLU, DM, and RG, as the pattern-matching capabilities of AIML and the associated templates are able to handle a wide range of user inputs and produce appropriate system responses. We have shown how the app developed in this chapter can be further extended, and we have reviewed some alternatives to AIML.

In the following chapters, we will examine more advanced technologies that are used to develop conversational interfaces, beginning with SLU in Chap. 8.

Further Reading
For readers interested in the Turing test, the collection of papers by an impressive range of scholars in Epstein et al. (2009) explores philosophical and methodological issues related to the quest for the thinking computer. The collection also includes Turing's 1950 paper "Computing machinery and intelligence." There is also an interesting paper by Levesque discussing the science of artificial intelligence in which the Turing test is criticized for relying too much on deception. A set of questions, known as the Winograd schema questions, is proposed as a more useful test of intelligence.[27]

References

Abu Shawar B, Atwell E, Roberts A (2005) FAQChat as an information retrieval system. In: Vetulani Z (ed) Human language technologies as a challenge. Proceedings of the 2nd language and technology conference, Wydawnictwo Poznanskie, Poznan, Poland, 21–23 April 2005: 274–278. http://eprints.whiterose.ac.uk/4663/. Accessed 20 Jan 2016

✳ Berg M, Isard A, Moore J (2013) An openCCG-based approach to question generation from concepts. In: Natural language processing and information systems. 18th international conference on applications of natural language to information systems, NLDB 2013, Lecture notes in computer science, vol 7934. Springer Berlin Heidelberg, Salford, UK, 19–21 June 2013, pp 38–52. doi:10.1007/978-3-642-38824-8_4

✳ Crocker MW, Pickering M, Clifton C Jr (1999) Architectures and mechanism for language processing, 1st edn. Cambridge University Press, Cambridge. doi:10.1017/cbo9780511527210

---

[27]http://www.cs.toronto.edu/∼hector/Papers/ijcai-13-paper.pdf. Accessed February 20, 2016.

De Gasperis G, Chiari I, Florio N (2013) AIML knowledge base construction from text corpora. In: Artificial intelligence, evolutionary computing and metaheuristics, vol 427. Studies in computational intelligence, pp 287–318. doi:10.1007/978-3-642-29694-9_12

Epstein R, Roberts G, Beber G (eds) (2009) Parsing the turing test: philosophical and methodological issues in the quest for the thinking computer. Springer, New York. doi:10. 1007/978-1-4020-6710-5

Klüwer T (2011) From chatbots to dialog systems. In: Perez-Marin D, Pascual-Nieto I (eds) Conversational agents and natural language interaction: techniques and effective practices. IGI Global Publishing Group, Hershey, Pennsylvania, pp 1–22. doi:10.4018/978-1-60960-617-6.ch001

Lundqvist KO, Pursey G, Williams S (2013) Design and implementation of conversational agents for harvesting feedback in eLearning systems. In: Hernandez-Leo D, Ley T, Klamma R, Harrer A (eds) Scaling up learning for sustained impact. Lecture notes in computer science, vol 8095, pp 617–618. doi:10.1007/978-3-642-40814-4_79

Wallace R (2003) The elements of AIML Style. ALICE A.I. Foundation, Inc. http://www.alicebot. org/style.pdf. Accessed 20 Jan 2016

Wallace R (2009) Anatomy of A.L.I.C.E. In: Epstein R, Roberts G, Beber G (eds) Parsing the turing test: philosophical and methodological issues in the quest for the thinking computer. Springer, New York, pp 81–210. doi:10.1007/978-1-4020-6710-5_13

Wallace R (2014a) AIML 2.0 working draft. https://docs.google.com/document/d/ 1wNT25hJRyupcG51aO89UcQEiG-HkXRXusukADpFnDs4/pub. Accessed 20 Jan 2016

Wallace R (2014b) AIML—sets and maps in AIML 2.0. https://docs.google.com/document/d/ 1DWHiOOcda58CflDZ0Wsm1CgP3Es6dpicb4MBbbpwzEk/pub. Accessed 20 Jan 2016

Wallace R (2014c) AIML 2.0—virtual assistant technology for a mobile era. In: Proceedings of the mobile voice conference 2014, San Francisco, 3–5 March http://wp.avios.org/wp-content/ uploads/2014/conference2014/35_mctear.pdf. Accessed 20 Jan 2016

Weizenbaum J (1966) ELIZA—a computer program for the study of natural language communication between man and machine. Commun ACM 9(1):36–45. doi:10.1145/ 365153.365168

Wilcox B (2011a) Beyond Façade: pattern matching for natural language applications. http://www. gamasutra.com/view/feature/134675/beyond_façade_pattern_matching_php. Accessed 20 Jan 2016

Wilcox B (2011b) Fresh perspectives—a Google talk on natural language processing http://www. gamasutra.com/blogs/BruceWilcox/20120104/90857/Fresh_Perspectives_A_Google_talk_on_ Natural_Language_Processing.php. Accessed 20 Jan 2016

Wu Y, Wang G, Li W, Li Z (2008) Automatic chatbot knowledge acquisition from online forum via rough set and ensemble learning. IEEE Network and Parallel Computing (NPC 2008). IFIP International Conference, pp 242–246. doi:10.1109/npc.2008.24

Web sites

Alice A.I. Foundation www.alicebot.org
AIML matching http://www.alicebot.org/documentation/matching.html
AIML tutorial http://www.pandorabots.com/pandora/pics/wallaceaimltutorial.html
AIML 2.0 Working Draft https://docs.google.com/document/d/1wNT25hJRyupcG51aO89UcQE iG-HkXRXusukADpFnDs4/pub
API.ai https://api.ai/
CallMom app http://callmom.pandorabots.com/static/callmombasic/features.html
Chatbots.org https://www.s.org/
Cleverbot http://www.cleverbot.com/
Jabberwacky http://www.jabberwacky.com/j2about
Kyle http://www.leeds-city-guide.com/kyle

Node.js http://nodejs.org/download/
Pandorabots http://www.pandorabots.com/
Pandorabots blog http://blog.pandorabots.com/
Pandorabots Command Line Interface (CLI) https://github.com/pandorabots/pb-cli
Pandorabots CLI instructions http://blog.pandorabots.com/introducing-the-pandorabots-cli/Node.js
Pandorabots Developer Portal https://developer.pandorabots.com/
Pandorabots Github https://github.com/pandorabots
Pandorabots Playground https://playground.pandorabots.com/en/
Pandorabots Playground tutorial https://playground.pandorabots.com/en/tutorial/
Pandorabots Rosie library https://github.com/pandorabots/rosie
Pandorabots Twitter https://twitter.com/pandorabots
Wit.ai https://wit.ai/

Chapter 8
Spoken Language Understanding

Abstract Spoken language understanding (SLU) involves taking the output of the speech recognition component and producing a representation of its meaning that can be used by the dialog manager (DM) to decide what to do next in the interaction. As systems have become more conversational, allowing the user to express their commands and queries in a more natural way, SLU has become a hot topic for the next generation of conversational interfaces. SLU embraces a wide range of technologies that can be used for various tasks involving the processing of text. In this chapter, we provide an overview of these technologies, focusing in particular on those that are relevant to the conversational interface.

8.1 Introduction

In a conversational interface, the spoken language understanding (SLU) component takes the output from the speech recognition (ASR) component and produces a representation of its meaning that it passes on to the dialog manager (DM) for further processing (see Chap. 10 for further details on the dialog management component).

In order to appreciate the issues involved in SLU, let us consider the following examples of the sorts of utterances that might be spoken to a conversational interface:

1. Set the alarm for 8 o'clock tomorrow morning.
2. Is there an Italian restaurant near here?
3. Tell me about spoken dialog systems.
4. Are you sad?

Analyzing these utterances informally, we might conclude that utterance 1 is a request to do an action (setting the alarm) in which the content of the request is a time value along the lines of "8:00 current date +1." Utterance 2 looks like a *yes/no* question, but often an utterance such as this is interpreted as a request—in this case, to find an Italian restaurant near the speaker's current location and return some answers, perhaps with addresses and a map (this is what usually happens when a question like this is put to virtual personal assistants such as Google Now). Utterance 3 has the form of a request but its purpose is to obtain information, so it is more like a question that would be sent to a question answering system or to an online encyclopedia such as Wikipedia. Finally, utterance 4 is also a question but unlike utterance 3 it is more like the sort of question that occurs in everyday conversation.

Note that we have mentioned several different concepts in this analysis of the meaning of these examples. We have referred to the function of the utterance—i.e., whether it is a question, request for action, statement, or some other type of dialog act. For conversational interfaces, knowing the function of the user's utterance—generally known as the *dialog act* being performed—is important in order to be able to recognize the speaker's intention and to decide what would be an appropriate type of response. So, for example, if the hearer interprets utterance 2 as a question, then it would be sufficient to answer "yes" or "no," whereas if it is interpreted as a request, then a more extensive answer is required, such as directions to the nearest Italian restaurant.

A conversational interface also needs to determine what the utterance is about. There can be two aspects to this: determining the *domain* of the utterance—is it about flight reservations, weather forecasts, stock quotes, and so on—and determining the user's *intent*—for example, within the flight reservations domain, is it to book a flight, change a booking, query a booking, and so on? In the examples above, there were utterances about setting alarms, finding restaurants, getting information about spoken dialog systems, and asking about personal feelings. In a multidomain system, it may be useful to first determine the domain and then determine the user's intent, although in many cases these two aspects are reduced to identifying the user's intent along with the relevant elements of the utterance—known as its *entities* (or *slots*). So, for example, in utterance 2 the intent might be something like `FindRestaurant` and the entities might be `cuisine: Italian` and `location:near`.

Various meaning representation languages have been used in semantic analysis to represent the content of utterances. In traditional semantic analysis, First Order Predicate Calculus (FOPC) has been used widely to represent literal content. FOPC has the advantage that it provides a sound basis for inference; that is, there are mechanisms that allow valid conclusions to be derived logically from the represented propositions. Generally, however, in practical conversational interfaces representations in the form of sets of attribute-value pairs have been used to capture the information in an utterance that is relevant to the application. Thus, in a representation of utterance 2 above the different elements of meaning that we have discussed—the dialog act type (request), the topic (FindRestaurant), and the attributes (cuisine, location)—may be represented as follows:

request(topic = FindRestaurant, cuisine = Italian, location = near)

One major issue for the SLU module of a conversational interface is that it has to deal with spoken language. Spoken language is less regular than the language of written texts. As discussed in Chap. 3, spoken utterances are often ill-formed in that they may contain characteristics of spontaneous speech, such as hesitation markers, self-corrections, and other types of disfluency. For example, a well-formed utterance such as "is there an Italian restaurant near here" might have been spoken more like:

is there uh is there—uh a Fre—I mean an Italian restaurant uh near here

Another problem is that the output of ASR may contain recognition errors that make the string to be analyzed more difficult to interpret.

Finally, the output from ASR may not be a single utterance but an *n*-best list, with potentially up to 100 hypotheses in the list. The simplest approach is to take the 1st-best hypothesis and pass it on to SLU, but this is not necessarily the optimal solution as some other hypothesis might be a closer transcription of what the user actually said. However, processing all the items in the *n*-best list in order to discover the most appropriate one could have a significant bearing on the real-time performance of the SLU component. An alternative that has been shown to work effectively is the use of word confusion networks, a special kind of lattice that encodes competing words in the same group (Hakkani-Tür et al. 2006).

8.2 Technologies for Spoken Language Understanding

Compared to ASR, SLU is much more diverse as it embraces a variety of different technologies and approaches. The choice of a particular approach will depend on the task to be performed by SLU—for example, it may be important to perform fairly low-level tasks such as normalizing the input before going on to higher-level tasks. Extracting the meaning may in some cases only require identifying keywords in the input while in other cases a deeper understanding may be required. In the following sections, we review a number of different technologies and approaches. These technologies and approaches are not necessarily mutually exclusive and it is often the case that several technologies are applied at different stages of the interpretation process.

8.3 Dialog Act Recognition

Dialog act recognition—also known as *intent determination* or *spoken utterance classification* (Tur and Deng 2011)—involves determining the function of an utterance in a dialog, for example, whether it is a question, suggestion, offer, and so on. As described in Chap. 3, there are a number of taxonomies of dialog acts that seek to account for the range of functions that utterances might have in a conversation.

In the 1980s and early 1990s, dialog act recognition involved the application of AI models of plan inference and reasoning (Allen 1995). An alternative approach using statistical methods emerged in the late 1990s in which dialog act interpretation is modeled as a supervised classification task.

Statistical dialog act recognition involves training classifiers on a corpus of dialog utterances where each utterance is annotated in terms of the dialog act it performs and features are identified that support the annotation. The Switchboard corpus (Godfrey et al. 1992) has been used in many studies, and tag sets for annotating a corpus include DAMSL (Allen and Core 1997), which provides a domain-independent set of dialog act types. In some cases, domain-specific tags have been employed, for example, in the Verbmobil dialog act tag set where the dialog acts are related specifically to planning schedules (Suzanne et al. 1995). Among the features often used for classification are the following:

- Words and phrases in the utterance, usually in the form of *n*-grams—for example, "please" is a good cue word for a request and the *n*-gram "are you" for a Yes–No Question (Webb et al. 2005).
- Prosody—for example, different prosodic features can help distinguish between the interpretation of "OK" as an Agreement or as a Back-channel act where the hearer indicates understanding of the preceding dialog act but does not attempt to take the floor (Shriberg et al. 1998).
- Syntactic and semantic information derived from deep linguistic processing (Klüwer et al. 2010).
- Dialog act sequences, in the form of dialog act *n*-grams (Nagata and Morimoto 1994).

Stolcke et al. (2000) used a combination of lexical, collocational, and prosodic features as well as dialog act sequences. They modeled dialogs as an HMM and dialog acts as observations from the model states. In other work, Bayesian networks have been used for classification (Keizer et al. 2002; Klüwer et al. 2010). For an overview of approaches to dialog act recognition, see Jurafsky and Martin (2009: 876–882) and Tur and Hakkani-Tür (2011).

Dialog state tracking (also known as belief tracking) is similar to dialog act recognition. Dialog state tracking involves estimating a user's goal in a spoken dialog system. In order to compare models and dialog state tracking algorithms, a series of dialog state tracking challenges have been organized in which a set of labeled dialogs is provided against which participants apply their algorithms (Black et al. 2011; Williams 2012; see also the Web page at Microsoft Research for Dialog State Tracking Challenge[1]). At the time of writing, the most recent challenge took place at the International Workshop on Spoken Dialog Systems (IWSDS) in 2016.[2]

8.4 Identifying Intent

Intent identification is similar to dialog act recognition as it also involves classifying the user's utterance. Dialog act recognition has been developed primarily in academic research where the aim has been to find a taxonomy of dialog acts that can classify the whole range of different functions that a user's utterance can take in a conversation—such as asking questions, making promises, acknowledging responses, and requesting clarification. In commercial work, on the other hand, for example, in domain-specific spoken dialog systems and in interactions with virtual personal assistants and social robots, a narrower range of utterance functions is used, such as asking a question, issuing a command, and providing information. These functions are combined with a domain or task and are known as the user's *intent*—for example, a query to book a flight, a statement to provide a departure time, or a command to set an alarm.

A precursor of current approaches to intent identification was the classification of calls to call centers in order to automatically route the caller to the appropriate agent. For example, in the HMIHY system described in Chap. 4 a caller's utterance was classified into one of a predefined set of categories (or call types). A number of different machine-learning techniques have been employed in call classification (Tur and Deng 2011).

Identifying the intent of an utterance can be seen as the first stage of the classification of the utterance before extracting the slot value pairs appropriate to that intent. For example, if the intent is identified as a hotel booking, a set of slots relevant to booking a hotel room is selected. Wu et al. (2010) found that this two-stage classification helped to constrain the semantic analysis of the content of the utterance, as the slots to be filled were restricted to those relevant to the identified intent. This would be important in transactions involving multiple domains, as in the ATIS task where utterances were mainly about flight reservations but there could also be questions about ground transportation and airplane specifications (Tur and Deng 2011).

There are a number of platforms including Api.ai,[3] Wit.ai,[4] and Microsoft's Language Understanding Intelligent Service (LUIS)[5] that enable developers to

---

specify and train the natural language understanding component of a conversational interface by defining the intents and entities associated with the utterances that users are likely to say (see Williams et al. 2015a, b for more detail on LUIS). In Chap. 9, we provide a tutorial on how to do this using the Api.ai platform.

8.5 Analyzing the Content of the User's Utterances

The main task for SLU is to analyze the content of the user's utterance and to provide a representation of its meaning. Most research in SLU has focused on this task. Determining the content of the user's utterance is essential when interpreting a search query and retrieving a ranked set of relevant documents; understanding a question and finding an answer from a collection of retrieved documents; extracting the key items of interest from a piece of text; or finding the required values to fill slots in a semantic frame. In the following sections, we review a number of different approaches that are used widely in SLU and that are relevant to applications involving conversational interfaces.

8.5.1 *Tokenization*

Tokenization is one of the first steps in language processing in which a text is broken up into units called tokens and normalized for further processing. Tokens may be words, punctuation marks, or other units such as numbers. Compared with languages such as Chinese where there are no explicit word boundaries, tokenization would appear to be relatively straightforward in languages such as English where words are generally separated by white space. However, there are some problematic cases. For example, should "New York" count as one or two tokens? Contracted items are also an issue. For example: "aren't" could be rendered as "are" + "n't," although there is still the issue that "n't" is not a word in English and would need to be normalized to "not" for further processing as a linguistically significant unit. Other problems in English involve abbreviations—for example, should "dr." be normalized as "doctor" or "drive" and is a period (or full stop) a signal of an abbreviation or indicating the end of a sentence?

8.5.2 *Bag of Words*

One of simplest ways to analyze the user's input is to collect the words as a set (or bag), ignoring any syntactic structure or word order information but counting the occurrences of each word. This approach is also known as the *vector space model*. Stop words—i.e., words that do not contribute to the content of the text—are often deleted. Words that are morphological variants—for example, "walk," "walks," "walking," "walked"—are represented as the base form of the word (in this case, "walk") through a process known as *lemmatization*. The bag-of-words method is applied widely in information retrieval to retrieve documents that are relevant to the user's query, and in document classification to train a classifier based on word frequencies. The bag-of-words approach implicitly defines the topic of a query by retrieving documents or text matching the words. One application for conversational interfaces is in question answering, where the bag-of-words approach can be used to identify documents in which the answer to a user's question can be found.

An advantage of the bag-of-words approach is that it does not require any linguistic knowledge. However, for some queries a more precise analysis might be required that takes into account linguistic information such as syntactic structure and semantic content. For example, the terms in the two sentences "John chased the dog" and "the dog chased John" would have the same vector representation but it is obvious that the different orderings of the words result in different meanings.

8.5.3 Latent Semantic Analysis

Latent Semantic Analysis (LSA), also known as Latent Semantic Indexing (LSI), is another technique that does not require specific linguistic knowledge such as syntactic structure or semantics. In LSA, documents are also represented as bags of words but instead of comparing the actual words LSA compares the meanings (or concepts) behind the words. Patterns of words that frequently occur together in documents are grouped under the assumption that words that are close in meaning will occur in similar pieces of text. The words in a document are listed in a matrix in which each row stands for a unique word and each column represents a document. Each cell in the matrix represents the frequency with which a particular word appears in the document that is represented by that cell's column. Since the matrix is often very large and sparse, a process known as singular value decomposition is applied to the matrix to transform it and reduce its dimensionality. As a result, a pattern of occurrences and relations is created that shows measures of word similarities by mapping together terms that occur frequently in the same context. LSA can also create mappings between terms that do not directly co-occur but that mutually co-occur with other terms.

LSA has been used for a number of natural language processing (NLP) tasks, for example, to predict query-document topic similarity judgments so that documents of similar topical meaning can be retrieved even though different words are used in the query and document.

8.5.4   Regular Expressions

Regular expressions provide a method for pattern matching of surface strings. As shown in Chap. 7, pattern matching is used in chatbot applications in which the user's input is matched at the surface level against a large set of stored patterns. In the simplest case, patterns take the form of a sequence of words but more powerful matching can be achieved through the use of wildcards that can represent an arbitrary string within the input. For example, in the pattern my name is *, the symbol * can stand for any words that follow the specific words my name is.

Patterns are a form of regular expression. Regular expressions are used widely in NLP tasks to specify text strings for applications such as Web search and information extraction. They are particularly useful for predicable stretches of text, such as dates, currencies, and post codes. Moreover, the processing of a regular expression can be implemented using a finite state automaton, thus avoiding the computational complexity of the types of parsing algorithm required for more complex NLP tasks.

8.5.5   Part-of-Speech Tagging

Part-of-speech (POS) tagging involves labeling each word in the input with a tag that indicates its syntactic role, for example, whether it is a noun, verb, adverb, preposition, and so on. POS tags may be useful in regular expressions to specify particular items, such as proper nouns that represent names, or to help disambiguate words that can be tagged with more than one POS—for example, the word "book" can be either a noun or a verb, but in a sentence such as "book a flight" it should be tagged as a verb, since a noun would not be syntactically appropriate in a string preceding a determiner and another noun. For this reason, words are often tagged with their parts of speech as a first stage in syntactic parsing. Figure 8.1 shows an example of POS tagging using the Stanford CoreNLP online demo.[6]

Part-of-speech taggers can be either rule-based or stochastic. In rule-based taggers, a large set of hand-written rules is created that help to disambiguate words such as "book" that can have more than one part of speech. So, for example, if "book" is preceded by a determiner, then it is a noun and not a verb. Stochastic part-of-speech taggers approach the problem of ambiguity by learning tag probabilities from a training corpus and treating tagging as a sequence classification problem using Hidden Markov models (HMMs).

---

[6]http://nlp.stanford.edu:8080/corenlp/process. Accessed February 20, 2016.

NNP VBD CD CD NNS IN NNP IN NNP
John acquired two thousand shares in Microsoft in July

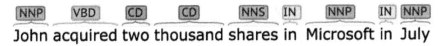

Fig. 8.1 Example of POS tagging using the Stanford CoreNLP online demo

PERSON NUMBER ORGANIZATION DATE
John acquired two thousand shares in Microsoft in July

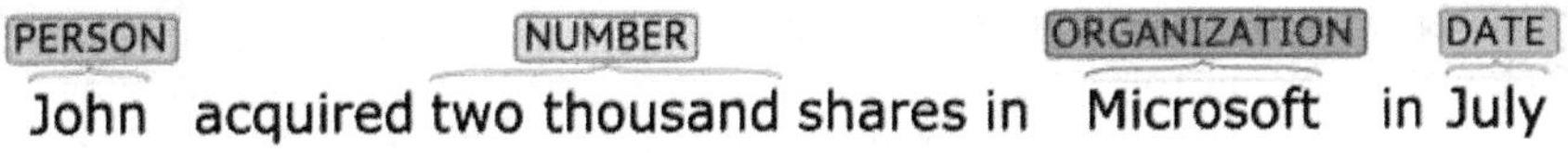

Fig. 8.2 Example of named entity recognition using the Stanford CoreNLP online demo

8.5.6 *Information Extraction*

Information extraction involves extracting information from texts to create machine-readable summaries or to create structured representations for knowledge bases and ontologies. In named entity recognition (NER), items such as persons, organizations, dates, and locations are extracted. Figure 8.2 shows an example of NER using the Stanford CoreNLP online demo.[7]

Relation extraction goes beyond NER by determining the relations among the named entities—for example, who did what to whom and when. Relations are used in a variety of applications, such as the Unified Medical Language System (UMLS), which consists of 134 entity types and 54 relations and is used to specify relations between items for medical purposes, such as "injury disrupts physiological function." Another example is Wikipedia's InfoBox relations in which a template is used to collect and present information about a topic across related articles.

Entities are used in platforms such as LUIS, Api.ai, and Wit.ai to specify the words and phrases relevant to a particular domain. For example, words such as "Italian," "Indian," and "Chinese" would be examples of the `cuisine` entity in the `Restaurant` domain.

8.5.7 *Semantic Role Labeling*

Semantic role labeling is a form of shallow semantic parsing in which the arguments of a verb are classified according to their semantic roles (Gildea and Jurafsky 2002). For example, in the string "book a flight to London on Friday," the object (sometimes known as the *theme*) of the verb "book" is "a flight," while "to London" is a location (or destination), and "on Friday" is a date. Semantic role labeling helps determine the meaning of a sentence independently of its syntactic structure. For example, in the sentences "obesity may cause diabetes" and "diabetes may be

[7]http://nlp.stanford.edu:8080/corenlp/process. Accessed February 20, 2016.

caused by obesity," the semantic roles of "diabetes" and "obesity" are the same even though their syntactic roles are different in the two sentences.

Automated semantic role labeling involves a series of steps. First the predicate of a sentence is identified and labeled, then the constituents that function as its arguments are found, and finally the semantic roles of those arguments are determined. Different predicates have different numbers of arguments. For example: "laugh" and "fall" are one-place predicates and "chase" and "kick" are two-place predicates. Automatic semantic role labellers are trained using supervised machine learning and resources such as FrameNet and PropBank. Often a syntactic analysis of the sentence is performed in order to identify the predicate and the constituents that function as its arguments, and then various features, such as the phrase type of the constituent, the constituent's headword, and the path in the parse tree from the constituent to the predicate, are used to classify the arguments (Gildea and Jurafsky 2002).

Semantic role labeling is useful for many NLP tasks. With reference to conversational interfaces, determining the semantic roles in an utterance is used in question answering systems where the semantic structures in the question are matched against similar structures in a collection of documents in order to find a potential answer to the question. Semantic role labeling can also be seen as a first step in creating a full semantic representation of an input in which a semantic grammar would then be used to specify the relations between the semantic roles.

8.6 Obtaining a Complete Semantic Interpretation of the Input

The methods described so far typically do not require a complete analysis of the user's utterance but rather extract from the utterance the parts that are relevant for a particular type of application. In this section, we consider the use of grammars to provide a semantic interpretation of the whole of the user's utterance. Obtaining the semantic interpretation can be done either directly using a semantic grammar or in a two-stage process in which the utterance is first parsed syntactically and then analyzed for its semantic interpretation. We discuss first the use of semantic grammars as they have been used widely to interpret the utterances spoken to conversational interfaces.

8.6.1   Semantic Grammar

The semantic grammar approach uses grammar rules in which the constituents of the utterance are classified in terms of their semantic roles. In the following example, the left-hand side of the rule specifies a non-terminal symbol that represents a semantic role, while the right-hand side specifies either terminal symbols that represent words in the input or another non-terminal symbol (e.g., CITY):

```
FLIGHT_BOOKING -> book | I would like | …
FLIGHT         -> (a) flight | flights
DESTINATION    -> to CITY
CITY           -> London | Paris | New York | …
DEPARTURE_DATE -> on DAY
DAY            -> Monday | Tuesday | …
```

Taking the utterance "book a flight to London on Friday" as input, the following is a possible output from the application of these rules:

```
FLIGHT_BOOKING:
    FLIGHT:
          DESTINATION: London
          DEPARTURE_DATE: Friday
```

The Phoenix spoken language system, which outperformed many other systems in the ATIS evaluations (Ward 1991), used manually constructed semantic grammars to detect keywords in an utterance and convert them to semantic tags.

Semantic grammars provide a shallow analysis and are not suitable for applications where finer distinctions need to be captured. For example, there is only one word that is different in the following sentences but the meaning representations should be different:

1. List all employees of the companies who are based in the city center
2. List all employees of the companies that are based in the city center

The interpretation of sentence 1 is that it is asking for a listing of employees who are based in the city center while the interpretation of sentence 2 is that it is asking for a listing of employees who are not necessarily based in the city center but who work for companies based there. This difference can only be picked up by an analysis that reflects the difference between the use of "who" and "that" in such sentences. Semantic grammars would not pick up such subtle distinctions.

Handcrafted semantic grammars have been used widely in commercial VoiceXML-based applications to output a semantic interpretation directly from the ASR output. Finite state grammars are defined as language models for recognition using SRGS (Speech Recognition Grammar Specification) (Hunt and McGlashan

2004), and the rules in the grammars are augmented with semantic interpretation tags based on Semantic Interpretation for Speech Recognition (SISR) (Van Tichelen and Burke 2007). Recognition is made less complex by specifying grammars for the predicted inputs at each step in a dialog. For example, if the system's prompt requires a "yes" or "no" response, then a grammar restricted to variants of "yes" and "no" is defined. A similar strategy is applied to other predictable inputs such as dates, times, and credit card details. However, this approach is not viable in more complex applications, for example in troubleshooting dialogs, where user inputs are more varied and less predictable, nor in user-directed systems where the user can issue a wide range of questions or commands to a conversational interface.

8.6.2 Syntax-Driven Semantic Analysis

In syntax-driven semantic analysis, it is assumed that the units retrieved using syntactic analysis can be mapped on to units of semantic analysis. Two grammar formalisms have been used widely for syntactic analysis: context-free grammar (CFG) and dependency grammar (DG).

8.6.2.1 Context-Free Grammar

CFG is based on the notion of constituency, where a constituent in a sentence is a group of words that forms a unit, for example, a noun phrase or a verb phrase. Figure 8.3 shows some CFG and lexical rules along with a parse tree based on application of the rule to the string "book a flight to London on Friday."

This is a syntactic representation of the input. To obtain a semantic interpretation, each of the constituents is analyzed using semantic attachment rules. Of course, a realistic grammar capable of parsing the diverse input to a conversational interface would require thousands of rules.

Adding more rules quickly gives rise to syntactic ambiguity, as the rules may combine with one another to produce multiple readings of a sentence. For example, in the rules in Fig. 8.3 PP is a constituent of NP. However, a PP can also be a constituent of VP as in the rule:

VP → Verb NP PP

Here, the PP modifies the action described, as in a sentence such as "book a flight with my credit card" where the meaning of the sentence is that the credit card

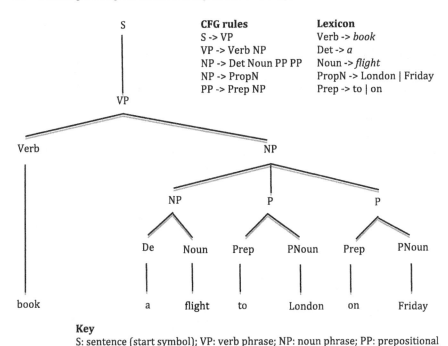

CFG rules
S -> VP
VP -> Verb NP
NP -> Det Noun PP PP
NP -> PropN
PP -> Prep NP

Lexicon
Verb -> *book*
Det -> *a*
Noun -> *flight*
PropN -> London | Friday
Prep -> to | on

Key
S: sentence (start symbol); VP: verb phrase; NP: noun phrase; PP: prepositional phrase; Det: determiner; PropN: proper noun; Prep: preposition

Fig. 8.3 CFG rules, lexical rules, and parse tree for the phrase "a flight to London on Friday"

is associated with the act of booking. The rule in this case is an example of VP attachment. On the other hand, in the sentence "book a flight to London on Friday," we would want the PP "on Friday" to be attached to "flight" and not to "book," as shown in Fig. 8.3, as otherwise we would have a reading in which the booking was to done on Friday. In fact, both readings would be syntactically possible given the rules and a parser would have no way of distinguishing between the two readings. This issue can quickly escalate. For example, Jurafsky and Martin (2009: 467) show how for the sentence "Show me the meal on Flight UA 386 from San Francisco to Denver" there are 14 alternative parse trees due to attachment ambiguities.

Finding all the parses for sentences with multiple readings could result in inefficient parsing of constituents, since the parser would repeatedly apply all the alternative rules in order to produce the subtrees associated with the different readings. For example, in the sentence "book a flight in the evening," the PP "in the evening" would be parsed as part of the VP rule and then again as part of the NP rule. Chart parsing has been proposed as a solution to this problem. With chart parsing, a data structure called a chart is used to store the parses of constituents as they are parsed so that they do not have to be parsed again later, and then the permissible combinations are constructed from the entries in the chart. Chart parsing is an example of dynamic programming in which solutions to subproblems are stored and then combined to construct an overall solution, resulting in efficiency gains.

Another problem is that where there are multiple parses for an utterance, there is no simple way of deciding which is the best parse. One solution to this problem is to use probabilistic parsers in which the CFG rules are assigned probability distributions, for example:

NP → Det Noun PP (0.2)

This rule states that the probability of this rule expansion is 0.2 in relation to the other NP rules in the grammar. The probabilities are learned from data such as a *treebank* of parsed sentences, for example, the Penn Treebank (Taylor et al. 2003). Thus, given appropriate probability assignments, the attachment of the PP *in the evening* in the sentence "book a flight in the evening" would return NP attachment as the most likely parse.

8.6.2.2 Dependency Grammar

DGs are an alternative to CFGs (Kübler et al. 2009). Here the focus is not on constituency as determined by CFG rules but on binary semantic or syntactic relations between the words in a sentence. Figure 8.4 shows a dependency representation for the sentence "book a flight":

Here the arrow labeled det from "flight" to "a" indicates that "flight" is the headword in the sequence "a flight" and "a" is a dependent word, in this case functioning as a modifier. The word "book" is the headword in a dependency relation with the word "flight," which functions as dobj (direct object). Figure 8.5 shows an analysis of the sentence "book a flight to London on Friday":

Here the pp relations represent the interpretation that the dependencies are between "flight" and "London," and "flight" and "Friday."

DGs have certain advantages over CFGs, particularly for parsing spontaneous spoken language (Béchet and Nasr 2009; Béchet et al. 2014). Another advantage is

Fig. 8.4 Dependency parse for the sentence "book a flight"

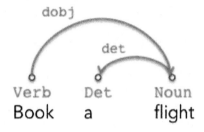

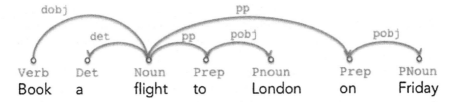

Fig. 8.5 Dependency parse for the sentence "book a flight to London on Friday"

that a single dependency structure can represent a number of surface level variations, whereas in a CFG different constituency structures would be required.

8.7 Statistical Approaches to Spoken Language Understanding

Up until the 1980s, NLP was dominated by the knowledge-based (or symbolic) paradigm in which grammars were handcrafted and meaning was represented using logic-based formalisms. At this time, the focus was on parsing written texts. In the 1990s, a paradigm shift occurred in which probabilistic and data-driven models that had already been deployed successfully in speech recognition were now applied to natural language understanding. The availability of large corpora of spoken and written text led to an increased use of machine-learning techniques so that the previously handcrafted rules of the knowledge-based paradigm could be learned automatically from labeled training data. At the same time, attention turned to SLU, for which statistical methods have proved to be more robust as they degrade gracefully with input that is previously unseen and potentially ill-formed.

Generally in statistically based semantic parsing, which is often referred to as *decoding*, the focus has been on extracting those aspects of meaning that are relevant to a particular task and domain rather than on developing a more general approach to semantic interpretation. Decoding is formalized as a pattern recognition problem in which the goal is to find a semantic representation of the meaning M of an utterance W that has the maximum a posteriori probability $P(M|W)$. Applying Bayes rule, this can be rewritten as the following decision rule (Wang et al. 2011):

$$\hat{M} = \arg\max_{M} P(M|W) = \arg\max_{M} P(W|M)P(M) \tag{8.1}$$

Here $P(M)$, the semantic prior, represents the probability of the meaning M, and $P(W|M)$, the lexicalization model, represents the probability of the utterance given M.

Two main types of statistical approach have been proposed in recent years to address the SLU task: generative and discriminative models (Raymond and

Riccardi 2007; Mairesse et al. 2009; Dinarelli 2010). The parameters of generative models refer to the joint probability of concepts and semantic constituents. Discriminative models learn a classification function based on conditional probabilities of concepts given words.

Generative models are robust to over-fitting and they are less affected by errors and noise. However, they cannot easily integrate complex structures. Discriminative models can easily integrate very complex features that can capture arbitrarily long-distance dependencies. On the other hand, they usually over-fit training data. These two statistical alternatives lead to very different and complex models depending on whether the SLU task involves dealing with simple classification problems or sequence labeling problems. Different models for SLU combine the strengths of both approaches (Dinarelli 2010).

8.7.1 Generative Models

The most representative generative models for language understanding are based on the Hidden Vector State model (HVS), Stochastic Finite State Transducers (SFST), and Dynamic Bayesian Networks (DBN).

8.7.1.1 The Hidden Vector State Model

The HVS model extends the discrete Markov model encoding the context of each state as a vector. State transitions are performed as stack shift operations followed by a push of a preterminal semantic category label as for a tree parser. As detailed in He and Young (2006), all the parameters of the model are denoted by λ and each state at time t is denoted by a vector of semantic concept labels.

$$c_t = [c_{1t}, c_{2t}, \ldots, c_{D,t}] \tag{8.2}$$

where c_{1t} is the preterminal concept and $c_{D,t}$ is the root concept.

The joint likelihood function is defined as:

$$L(\lambda) = \log P(W, C, N|\lambda) \tag{8.3}$$

where W is the word sequence, C is the concept vector sequence, and N is the sequence of stack pop operations.

The auxiliary function Q is defined to apply the Expectation-Maximization (EM) technique to maximize the expectation of $L(\lambda)$ given the observed data and current estimates:

$$Q(\lambda|\lambda^k) = E\left[\log P(W,C,N|\lambda)|W,\lambda^k\right] \sum_{C,N} P(C,N|W,\lambda) \log P(W,C,N|\lambda^k) \quad (8.4)$$

The term $P(W, C, N)$ is decomposed as follows:

$$P(W,C,N) = \prod_{t=1}^{T} P(n_t|W_1^{t-1}, C_1^{t-1}) \cdot P\left(c_t[1]|W_1^{t-1}, C_1^{t-1}, n_t\right) \cdot P(w_t|W_1^{t-1}, C_1^t)$$

$$(8.5)$$

These three terms are approximated in He and Young (2006) to solve this equation choosing a value for n_t, then selecting a preterminal concept tag $c_t[1]$, and finally selecting a word w_t.

8.7.1.2 Stochastic Finite State Transducers Models

SFSTs model the SLU task as a translation process from words to concepts, using Finite State Machines (FSM) to implement the stochastic language models (Raymond and Riccardi 2007). An FSM is defined for each elementary concept. They can be manually designed or learned using an annotated corpus. Each transducer takes words as input and outputs the concept tag conveyed by the accepted sentence.

All these transducers are grouped together into a single transducer, called λ_{W2C}, which is the union of all of them. A stochastic conceptual language model is computed as the joint probability $P(W, C)$:

$$P(W,C) = \prod_{i=1}^{k} P(w_i c_i|h_i) \quad (8.6)$$

where
$h_i = \{w_{i-1}c_{i-1}\ldots w_1 c_1\}$ This term is usually approximated by $\{w_{i-1}c_{i-1}, w_{i-2}c_{i-2}\}$ as a 3-gram model.
$C = \{c_1, c_2, \ldots, c_k\}$ is the sequence of concepts.
$W = \{w_1, w_2, \ldots, w_k\}$ is the sequence of words.

This model, called λ_{SLM}, is also encoded as an FSM. Given a new sentence W and its FSM representation λ_W, the translation process is to find the best path of the transducer resulting in the next composition:

$$\lambda_{SLU} = \lambda_W \circ \lambda_{W2C} \circ \lambda_{SLM} \quad (8.7)$$

In the SFST model, the best sentence segmentation (concept boundaries and labels) is computed over all possible hypotheses in λ_{SLU}.

8.7.1.3 Dynamic Bayesian Networks Models

Two interesting studies of DBNs for SLU are described in Lefèvre (2006, 2007). These models integrate automatic concept labeling and attribute-value extraction into a single model, thus allowing a complete stochastic modeling.

The concept of decoding is reformulated in this approach to combine the concept sequence with the value sequence as follows:

$$\hat{c}_1^N, \hat{v}_1^N = \arg\max_{c_1^N, v_1^N} p(c_1^N, v_1^N | w_1^T) \arg\max_{c_1^N, v_1^N} p(w_1^T | c_1^N, v_1^N) p(v_1^N | c_1^N) p(c_1^N) \quad (8.8)$$

where the concepts c_1^N are hypothesized by means of:

$$\hat{c}_1^N = \arg\max_{c_1^N} \sum_{v_1^N} \arg\max_{c_1^N, v_1^N} p(w_1^T | c_1^N, v_1^N) p(v_1^N | c_1^N) p(c_1^N) \quad (8.9)$$

Factored language models (FLM) can be used to improve the estimates of the terms in the previous equation (Bilmes and Kirchhoff 2003).

8.7.2 Discriminative Models

The most representative discriminative models for language understanding are based on support vector machines (SVMs) (Raymond and Riccardi 2007) and conditional random fields (CRFs) (Lafferty et al. 2001).

8.7.2.1 Support Vector Machines Models

SVMs are machine-learning algorithms included into the class of linear classifiers. They learn a hyperplane $H(\vec{x}) = \vec{w}\vec{x} + b = 0$ that divides training examples with maximum margin, where the learned parameters are given as follows:

- $\vec{x}$, the feature vector representation of a classifying object o,
- $\vec{w} \in R$,
- $b \in R$ (Vapnik 1998).

Applying the lagrangian optimization theory, the hyperplane can be represented in the following dual form:

$$\sum_{i=1...l} y_i \alpha_i \vec{x_i}\vec{x} + b = 0 \quad (8.10)$$

where $\vec{x_i}$ are the training examples, y_i is the label associated with $\vec{x_i}$ (+1 or −1), and α_i are the lagrange multipliers.

SVMs are applied to the SLU task to solve the concept-labeling problem as a sequence of classification problems using binary classifiers that can be trained taking into account non-local features and deciding the current concept in the sequence locally without using decisions made at previous steps.

8.7.2.2 Conditional Random Fields Models

CRFs are log-linear models that train conditional probabilities, taking into account features of the input sequence. Conditional dependence is captured using feature functions and a factor for probability normalization. A sequence level normalization leads to a linear chain CRF and a positional level normalization leads to the maximum entropy model (Bender et al. 2003).

The conditional probabilities of the concept sequences $c_N^1 = c_1, \ldots, c_N$ given the word sequences $w_N^1 = w_1, \ldots, w_N$ are calculated by means of:

$$p(c_1^N | w_1^N) = \frac{1}{Z} \prod_{n=1}^{N} \exp\left(\sum_{m=1}^{M} \lambda_m \cdot h_m(c_{n-1}, c_n, w_{n-2}^{n+2}) \right) \tag{8.11}$$

where λ_m is the vector of parameters to be trained, $h_m(c_{n-1}, c_n, w_{n-2}^{n+2})$ are the feature functions used to capture dependencies between input features (words and other features that can be associated with words in a certain window around the current word to be labeled) and the output concept (Macherey et al. 2009). Lexical, Prefix and Suffix, Capitalization, Transition, Prior, and Compound Features are usually used (Dinarelli 2010).

One of the problems for statistical SLU is the need for large amounts of annotated data to train the models. Increasingly, resources are becoming available in the public domain, for example, the ATIS corpus (Dahl et al. 1994), the DARPA Communicator corpus (Walker et al. 2002), and the LUNA corpus (Dinarelli et al. 2009; Hahn et al. 2011).

8.7.3 Deep Learning for Natural and Spoken Language Understanding

Since around 2006 deep learning methods have begun to outperform other machine-learning methods in areas such as speech recognition and image processing, and more recently also in NLP and SLU.

In deep learning for NLP (or *deep NLP*), syntactic categories such as N, V, NP, VP, and S, which are traditionally treated as discrete atomic categories, are represented as vector representations. At the level of individual words, word embeddings are used to encode among other things syntactic and semantic relationships of words to neighboring words, typically using a window of 5 words on each side of the word being represented. This allows semantically similar words to be mapped to neighboring points in a continuous vector space (Mikolov et al. 2013a).

Deep learning has been used successfully to perform many of the NLP tasks discussed in this chapter, including part-of-speech tagging, chunking (identifying phrases such as NP and VP), NER, and semantic role labeling (Collobert et al. 2011). For syntactic processing, a compositional approach is used in which a recursive neural network (RNN) combines two word vectors into a single vector representing a phrase and then combines the phrases into a higher-level category such as a sentence, working bottom-up from individual words to the higher-level categories (Mikolov et al. 2013b).

The deep NLP approach allows richer information to be captured compared with traditional vector-based approaches such as bag of words in which structural relations between words and phrases are not captured. In recent work, a compositional vector grammar (CVG) parser was used to jointly find syntactic structure and compositional semantic information by combining a probabilistic CFG with an RNN (Socher et al. 2013a). The grammar finds discrete syntactic categories such as NP and VP in the input while the RNN captures compositional semantic information. One advantage of this approach is that sentences that are syntactically ambiguous can be resolved using the more fine-grained semantic information—for example, in sentences involving PP-attachment, as in "book a flight on Friday" (where the flight is on Friday), as opposed to "book a flight with my credit card" (where the booking is made with a credit card).

Deep learning has been used for slot filling tasks, in which the items of information are extracted from the input that are required to fill slots in a semantic frame, for example, in a domain such as flight reservations to fill the slots for the departure and arrival cities. Mesnil et al. (2015) compared the use of RNNs with CRF approaches and found that the RNN-based models outperformed the CRF baseline in terms of error reduction on the Airline Travel Information Service (ATIS) benchmark. Other uses of deep learning in NLP include: logical semantics involving learning to identify logical relationship such as entailment and contradiction (Bowman et al. 2015), question answering (Kumar et al. 2015), and sentiment analysis (Socher et al. 2013b).

8.8 Summary

In this chapter, we have provided an overview of the various technologies of SLU, with particular reference to those technologies that are relevant to the extraction of the meaning of a user's utterance in an interaction with a conversational interface. We looked at various tasks, in particular dialog act recognition, topic identification, intent recognition, entity extraction, and analysis of the content of the user's utterance. A wide range of techniques and approaches was reviewed. Some of these involve low-level tasks such as tokenization that are used to analyze the text for higher-level tasks such as semantic analysis. There is no single method in SLU that is applicable in all tasks and in all types of application. On the whole, statistical techniques dominate the research literature, but there are still many proponents of handcrafted approaches, particularly in industry where designers wish to have greater control over the output of their systems and how this output is obtained.

In the next chapter, we introduce the Api.ai platform and provide a tutorial introduction with exercises that will enable readers to specify and extract semantic information from utterances. We will also describe some other open source tools that have been used widely for SLU.

Further Reading

In addition to Jurafsky and Martin (2009), which covers formal as well as stochastic approaches to SLU, and Huang et al. (2001), which focuses mainly on stochastic approaches, Allen (1995) provides a comprehensive overview of the knowledge-based tradition in NLP. Manning and Schütze (1999) is a thorough introduction to statistical approaches to natural language understanding with a useful companion Web site.[8] Tur and de Mori (2011) is a collection of chapters on statistical SLU. The chapter by Wang et al. (2011) in this collection is a detailed tutorial on semantic frame-based SLU. Henderson and Jurčíček (2012) describe three different parsers developed within the EU CLASSIC project (Lemon and Pietquin 2012) and evaluate their performance against several other parsers, including Phoenix. Reese (2015) is a practical guide to NLP with Java, focusing in particular on various approaches to organize and extract useful text from unstructured data. Dahl (2013) provides an excellent tutorial overview of approaches to NLP and of its application in areas including mobile personal assistants, dialog systems, and question answering.

Deep learning is a new approach to NLP. Online materials include: lectures 14, 15, and 16 in Chris Manning's Natural Language Processing course (CS 224 N/Ling 284) at Stanford[9]; Richard Socher's Deep Learning for Natural Language Understanding (CS224d) at Stanford[10]; Goldberg's Primer on neural network models for NLP[11]; and a series of videos by Chris Potts.[12]

SLU is a main topic at several academic conferences, in particular, ACL,[13] INTERSPEECH,[14] LREC,[15] COLING,[16] CoNLL,[17] NAACL,[18] and EACL.[19] Natural language is also a topic at commercially oriented conferences such as SpeechTEK[20]

---

[8]http://nlp.stanford.edu/fsnlp/. Accessed February 20, 2016.

[9]http://web.stanford.edu/class/cs224n/. Accessed February 20, 2016.

[10]http://cs224d.stanford.edu/. Accessed February 20, 2016.

[11]http://u.cs.biu.ac.il/~yogo/nnlp.pdf. Accessed February 20, 2016.

[12]https://www.youtube.com/playlist?list=PLfmUaIBTH8exY7fZnJss508Bp8k1R8ASG. Accessed February 20, 2016.

[13]https://www.aclweb.org/. Accessed February 20, 2016.

[14]http://www.interspeech2016.org/. Accessed February 20, 2016.

[15]http://lrec-conf.org/. Accessed February 20, 2016.

[16]http://nlp.shef.ac.uk/iccl/. Accessed February 20, 2016.

[17]http://ifarm.nl/signll/conll/. Accessed February 20, 2016.

[18]http://naacl.org/. Accessed February 20, 2016.

[19]http://www.eacl.org/. Accessed February 20, 2016.

[20]http://www.speechtek.com/. Accessed February 20, 2016.

and MobileVoice.[21] Key journals are: IEEE Transactions on Audio, Speech, and Language Processing (since 2014 renamed as IEEE/ACM Transactions on Audio, Speech, and Language Processing); Speech Communication; Computer Speech and Language; Computational Linguistics; and Natural Language Engineering.

References

Allen JF (1995) Natural language understanding, 2nd edn. Benjamin Cummings Publishing Company Inc., Redwood

Allen JF, Core M (1997) Draft of DAMSL: dialog act markup in several layers. The Multiparty Discourse Group, University of Rochester, Rochester. http://www.cs.rochester.edu/research/cisd/resources/damsl/RevisedManual/. Accessed 20 Jan 2016

Béchet F, Nasr A (2009) Robust dependency parsing for spoken language understanding of spontaneous speech. In: Proceedings of the 10th annual conference of the international speech communication association (Interspeech2009), Brighton, UK, 6–10 Sept 2009, pp 1027–1030. http://www.isca-speech.org/archive/archive_papers/interspeech_2009/papers/i09_1039.pdf. Accessed 21 Jan 2016

Béchet F, Nasr A, Favre B (2014) Adapting dependency parsing to spontaneous speech for open domain language understanding. In: Proceedings of the 15th annual conference of the international speech communication association (Interspeech2014), Singapore, 14–18 Sept 2014, pp 135–139. http://www.isca-speech.org/archive/archive_papers/interspeech_2014/i14_0135.pdf. Accessed 21 Jan 2016

Bender O, Macherey K, Och F-J, Ney H (2003) Comparison of alignment templates and maximum entropy models for natural language understanding. In: Proceedings of the 10th conference of the European chapter of the association for computational linguistics, Budapest, Hungary, 12–17 Apr 2003, pp 11–18. doi:10.3115/1067807.1067811

Bilmes JA, Kirchhoff K (2003) Factored language models and generalized parallel backoff. In: Proceedings of the 2003 conference of the North American chapter of the association for computational linguistics on human language technology (HLT-NAACL 2003), Edmonton, Canada, 27 May–1 June 2003, pp 4–6. doi:10.3115/1073483.1073485

Black AW, Burger S, Conkie A, Hastie H, Keizer S, Lemon O, Merigaud N, Parent G, Schubiner G, Thomson B, Williams JD, Yu K, Young S, Eskenazi M (2011) Spoken dialogue challenge 2010: comparison of live and control test results. In: Chai JY, Moore JD, Passonneau RJ, Traum DR (eds) Proceedings of the SIGDial 2011 conference, Portland, Oregon, June 2011. http://www.aclweb.org/anthology/W/W11/W11-2002.pdf. Accessed 23 Jan 2016

Bowman SR, Potts C, Manning CD (2015) Recursive neural networks can learn logical semantics. In: Proceedings of the 3rd workshop on continuous vector space models and their compositionality (CVSC), Beijing, China, 26–31 July 2015, pp 12–21. doi:10.18653/v1/w15-4002

Collobert R, Weston J, Bottou L, Karlen M, Kavukcuoglu K, Kuksa P (2011) Natural language processing (almost) from scratch. J Mach Learn Res 12:2493–3537. http://arxiv.org/pdf/1103.0398.pdf

Dahl DA (2013) Natural language processing: past, present and future. In: Neustein A, Markowitz JA (eds), Mobile speech and advanced natural language solutions. Springer Science +Business Media, New York, pp 49–73. doi:10.1007/978-1-4614-6018-3_4

Dahl DA, Bates M, Brown M, Fisher W, Hunicke-Smith K, Pallett D, Pao C, Rudnicky A, Shriberg E (1994) Expanding the scope of the ATIS talk: the ATIS-3 corpus. In: Proceedings

---

[21]http://www.mobilevoiceconference.com/. Accessed February 20, 2016.

of the workshop on human language technology (HLT'94), Association for computational linguistics, Stroudsburg, pp 43–48. doi:10.3115/1075812.1075823

Dinarelli M (2010) Spoken language understanding: from spoken utterances to semantic structures. Dissertation, University of Trento, 2010. http://eprints-phd.biblio.unitn.it/280/

Dinarelli M, Quarteroni S, Tonelli S, Moschitti A, Riccardi G (2009) Annotating spoken dialogs: from speech segments to dialog acts and frame semantics. In: Proceedings of SRSL 2009, the 2nd workshop on semantic representation of spoken language, Association for computational linguistics, Athens, Greece, March, pp 34–41. doi:10.3115/1626296.1626301

Gildea D, Jurafsky D (2002) Automatic labeling of semantic roles. Comp Linguist 28(3):245–288. doi:10.1162/089120102760275983

Godfrey JJ, Holliman EC, McDaniel J (1992) Switchboard: telephone speech corpus for research and development. In: Proceedings of the international conference on acoustics, speech, and signal processing (ICASSP-92), vol 1. San Francisco, 23–26 March, pp 517–520. doi:10.1109/icassp.1992.225858

Hahn S, Dinarelli M, Raymond C, Lefevre F, Lehnen P. De Mori R, Moschitti A, Ney H, Riccardi G (2011) Comparing stochastic approaches to spoken language understanding in multiple languages. IEEE Trans Speech Audio Proc 19(6):1569–1583. doi:10.1109/tasl.2010.2093520

Hakkani-Tür D, Béchet F, Riccardi G, Tur G (2006) Beyond ASDR 1-best: using word confusion networks in spoken language understanding. Comp Speech Lang 20(4):495–514. doi:10.1016/j.csl.2005.07.005

He Y, Young S (2006) Spoken language understanding using the hidden vector state model. Speech Commun 48(3–4):262–275. doi:10.1016/j.specom.2005.06.002

Henderson J, Jurčíček F (2012) Data-driven methods for spoken language understanding. In: Lemon O, Pietquin O (eds) Data-driven methods for adaptive spoken dialogue systems: computational learning for conversational interfaces. Springer, New York, pp 19–38. doi:10.1007/978-1-4614-4803-7_3

Huang X, Acero A, Hon H-W (2001) Spoken language processing: a guide to theory, algorithm, and system development. Prentice Hall, Upper Saddle River

Hunt A, McGlashan S (2004) Speech recognition grammar specification version 1.0. http://www.w3.org/TR/speech-grammar/. Accessed 21 Jan 2016

Jurafsky D, Martin JH (2009) Speech and language processing: an introduction to natural language processing, computational linguistics, and speech recognition, 2nd edn. Prentice Hall, Upper Saddle River

Keizer S, op den Akker R, Nijholt A (2002) Dialogue act recognition with Bayesian networks for Dutch dialogues. In: Proceedings of the 3rd SIGdial workshop on discourse and dialogue, Philadelphia, PA, pp 88–94. doi: 10.3115/1118121.1118134

Klüwer T, Uszkoreit H, Xu F (2010) Using syntactic and semantic based relations for dialog act recognition. In: Proceedings of the 23rd international conference on computational linguistics (COLING'10), Association for computational linguistics, Stroudsburg, pp 570–578. http://www.aclweb.org/anthology/C10-2065.pdf. Accessed 21 Jan 2016

Kübler S, McDonald R, Nivre J (2009) Dependency parsing. Synthesis lectures on human language technologies. Morgan and Claypool Publishers, San Rafael. doi:10.2200/S00169ED1V01Y200901HLT002

Kumar A, Irsoy O, Ondruska P, Iyyer M, Bradbury J, Gulrajani I, Socher R (2015) Ask me anything: dynamic memory networks for natural language processing. arXiv: http://arxiv.org/abs/1506.07285. Accessed 21 Jan 2016

Lafferty JD, McCallum A, Pereira FCN (2001) Conditional random fields: probabilistic models for segmenting and labeling sequence data. In: Proceedings of the 18th international conference on machine learning (ICML'01), Williamstown, MA, USA, 28 June–1 July 2001, pp 282–289. http://dl.acm.org/citation.cfm?id=655813

Lefèvre F (2006) A DBN-based multi-level stochastic spoken language understanding system. In IEEE spoken language technology workshop, Palm Beach, Aruba, 10–13 Dec 2006, pp 82–85. doi:10.1109/slt.2006.326822

Lefèvre F (2007) Dynamic bayesian networks and discriminative classifiers for multistage semantic interpretation. In: Proceedings of the IEEE international conference on acoustics, speech and signal processing (ICASSP'07), vol 4. Honolulu, HI, USA, 15–20 Apr 2007, pp 13–16. doi:10.1109/ICASSP.2007.367151

Lemon O, Pietquin O (eds) (2012) Data-driven methods for adaptive spoken dialogue systems: computational learning for conversational interfaces. Springer, New York. doi:10.1007/978-1-4614-4803-7

Macherey K, Bender O, Ney H (2009) Applications of statistical machine translation approaches to spoken language understanding. IEEE Trans Speech Audio Proc 17(4):803–818. doi:10.1109/tasl.2009.2014262

Mairesse F, Gašić M, Jurčíček F, Keizer S, Thomson B, Yu K, Young S (2009) Spoken language understanding from unaligned data using discriminative classification models. In: Proceedings of the IEEE international conference on acoustics, speech and signal processing (ICASSP'09), Taipei, Taiwan, 19–24 Apr 2009, pp 4749–4752. doi:10.1109/icassp.2009.4960692

✳ Manning CD, Schütze H (1999) Foundations of statistical natural language processing. MIT Press, Cambridge

Mesnil G, Dauphin Y, Yao K, Bengio Y, Deng L, Hakkani-Tur D, He X, Heck L, Tur G, Yu D, Zweig G (2015) Using recurrent neural networks for slot filling in spoken language understanding. IEEE/ACM Trans Speech Audio Proc 23(3):530–539. doi:10.1109/taslp.2014.2383614

Mikolov T, Chen K, Corrado GS, Dean J (2013a) Efficient representation of word representations in vector space. In: Proceedings of the international workshop on learning representations (ICLR) 2013, Scottsdale, AZ, USA, 2–4 May 2013. http://arxiv.org/pdf/1301.3781.pdf. Accessed 21 Jan 2016

Mikolov T, Sutskever I, Chen K, Corrado GS, Dean J (2013b) Distributed representations of words and phrases and their compositionality. In: Proceedings of the twenty-seventh conference on neural information processing systems 26 (NIPS 2013), Lake Tahoe, 5–10 Dec 2013. http://papers.nips.cc/paper/5021-distributed-representations-of-words-and-phrases-and-their-compositionality.pdf. Accessed 21 Jan 2016

Nagata M, Morimoto T (1994) First steps toward statistical modeling of dialogue to predict the speech act type of the next utterance. Speech Commun 15:193–203. doi:10.1016/0167-6393(94)90071-x

Raymond C, Riccardi G (2007) Generative and discriminative algorithms for spoken language understanding. In: Proceedings of the 8th annual conference of the international speech communication association (Interspeech 2007), Antwerp, Belgium, 27–31 Aug, pp 1605–1608. http://www.isca-speech.org/archive/archive_papers/interspeech_2007/i07_1605.pdf. Accessed 21 Jan 2016

✳ Reese RM (2015) Natural language processing with Java. Packt Publishing Ltd., Birmingham

Shriberg E, Bates R, Stolcke A, Taylor P, Jurafsky D, Ries K, Coccaro N, Martin R, Meteer M, Ess-Dykema CV (1998) Can prosody aid the automatic classification of dialog acts in conversational speech? Lang Speech 41(3–4):439–487. http://www.ncbi.nlm.nih.gov/pubmed/10746366. Accessed 21 Jan 2016

Socher R, Bauer J, Manning CD, Ng AY (2013a) Parsing with compositional vector grammars. In: Proceedings of the 51st meeting of the association for computational linguistics (ACL) 2013, Sofia, Bulgaria, 4–9 Aug. http://www.aclweb.org/anthology/P/P13/P13-1045.pdf. Accessed 21 Jan 2016

Socher R, Perelygin A, Wu JY, Chuang J, Manning CD, Ng AY, Potts C (2013b) Recursive deep models for semantic compositionality over a sentiment treebank. In: Proceedings of the 2013 conference on empirical methods in natural language processing (EMNLP 2013), Seattle, Washington, USA, 18–21 Oct 2013, pp 1631–1642. http://www.aclweb.org/anthology/D/D13/D13-1170.pdf. Accessed 21 Jan 2016

Stolcke A, Ries K, Coccaro N, Shriberg E, Bates R, Jurafsky D, Taylor P, Martin R, Meteer M, Van Dykema C (2000) Dialogue act modelling for automatic tagging and recognition of conversational speech. Comp Linguist 26(3):339–371. doi:10.1162/089120100561737

Suzanne J, Klein A, Maier E, Maleck I, Mast M, Quantz J (1995) Dialogue acts in Verbmobil. Report 65, University of Hamburg, DFKI GmbH, University of Erlangen, TU Berlin

Taylor A, Marcus M, Santorini B (2003) The penn treebank: an overview. In: Abeillé A (ed) Treebanks: building and using parsed corpora. Kluwer Academic Publishers, Dordrecht, pp 5–22. doi:10.1007/978-94-010-0201-1_1

Tur G, de Mori R (eds) (2011) Spoken language understanding: systems for extracting semantic information from speech. Wiley, Chichester. doi:10.1002/9781119992691

Tur G, Deng L (2011) Intent determination and spoken utterance classification. In: Tur G, de Mori R (eds) Spoken language understanding: systems for extracting semantic information from speech. Wiley, Chichester, pp 93–118. doi:10.1002/9781119992691.ch4

Tur G, Hakkani-Tür D (2011) Human/human conversation understanding. In: Tur G, de Mori R (eds) Spoken language understanding: systems for extracting semantic information from speech. Wiley, Chichester, pp 225–255. doi:10.1002/9781119992691.ch9

Van Tichelen L, Burke D (2007) Semantic interpretation for speech recognition (SISR) version 1.0. http://www.w3.org/TR/semantic-interpretation/. Accessed 21 Jan 2016

Vapnik VN (1998) Statistical learning theory. Wiley, Chichester

Walker MA, Rudnicky A, Prasad R, Aberdeen J, Bratt EO, Garofolo J, Hastie H, Le A, Pellom B, Potamianos A, Passonneau R, Roukos S, Sanders G, Seneff S, Stallard D (2002) DARPA communicator: cross-system results for the 2001 evaluation. In: Proceedings of the 7th international conference on spoken language processing (ICSLP2002), vol 1. Denver, Colorado, pp 273–276. http://www.isca-speech.org/archive/archive_papers/icslp_2002/i02_0269.pdf. Accessed 21 Jan 2016

Wang YY, Deng L, Acero A (2011) Semantic frame-based spoken language understanding. In: Tur G, de Mori R (eds) Spoken language understanding: systems for extracting semantic information from speech. Wiley, Chichester, pp 41–91. doi:10.1002/9781119992691.ch3

Ward W (1991) Understanding spontaneous speech: the Phoenix system. In: Proceedings of the IEEE international conference on acoustics, speech, and signal processing (ICASSP-91), Toronto, Canada, 14–17 Apr, pp 365–367. doi:10.1109/icassp.1991.150352

Webb N, Hepple M, Wilks Y (2005) Dialogue act classification using intra-utterance features. In: Proceedings of the AAAI workshop on spoken language understanding, Pittsburgh, PA, pp 451–458. http://staffwww.dcs.shef.ac.uk/people/Y.Wilks/papers/AAAI05_A.pdf. Accessed 21 Jan 2016

Williams JD (2012) A belief tracking challenge task for spoken dialog systems. In: NAACL-HLT Workshop on future directions and needs in the spoken dialog community: tools and data. NAACL 2012, Montreal, 7 June, 2012, 23–24. http://www.aclweb.org/anthology/W12-1812. Accessed 23 Jan 2016

Williams JD, Kamal E, Ashour M, Amr H, Miller J, Zweig G (2015a) Fast and easy language understanding for dialog systems with Microsoft Language Understanding Intelligent Service (LUIS). In: Proceedings of the SIGDIAL 2015 conference, Prague, Czech Republic, 2–4 Sept 2015, pp 159–161. doi:10.18653/v1/w15-4622

Williams JD, Niraula NB, Dasigi P, Lakshmiratan A, Suarez CGJ, Reddy M, Zweig G (2015b) Rapidly scaling dialog systems with interactive learning. In: Lee GG, Kim HK, Jeong M, Kim J-H (eds) Natural language dialog systems and intelligent assistants. Springer, New York, pp 1–12. doi:10.1007/978-3-319-19291-8_1

Wu W-L, Lu R-Z, Duan J-Y, Liu H, Gao F, Chen Y-Q (2010) Spoken language understanding using weakly supervised learning. Comp Speech Lang 24(2):358–382. doi:10.1016/j.csl.2009.05.002

Chapter 9
Implementing Spoken Language Understanding

Abstract There is a wide range of tools that support various tasks in spoken language, some of which are particularly relevant for processing spoken language understanding in conversational interfaces. Here, the main task is to detect the user's intent and to extract any further information that is required to understand the utterance. This chapter provides a tutorial on the Api.ai platform that has been widely used to support the development of mobile and wearable devices as well as applications for smart homes and automobiles. The chapter also reviews some similar tools provided by Wit.ai, Amazon Alexa, and Microsoft LUIS, and looks briefly at other tools that have been widely used in natural language processing and that are potentially relevant for conversational interfaces.

9.1 Introduction

As we saw in Chap. 8, spoken language understanding is not a uniform technology. Some tasks such as tokenization and part-of-speech tagging are used for low-level processing that will contribute to subsequent stages of analysis, while others perform more high-level tasks such as providing a semantic interpretation of an utterance. For conversational interfaces, a widely used approach involves detecting the intent behind the user's utterance and extracting the relevant entities. The intent might be some action such as setting an alarm, scheduling a meeting, sending a text message, or booking a table at a restaurant. The entities are those elements of meaning that are essential to the execution of the action, such as the time for the alarm or the meeting, the recipient of the text message and its content, or the number of people for the restaurant booking.

A number of spoken language understanding platforms take the approach of intent recognition and entity extraction, including Api.ai, Wit.ai, Amazon Alexa, and Microsoft LUIS. In this chapter, we will focus mainly on the Api.ai platform. In the final part of the chapter, we will provide an overview of some other tools that are used widely for spoken language understanding.

© Springer International Publishing Switzerland 2016
M. McTear et al., *The Conversational Interface*,
DOI 10.1007/978-3-319-32967-3_9

9.2   Getting Started with the Api.ai Platform

In 2010, the company Api.ai released the highly rated and widely used virtual personal assistant known as Assistant or Assistant.ai (previously known as Speaktoit Assistant). More recently, the platform on which Assistant is based has been made available to developers.[1] The Api.ai platform supports the development of mobile apps as well as apps involving wearables, robots, motor vehicles, smart homes, and smart TV. Currently, 15 languages are available, including English, German, French, Chinese, Korean, Portuguese, and Spanish, on a range of platforms and coding languages, including iOS, Apple Watch, Android, Cordova, Python, C#, Xamarin, Windows Phone, and Unity.

The following is an example of an interaction with the Api.ai platform:

> User input: *Where is the nearest Italian restaurant*
> System output (simplified):
> *action: maps.places,*
> *parameters:*
> *"cuisine": "Italian",*
> *"request_type": "address",*
> *"sort": ["nearest"],*
> *"venue_type": "restaurant".*

The system output displays information associated with the intent that has been identified from the user's input. Intents are one of the key concepts in Api.ai. An intent represents a mapping between the user's input and the action to be taken by the app—in this case, a map search for a place. The parameters specify further details about the action that are extracted as *entities* from the user's input. We will describe how to specify Intents and entities later in the chapter.

9.2.1   Exercise 9.1 Creating an Agent in Api.ai

On the Api.ai platform, an app is known as an *agent*. There is a range of documents and videos on the Api.ai Web site that explain how to develop an agent using the Api.ai platform.[2] Here for the sake of exposition, we provide a simplified set of instructions.

---

[1]https://api.ai/. Accessed February 21, 2016.
[2]https://docs.api.ai/v3/docs/get-started. Accessed February 21, 2016.

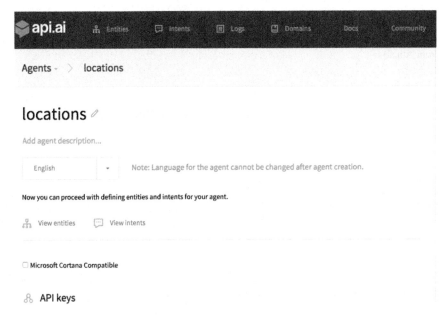

Fig. 9.1 Creating an agent in Api.ai (reproduced with permission from Api.ai)

1. Go to the platform Web site and sign up for a free account.
2. Create an agent called `locations`. This will result in the screen as shown in Fig. 9.1.
3. Choose a language for your agent (the default is English).
4. Scroll down to the section API keys. Here, you will see three keys listed. For the Android app that we will create shortly, you will need the subscription key and the client access token. Take a note of these and where to find them when you need them later for your app.

The developer access token will also be required if you wish to create intents and entities using `curl` commands.

9.2.2   Exercise 9.2 Testing the Agent

You can test your agent by typing or speaking queries into the test console, which is in the top right corner of the Api.ai developer console, as shown in Fig. 9.2.

Fig. 9.2 The Api.ai
developer console
(reproduced with permission
from Api.ai)

ⓘ Please use test console above to try
a sentence.

Table 9.1 Parameters and
values for the input "where is
the nearest Italian restaurant"

Parameter	Value
Cuisine	Italian
Request_type	Address
Sort	["Nearest"]
Venue_type	Restaurant

1. Test and train your agent by typing in queries such as:

> Where is the nearest Italian restaurant
> Find me an Italian restaurant
> Show me directions to an Italian restaurant
> I am hungry

You will see that all of these queries will return the action: maps.places, but there will be differences in the parameters and their values. For example, for the input "where is the nearest Italian restaurant," the parameters, shown in Table 9.1, are returned.

At this point, you might be wondering how the system is able to process your queries and return the responses, particularly as you have not yet specified any spoken language understanding rules for your agent. The answer is that Api.ai provides a range of built-in domains (or predefined knowledge packages).[3] When you make a request in the test console, the responses that you have specified for your agent will be displayed along with those produced from the Domains Knowledge Base. Currently as we do not have our own specified inputs, only those from the Domains Knowledge Base will be displayed. When we make our query using an app, the response that we have specified for the agent will be displayed, unless none is available, in which case a domain response will be returned.

---

[3]http://api.ai/docs/domains/. Accessed February 21, 2016.

You may find that you do not need to specify your own intents and entities for your app since the Domains Knowledge Base is quite extensive and is continually being expanded. However, if you find that not all inputs receive appropriate responses or if you are creating an app for a domain that is not covered in the Domains Knowledge Base, then you will need to specify the Intents and entities. We will show how to do this later. First, we will show how to create an Android app that allows you to interact with your agent.

9.3 Creating an Android App for an Agent

For this app, we will use the `VoiceActivity` class that was introduced in Chap. 6 and that we used to interact with the Pandorabots platform in Chap. 7. As before, we will process a speech recognition result, but this time we will send it to Api.ai's `AIDataService` for processing. An alternative would be to use Api.ai's integrated speech recognition available with their Android SDK. The SDK and sample code and a tutorial showing how to do this can be found here.[4] There is also a tutorial on how to use your own speech recognition in conjunction with the SDK library.

In this app, we will take a recognized string and call the `AIDataService` to process the string and produce a semantic parse that will be displayed on the device as the recognized query, the recognized action (i.e., intent), and the parameters of the intent.

9.3.1 Exercise 9.3 Producing a Semantic Parse

1. Create an Android app and call it Understand.
2. Modify the `MainActivity` class that we used in Chap. 6 and adapt it for use with the Api.ai platform.

Note: you can download the following code from GitHub, in the folder chapter9 of the ConversationalInterface[5] repository. To run the code, you will need to insert the subscription key and the client access token that were provided when you created your agent. In the following, we explain the modifications to `MainActivity` and other changes that are required to run the app.

Modifications to MainActivity

1. Add the following additional imports (Code 9.1).
2. Change the value of the `LOGTAG` to `Understand`
3. Add a `TextView` to the `main.xml` file to display the results of the spoken language understanding analysis (Code 9.2).

---

[4]https://github.com/api-ai/api-ai-android-sdk#android-sdk-for-apiai. Accessed February 21, 2016.
[5]http://zoraidacallejas.github.io/ConversationalInterface/. Accessed March 2, 2016.

```
import ai.api.AIConfiguration;
import ai.api.AIService;
import ai.api.AIDataService;
import ai.api.AIServiceException;
import ai.api.model.AIRequest;
import ai.api.model.AIResponse;
import ai.api.model.Result;

import android.widget.TextView;

import com.google.gson.JsonElement;
import java.util.Map;
```

Code 9.1   Additional imports in `MainActivity`

```
<TextView
    android:layout_width="wrap_content"
    android:layout_height="wrap_content"
    android:id="@+id/resultTextView"
    android:layout_alignParentBottom="true"
    android:layout_alignParentLeft="true"
    android:layout_alignParentStart="true"
    android:layout_alignParentRight="true"
    android:layout_alignParentEnd="true" />
```

Code 9.2   The `TextView` for the `Understand` app

```
resultTextView = (TextView) findView-
ById(R.id.resultTextView);
```

Code 9.3   Reference to the `TextView` in the `onCreate` method

4. In the `onCreate` method, add a reference to the `TextView` to display the results of the spoken language processing (Code 9.3).
5. Modify the method `processAsrResults`, as follows:

   – Remove the section containing the query to Pandorabots and the processing of the Pandorabots response.
   – Add the method: `apiSLU(userQuery)`. The main work will be done in this method.
   – Remove the methods `processBotErrors`, `processBotResults`, and `removeTags`, which are not required for this application.

6. In the method `apiSLU()`:

   – Set up the configuration of the `AIDataService` using the API keys `Client Access Token` and `Subscription Key` that can be found if you look at the settings for your agent (Code 9.4).

```
final AIConfiguration config = new AIConfiguration("Client
access token", "Subscription key", AIConfigur-
tion.SupportedLanguages.English,
AIConfiguration.RecognitionEngine.System);
final AIDataService aiDataService =

    new AIDataService(this, config);
```

Code 9.4 Configuration of the AIDataService

```
new AsyncTask<String,Void,AIResponse>() {
  @Override
  protected AIResponse doInBackground(String... strings) {
      final String request = strings[0];
      try {
          final AIRequest aiRequest

                = new AIRequest(request);
          final AIResponse response

                = aiDataService.request(aiRequest);
          return response;
      } catch (AIServiceException e) {

          //Manage exception
      }
      return null;
  }
  @Override
  protected void onPostExecute(AIResponse aiResponse) {
      if (aiResponse != null)
          //Process the response
  }
}.execute(userQuery);
```

Code 9.5 Sending a request to the AIService and getting a response

7. Send the request to the AIService and get a response using a background thread (Code 9.5).
8. To process the response, we will extract the intent's action and its parameters and show them as a formatted string in the GUI by means of resultTextView (Code 9.6).
   To use the AIDataService, we need to declare the Api.ai sdk as a dependency in the build.gradle (Module:app) file along with references to gson and commons:io.

```
Result result = aiResponse.getResult();
// Get the parameters
String parameterString = "";
if (result.getParameters() != null

    && !result.getParameters().isEmpty()) {


    for (final Map.Entry<String, JsonElement> entry :

      result.getParameters().entrySet()) {
          parameterString += "(" +

          entry.getKey() + ", " + entry.getValue() + ") ";
    }
}
// Show the results in the TextView

resultTextView.setText("Query:" +

  result.getResolvedQuery() +
  "\nAction: " + result.getAction() +
  "\nParameters: " + parameterString);
```

Code 9.6 Extracting and displaying the intent and its parameters

Fig. 9.3 Output from the query "where is the nearest Italian restaurant"

Understand

Press to speak

Query:where is the nearest Italian restaurant
Action: maps.places
Parameters: (sort, ["nearest"])
(request_type, "address") (cuisine,
"Italian") (venue_type, "restaurant")

9.3.2 Exercise 9.4 Testing the App

Test the app with a variety of relevant inputs. For the input "where is the nearest Italian restaurant" you should receive an output as displayed in Fig. 9.3.

Table 9.2 Additional values for the `venue_type` parameter

Location	Venue_type
Petrol station	*Gas station*
Chemist	*Drugstore*
Pub	*Pub*
Golf course	*Golf*

9.4 Specifying Your Own Entities and Intents

So far we have made use of the built-in entities and Intents provided in the Domains Knowledge Base. However, we may need to specify Intents and entities for our agents that are not provided. The documents and videos at the Api.ai Web site show how to do this. Here, we will provide a basic overview. More details can be found by consulting the Api.ai documentation as well as the tutorials.

If we try a number of queries such as "where is the nearest Italian restaurant" we will see that the parameter `venue_type` is returned with the value `restaurant`. Similarly, by substituting other locations, we can return different values for `venue_type`, as shown, for example, in Table 9.2:

Before we create new entities and intents, we should check how similar entities and Intents are represented in the Domains Knowledge Base in order to maintain consistency. If we look at the section Maps: Places in the Domains Knowledge Base, we can see from the extract shown in Fig. 9.4 that the relevant action for matching expressions about locations is `maps.places`. This is followed by a list of parameters under the heading Data, along with notes and examples. For the complete listing, see the documentation.[6] Note that there are also built-in system entities for concepts such as color, date and time, email, geography, names (common given names), music (artists and genres), numbers, and phones.[7]

We will focus on the `venue_type` parameter. The following are some entities that are not currently included:

Discotheque,
Football stadium,
Running club,
Tea shop (Note: coffee shop is included).

9.4.1 Exercise 9.5 Creating Entities

1. Click on the tab `Create new entity`.
2. Name the new entity `venue_type`.

---

[6]https://docs.api.ai/docs/maps-and-points-of-interest#maps-places. Accessed February 21, 2016.
[7]http://api.ai/docs/getting-started/entity-overview.html#system-entities. Accessed February 21, 2016.

Maps: Places

Action

Name	Description
maps.places	Requests to search for different types of venues

Data

Name	Parent	Data type	Required / Optional	Description	Request examples
request_type	data	string	required	Defines the type of request: *search, address, postalcode, phone*	Address of the nearest cafe (request_type=address) Find hotel nearby (request_type=search)
venue_type	data	string	optional	Defines the type of requested venue.	Find a restaurant (venue_type=restaurant)
venue_title	data	string	optional	Defines the title of the venue.	Find Seven Glaciers restaurant (venue_title=Seven Glaciers)
venue_chain	data	string	optional	Defines the chain of the venue.	Find FamilyMart (venue_chain=FamilyMart)
location	data	string	optional	Defines the location.	Find a restaurant in new york (location=new york)
sort	data	list	optional	Specifies sort order: *nearest, best, cheapest*	Find a restaurant nearby (sort=[nearest]) Find a good restaurant (sort=[best]) Find the cheapest supermarket nearby (sort=[nearest, cheapest])
stars	data	digit	optional	Defines the class of the venue: *1, 2, 3, 4, 5*	Find 3 stars restaurant (stars=3) Find 5 star hotel (stars=5)

Fig. 9.4 Parameters for the `maps.places` domain in the Domains Knowledge Base (reproduced with permission from Api.ai)

3. Enter the locations listed above ("discotheque," "football stadium," etc.) in the rows labeled `Enter synonym`.
4. As there can be more than one way to express an entity, you can add further synonyms on the same row as the entity. For example, "discotheque" can have the synonym "disco."
5. When you have added the entities and their synonyms, click `Save`. Your list of entities should look like this (Fig. 9.5).

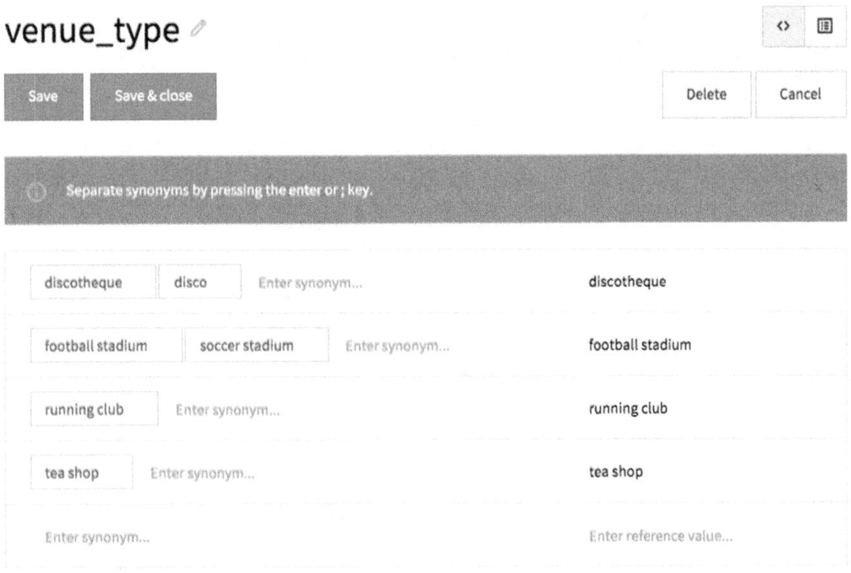

venue_type ✎

‹› ▦

| Save | Save & close | | Delete | Cancel |

ⓘ Separate synonyms by pressing the enter or ; key.   ✕

discotheque	disco	Enter synonym...		discotheque
football stadium	soccer stadium	Enter synonym...		football stadium
running club	Enter synonym...		running club
tea shop	Enter synonym...		tea shop
Enter synonym...			Enter reference value...

Fig. 9.5 New entities for the parameter venue_type (reproduced with permission from Api.ai)

Now that we have created some entities, we need to associate these entities with the expressions in an intent that the user might use to make requests. Note that to refer to an entity within an intent, you need to prefix it with the symbol @, for example, @venue_type.

9.4.2 Exercise 9.6 Creating an Intent

1. Click on the tab Create an intent.
2. Name your intent that makes reference to your entities: @venue_type.
3. In the section entitled User says, enter the expressions that the user can say.

The following is a list of some of the expressions that a user might say when looking for a location:

> Where is the nearest @venue_type
> I am looking for the nearest @venue_type
> Find me the nearest @venue_type
> Find me a @venue_type [nearby, around here, in this neighborhood, in this area]
> Is there a @venue_type [nearby, around here, in this neighborhood, in this area]

(Note: The items in square brackets are alternatives)

Normally, there would be a much larger set of expressions to cover all of the different ways in which the user might express Intents, but these examples will suffice for the purposes of illustration.

You will notice that the entity reference is automatically expanded e.g., @venue_type:venue_type. The additional part is the *alias*, which acts like a variable name that can be referenced later to specify the parameter name as $alias—for example, in this case, it is venue_type. We will explain aliases in more detail later.

You now need to add an action to your intent to specify the type of intent—in this case, following the action types specified for similar queries in the Domains Knowledge Base, we will call the action maps.place.

---

where is the nearest @venue_type ✎

⌄ Set priority

( ) **Contexts**

IN   Add input context...

OUT | location | Add output context...

▢ **User says**   Machine learning ON

| where is the nearest @venue_type:venue_type |
| i am looking for the nearest @venue_type:venue_type |
| find me the nearest @venue_type:venue_type |
| is there a @venue_type:venue_type nearby, around here, in this neighbourhood, in this area |
| find me a @venue_type:venue_type nearby, around here, in this neighbourhood, in this area |
| Add user expression... |

+ Add

⋇ **Action**

| maps.places |

PARAMETER NAME	VALUE	DEFAULT VALUE
venue_type	$venue_type	Enter default value...
request_type	Enter value...	address
sort	Enter value...	nearest

Fig. 9.6 The intent page (reproduced with permission from Api.ai)

Parameter	Value
Request_type	Address
Sort	Nearest
Venue_type	Football stadium

Table 9.3 Parameters and values for the input "where is the nearest football stadium"

One parameter has already been added automatically: `venue_type` with the value `$venue_type`. You will need to add further parameters, for which the values should be inserted in the `default value` cell:

request_type	address
sort	nearest

At this point, your intent page for the expressions listed above should look like this (Fig. 9.6).

9.4.3 Exercise 9.7 Testing the Custom Entities and Intents

You can now test your agent in the test console using various combinations of the sentences listed in Fig. 9.3. For the query "where is the nearest football stadium" the action: `maps.places` is returned along with the parameters and values shown in Table 9.3.

9.5 Using Aliases

Aliases act like variable names. They are required since an expression might use an entity more than once and the alias helps to keep them distinct. For example, a travel app might need to reference a source city as well as a destination city. To do this, the `@city` entity should have different aliases to represent the source and the departure, e.g., `@city:fromCity` and `@city:toCity`. These can then be referred to as `$fromCity` and `$toCity`, respectively.

9.6 Using Context

Contexts represent the current context of a user expression. This is useful for dealing with follow-up queries on the same topic. For example, the first query might be:

Where is the nearest running club

and the next query might be:

and the nearest football stadium

In order to make this possible, we need to set contexts for the queries—an output context for the initial query and an input context for the follow-up query.

9.6.1 Exercise 9.8 Defining Contexts

1. Open the intent: Where is the nearest @venue_type.
2. Click on "Define contexts."
3. Add an output context, e.g., location.
4. Add a new intent for the expression: and the nearest @venue_type.
5. For this intent, to make it a follow-up query, add the input context: location.
6. Also, in case of another follow-up query, add the output context: location.

Variables can be collected in a context for later reference. To retrieve a parameter, type: #context-name.parameter-name.

Contexts expire after 5 queries or 5 min, unless they are set again. For more information on context, see here.[8]

9.7 Creating a Slot Filling Dialog

In the apps created so far, the interaction has taken the form of a one-shot dialog in which the user specified everything required in the input and the system extracted the user's intent and its parameters. However, in many interactions, the user may not specify all the required parameters in a single utterance and the app has to collect the remaining parameters in a subsequent dialog. In VoiceXML, each field represents a parameter value to be elicited from the user (see Chap. 11). Api.ai provides a different approach in which the possible user inputs are listed and the parameters in the inputs are marked as required and associated with prompts that the system uses to elicit any missing parameters.[9,10] In the following, we provide a simple example for setting an alarm.

---

[8]http://api.ai/docs/getting-started/quick-start-contexts.html. Accessed February 21, 2016.
[9]https://api.ai/blog/2015/11/09/SlotFilling/. Accessed February 21, 2016.
[10]https://docs.api.ai/docs/dialogs. Accessed February 21, 2016.

Let us assume that to set an alarm, the system needs to know the time and the day, but that the user may not necessarily specify values for both these parameters in the initial input. Examples of complete inputs would be:

Set the alarm 7 tomorrow.
Wake me 7 tomorrow.

Examples of incomplete inputs might be:

Set the alarm.
Set the alarm 7.
Wake me tomorrow.
Wake me 7.

We can create a single intent to handle a range of inputs such as these as follows. Note that you will have to experiment with wildcards if you want to accept input such as "set the alarm for seven on Friday."

9.7.1 Exercise 9.9 Creating a Slot-Filling Dialog

1. Create an agent called "alarm."
2. Create an intent called "set alarm."
3. List examples of the user inputs, but replace the time expression with the variable @sys.any:time and the day expression with the variable @sys.any:day.
4. Scroll down to the list of parameters—in this case, you should see the parameters time and day listed.
5. Mark these parameters as required and provide prompts to elicit their values. The resulting parameter list should look like Fig. 9.7.
6. Finally, create the speech response in which the elicited values are spoken back using the variables $time and $day, for example, Setting the alarm for $time $day.

REQUIRED ❷	PARAMETER NAME ❷	ENTITY ❷	VALUE	PROMPTS ❷
☑	time	@sys.any	$time	what time? [1]
☑	day	@sys.any	$day	what day? [1]

Fig. 9.7 Parameters for setting an alarm (reproduced with permission from Api.ai)

Note that currently the Understand app only provides a semantic interpretation for your initial input so you will have to test these additional functionalities using the Api.ai developer console. To include these functionalities in the app, you would need to extend it to capture the prompts from the system, the user responses, and the system feedback.

You can take this example further by allowing the user to confirm these values or to edit and correct them. This involves creating additional Intents and also specifying input and output contexts to link the Intents together. An example of how to do this is provided here.[11] Again, you can test these additions in the developer console or you can extend the Understand app to include them.

9.7.2 Exercise 9.10 Additional Exercises

1. Add further intents to allow the user to confirm or correct the time and date values. Clear the context so that each new interaction begins with a new context. You can see an example of how to do this here.[12]
2. Implement some other examples that might require a parameter collection dialog —for example, booking a table at a restaurant.
3. Compare this approach to the approach used in VoiceXML for slot filling (see Chap. 11).

9.8 Overview of Some Other Spoken Language Understanding Tools

There is a wide range of tools available to support various tasks in spoken language understanding. Some of these are intended for processing spoken language understanding in conversational interfaces where the main task is to detect the user's intent (what action they wish to have performed) along with any further information that is required to complete the action (the entities or slots associated with the intent). Other tools provide different functions, ranging from lower-level tasks such as tokenization and part-of-speech tagging to tasks such as a deep semantic analysis of the input. In the following sections, we provide a brief overview of a selection of these tools.

---

[11]https://api.ai/blog/2015/11/09/SlotFilling/. Accessed February 21, 2016.
[12]https://api.ai/blog/2015/11/23/Contexts/. Accessed February 21, 2016.

9.8.1   Tools Using Intents and Entities

Intent detection and entity extraction are used by a number of providers, including Wit.ai, Amazon Alexa, and Microsoft LUIS. In the following, we provide a brief overview of these platforms.

9.8.1.1   Wit.ai

Wit.ai,[13] which was acquired in January 2015 by Facebook, provides an API for creating voice-activated interfaces and, in particular, for extracting a semantic representation from the user's utterances. Its approach is similar to that of Api.ai. The developer begins by listing some utterances that the app should understand. Next, an existing intent can be selected from the Wit.ai community or a custom intent can be created. Entities can be highlighted in the input utterances, and built-in entities can be selected or custom entities can be specified. Once a sufficient set of sample utterances has been collected and labeled, they can be tested and validated in the console. Machine learning is then applied to the utterances that have been validated and to similar utterances from the community to generate new classifiers.

Wit.ai has a feature called `roles` that is similar to the `alias` feature in Api.ai. For example, it might be necessary to label a location as either an origin or a destination in an utterance such as "I want to go from Belfast to London." The built-in entity `wit/location` would capture both "Belfast" and "London" as locations but by specifying their roles in the utterance "Belfast" would be assigned the role `origin` and "London" the role `destination`.

Normally, entities are declared by specifying a word or phrase, known as a *span*, in the utterance to which the entity refers. However, there are some entities that cannot be specified in this way, as their values cannot be inferred from a particular span in the utterance but only from the utterance as a whole. These entities are known as *spanless entities*. The following example from the Wit.ai documentation illustrates the following[14]:

1. It's too hot here.
2. Can you make it more hot.

The entity that should be extracted from 1 should be `temperature = down`, while for 2, the entity should be `temperature = up`. However, the single word "hot" is not sufficient and the meaning has to be inferred from the utterance as a whole.

---

[13]https://wit.ai/. Accessed February 21, 2016.

[14]https://wit.ai/docs/console/complete-guide#advanced-topics-link. Accessed February 21, 2016.

There is also a feature called `state` that preserves context, as in the following example:

> System: Do you prefer email or SMS?
> User: SMS.

To specify that the system should be in a specific state when it expects responses such as "email" or "SMS," additional intents for `email_answer` and `sms_answer` are created that will only be activated in certain defined states.

Wit.ai is available on several platforms, including iOS, Android,[15] Windows, and Raspberry Pi. Further details about the Wit.ai platform can be found at the documentation page.[16]

9.8.1.2   Amazon Alexa

Amazon Alexa is the voice service that powers Amazon Echo and other devices such as Amazon Fire and Amazon TV. Alexa consists of a set of *skills* that represent tasks such as performing an action and searching for information. The Alexa Skills Kit (ASK) provides a number of APIs and tools for adding skills to Alexa. There is a wide range of built-in skills, including playing music from various providers, answering general knowledge questions, and carrying out actions such as setting alarms.[17] Designing a skill involves defining a voice interface[18] that specifies a mapping between the user's utterances and the intents that the system can handle. This mapping consists of two inputs:

1. An intent schema in the form of a JSON structure.
2. The spoken input data, consisting of sample utterances and custom values.

As with other similar approaches, an Intent represents an action that the user wishes to have carried out. Intents can also have slots. For example, an Intent called `SetAlarm` would have a slot for the time. ASK provides a number of built-in Intents as well as support for slot types such as `AMAZON.NUMBER`, `AMAZON.DATE`, `AMAZON.TIME`, `AMAZON.US_CITY`, and others. Developers can create their own custom slot types in which the values can be anything that can be spoken by the user that is supported by the skill.

---

[15]https://wit.ai/docs/android/3.1.0/quickstart. Accessed February 21, 2016.

[16]https://wit.ai/docs. Accessed February 21, 2016.

[17]https://developer.amazon.com/public/solutions/alexa/alexa-skills-kit/getting-started-guide. Accessed February 21, 2016.

[18]https://developer.amazon.com/public/solutions/alexa/alexa-skills-kit/docs/defining-the-voice-interface. Accessed February 21, 2016.

The Sample Utterances File contains all the possible utterances for a particular intent with the slots specified using curly brackets, as in:

What {TIME} do you want to set the alarm for?

There are three types of intent in Alexa:

1. Full intents,
2. Partial intents,
3. No intent.

With a full Intent, the user says everything in a single utterance that is required to fulfill their request, as in a one-shot query. With a partial intent, one or more slots are missing and the system has to prompt for the missing values. This is similar to a slot-filling dialog (see further Chap. 11). A no intent is where the user's intent is unclear and the system has to request clarification by presenting a short list of options to choose from.[19]

Users interact with Alexa either by asking a question or by telling Alexa to do something. Requests are sent to the Alexa service in the cloud and routed to the specific service that provides the logic and a response. The response can be a spoken message as well as a card that is displayed in the Alexa app. The text to be spoken can be marked up using the Speech Synthesis Markup Language (SSML).

9.8.1.3 Microsoft Language Understanding Intelligent Service (LUIS)

The Microsoft Language Understanding Intelligent Service (LUIS) is part of Microsoft's Project Oxford[20] that provides APIs for several areas of artificial intelligence, including vision, speech, and language understanding. For language understanding, LUIS enables developers to apply machine learning techniques to create custom Intents and entities or use preexisting models from Bing and Cortana.

The process of creating Intents and entities in LUIS involves first specifying the Intents and entities relevant to a particular domain and then listing some sample utterances and labeling them according to their intents and the entities. A model is built using machine learning that is then trained and retrained through a process known as *active learning*. In active learning, an initial classifier is trained from a small seed set of labeled examples; then, this classifier is applied to a larger set of unseen examples and problematic classifications are identified. These examples are

---

[19]https://developer.amazon.com/public/solutions/alexa/alexa-skills-kit/docs/alexa-skills-kit-voice-design-handbook. Accessed February 21, 2016.

[20]https://www.projectoxford.ai/. Accessed February 21, 2016.

labeled and the classifier is retrained. The process continues until a satisfactory level of performance is achieved.

As with other platforms, a large set of prebuilt entities is provided. To improve the performance of the classifier, the developer can specify features to be used by the machine learning algorithm in the identification of entities. For example, a feature such as `SearchPhrase` might include a list of many words and phrases that could be used in a spoken search query, such as "tell me about" or "information about". There are also several tools to assist with the development process, such as the Interactive Classification and Extraction (ICE) tool. For further details, see the papers by Williams et al. (2015a, b) as well as the video demo at the LUIS Web site.[21]

9.8.2 Toolkits for various other NLP Tasks

Stanford CoreNLP[22] is a suite of natural language processing tools for a variety of tasks including tokenization, sentence splitting, part-of-speech (POS) tagging, morphological analysis, named entity recognition, syntactic parsing, coreference resolution, and sentiment analysis (Manning et al. 2014). Several languages are supported. The output can also be formatted as XML, JSON, and CoNLL, or displayed using a pretty print format.

The Apache OpenNLP tools[23] support various natural language processing tasks including tokenization, sentence segmentation, part-of-speech tagging, named entity extraction, chunking, parsing, and coreference resolution. There is a tutorial on how to use Apache OpenNLP through a set of simple examples here.[24]

GATE[25] is open-source free software that has been developed over more than 15 years at the University of Sheffield, England. GATE supports a variety of text processing tasks, including an information extraction system (ANNIE), parsing, morphological analysis, tagging, and semantic annotation.

LingPipe[26] is a toolkit for processing text involving tasks such as part-of-speech tagging, named entity recognition, topic classification, spelling correction, and sentiment analysis (Baldwin and Dayanidhi 2014).

Natural Language Toolkit (NLTK)[27] is a platform for building Python programs to analyse natural language data, including a suite of libraries for

---

[21]https://www.luis.ai/. Accessed February 21, 2016.

[22]http://nlp.stanford.edu/software/corenlp.shtml. Accessed February 21, 2016.

[23]http://opennlp.apache.org/. Accessed February 21, 2016.

[24]http://www.programcreek.com/2012/05/opennlp-tutorial/. Accessed February 21, 2016.

[25]https://gate.ac.uk/. Accessed February 21, 2016.

[26]http://aliasi.com/lingpipe/index.html. Accessed February 21, 2016.

[27]http://www.nltk.org/. Accessed February 21, 2016.

classification, tokenization, stemming, tagging, parsing, and semantic reasoning. The Natural Language Toolkit is described in an online book.[28]

AlchemyAPI[29] provides tools for a number of text processing tasks including entity extraction, sentiment analysis, keyword extraction, concept tagging, relation extraction, taxonomy classification, author extraction, language detection, text extraction, and feed detection. AlchemyAPI was acquired in March 2015 by IBM and is being integrated into IBM's Watson platform to provide deep learning technology for next-generation cognitive computing applications.

9.9 Summary

There is a wide range of tools that can be used to process the input to a conversational interface and produce a semantic analysis. Many of these tools treat the semantic analysis as the identification of the intent behind the user's utterance along with the parameters that are related to this intent. We have provided a tutorial and a set of exercises showing how to produce a semantic analysis using the Api.ai platform. Other platforms such as Wit.ai, Amazon Alexa, and Microsoft LUIS take a similar approach but differ in the tools that they provide. The final part of the chapter looked briefly at a number of other tools that may be used for spoken language processing tasks.

The next stage for the conversational interface is to take the semantic interpretation of the user's input and decide how to respond. This is the task for the dialog manager. Chapter 10 reviews current approaches to dialog management, while Chap. 11 provides tutorials showing how dialog management can be implemented.

Further Reading
Most of the Web sites referenced above provide online demos. There is also a collection of online demos at the Conversational Technologies Web site.[30] Some open-source NLP tools are described here.[31] This page provides a list of NLP tools and resources.[32]

---

[28]http://www.nltk.org/book/. Accessed February 21, 2016.

[29]http://www.alchemyapi.com/. Accessed February 21, 2016.

[30]http://www.conversational-technologies.com/nldemos/nlDemos.html. Accessed February 21, 2016.

[31]https://opensource.com/business/15/7/five-open-source-nlp-tools. Accessed February 21, 2016.

[32]http://ils.unc.edu/~stephani/nlpsp08/resources.html#tools. Accessed February 21, 2016.

References

Baldwin B, Dayanidhi K (2014) Natural language processing with Java and LingPipe Cookbook. Packt Publishing, Birmingham, UK

Manning CD, Surdeanu M, Bauer J, Finkel J, Bethard SJ, McClosky D (2014) The stanford CoreNLP natural language processing toolkit. In: Proceedings of the 52nd annual meeting of the association for computational linguistics: system demonstrations, Baltimore, 23–25 June 2014, pp 55–60. doi:10.3115/v1/p14-5010

Williams JD, Kamal E, Ashour M, Amr H, Miller J, Zweig G (2015a) Fast and easy language understanding for dialog systems with Microsoft language understanding intelligent service (LUIS). In: Proceedings of the SIGDIAL 2015 conference, Prague, Czech Republic, 2–4 Sept 2015, pp 159–161. doi:10.18653/v1/w15-4622

Williams JD, Niraula NB, Dasigi P, Lakshmiratan A, Suarez CGJ, Reddy M, Zweig G (2015b) Rapidly scaling dialog systems with interactive learning. In Lee GG, Kim HK, Jeong M, Kim J-H (eds) Natural language dialog systems and intelligent assistants. Springer, New York, pp 1–12. doi:10.1007/978-3-319-19291-8_1

Chapter 10
Dialog Management

Abstract One of the core aspects in the development of conversational interfaces is to design the dialog management strategy. The dialog management strategy defines the system's conversational behaviors in response to user utterances and environmental states. The design of this strategy is usually carried out in industry by handcrafting dialog strategies that are tightly coupled to the application domain in order to optimize the behavior of the conversational interface in that context. More recently, the research community has proposed ways of automating the design of dialog strategies by using statistical models trained with real conversations. This chapter describes the main challenges and tasks in dialog management. We also analyze the main approaches that have been proposed for developing dialog managers and the most important methodologies and standards that can be used for the practical implementation of this important component of a conversational interface.

10.1 Introduction

This chapter describes the main aspects, tasks, and approaches involved in the dialog management (DM) process. Section 10.2 defines the DM process and the tasks involved. To illustrate the complexity of dialog strategy design, this section analyzes two frequently arising design issues: the interaction strategy and the choice of a confirmation strategy.

Dialog management can be classified into handcrafted approaches using rules, which are described in Sect. 10.3, and statistical approaches using machine learning methodologies, which are described in Sect. 10.4. Statistical approaches have been proposed to model the variability in user behaviors and to allow the exploration of a wider range of strategies. This section provides two detailed examples of the practical application of reinforcement learning and corpus-based supervised learning for the development of statistical dialog managers.

© Springer International Publishing Switzerland 2016   209
M. McTear et al., *The Conversational Interface*,
DOI 10.1007/978-3-319-32967-3_10

10.2   Defining the Dialog Management Task

As has been described in previous chapters, different modules and processes must cooperate to achieve the main goal of a conversational interface. Automatic speech recognition (ASR) is the process of obtaining the text string corresponding to an acoustic input (see Chap. 5). Once the speech recognition component has recognized what the user uttered, it is necessary to understand what was said. Spoken language understanding (SLU) is the process of obtaining a semantic interpretation of a text string (see Chap. 8). This generally involves morphological, lexical, syntactical, semantic, discourse, and pragmatic knowledge.

DM relies on the fundamental task of deciding what action or response a system should take in response to the user's input. There is no universally agreed definition of the tasks that this component has to carry out to make this decision. Traum and Larsson (2003) state that DM involves four main tasks:

1. Updating the dialog context.
2. Providing a context for interpretation.
3. Coordinating other modules.
4. Deciding the information to convey and when to do it.

Thus, the dialog manager has to deal with different sources of information such as the SLU results, results of database queries, application domain knowledge, and knowledge about the users and the previous dialog history. The complexity of DM depends on the task, the extent to which the dialog is flexible, and who has the initiative in the dialog, the system, the user, or both.

Although DM is only one part of the information flow of a conversational interface, it can be seen as one of the most important tasks given that this component encapsulates the logic of the speech application. The selection of a particular action depends on multiple factors, such as the output of ASR (e.g., measures that define the reliability of the recognized information), the dialog interaction (e.g., the number of repairs carried out so far), the application domain (e.g., guidelines for customer service), and the responses and status of external back-ends, devices, and data repositories. Given that the actions of the system directly impact users, the dialog manager is largely responsible for user satisfaction. Because of these factors, the design of an appropriate DM strategy is at the core of conversational interface engineering.

ASR is not perfect, so one of the most critical aspects of the design of the dialog manager involves error handling. The ASR and SLU components make errors, and so conversational interfaces are generally less accurate than humans. For all of this technology to work, severe limitations need to be imposed on the scope of the applications and this requires a great amount of manual work for designers. One common way to alleviate errors is to use techniques aimed at establishing a confidence level for the ASR result and to use that to decide when to ask the user for confirmation, or whether to reject the hypothesis completely and reprompt the user. Too many confirmations as well as too many reprompts could annoy users. So it is important to reduce the number of confirmations and rejections to a minimum while at the same time preserving a reasonable level of accuracy.

In order to complete the tasks described above and to decide "what to say" and "what to do," the dialog manager needs to track the dialog history and update its representation of the current state of the dialog. In addition, the dialog manager needs a dialog strategy that defines the conversational behavior of the system, for example, when to take the initiative in a dialog or when to establish common ground. The performance of a conversational interface is highly dependent on the quality of its DM strategy.

Unfortunately, the design of a good strategy is far from trivial since there is no clear agreement as to what constitutes a good strategy. User populations are often diverse, thus making it difficult to foresee which form of system behavior will lead to a quick and successful completion of the dialog. In addition, the omnipresence of ASR errors leads to constant uncertainty about the true intentions of the user. As applications become more complex, the process of designing robust dialog strategies becomes very time-consuming and expensive.

To further illustrate the complexity involved in dialog strategy design, the following subsections briefly discuss two frequently arising design issues: interaction strategies and the choice of a confirmation strategy.

10.2.1 Interaction Strategies

The interaction strategy of a conversational interface determines who takes the initiative in the dialog—the system, the user, or both. Three types of interaction strategies are often distinguished in the literature: user-directed, system-directed, and mixed-initiative.

When user-directed initiative is used, the user always has the initiative in the dialog, and the system just responds to the user's queries and commands. The main problem with this strategy is that the user may think that they are free to say whatever they want, which tends to cause ASR and SLU errors. The following example shows a user-directed dialog with the EDECAN dialog system for booking sports facilities at the Technical University of Valencia, Spain (Griol et al. 2014).

System: Welcome to the sports service. How can I help you?
User: I want to book a basketball court.
System: You can see the available courts on the screen.
User: I want a court for November 12th in the afternoon.
System: Sorry, do you want a court from 15 to 21 hours?
User: Yes.
System: There are several courts available on Friday November 12th. Please select one of them.
User: The one at 7 pm.

System: There is only one court with the required features. Do you want to book it?
User: Yes.
System: The basketball court has been booked. Anything else?
User: Yes, I want to know my list of court bookings.
System: You can see the courts on the screen. Anything else?
User: Yes, cancel the one at 12 am.
System: The specified court has been released. Anything else?
User: No, thank you.
System: Thank you for using the sports service. Goodbye.

When system-directed initiative is used, the system has the initiative in the dialog, and the user just answers its queries. The advantage of this strategy is that it helps to constrain the user's input, leading to more efficient dialogs. The disadvantage is a lack of flexibility, as the user is restricted to behaving according to the system's expectations, providing the necessary data to perform some action in the order specified by the system. The following example shows a system-directed dialog with the LUNA dialog system that supports problem-solving tasks in the domain of software/hardware repair (Griol et al. 2014).

System: Help Desk, I am Paola! How may I help you?
User: Hello I have a problem with the printer. It is not printing.
System: Where do you work?
User: For CSI Piemonte.
System: Tell me your name and surname.
User: My name is Ugo Foscolo.
System: I didn't hear you very well. Please repeat and speak louder.
User: Ugo Foscolo.
System: Can you tell me your phone number?
User: Zero eleven forty fifty sixty-three.
System: Can you tell me the brand and model?
User: Epson EPL nine thousand.
System: Is the printer on now?
User: It is on.
System: Can you tell me the message on the printer display?
User: It says No network connection.
System: Can you see the printer from the control panel of your PC?
User: It is active.
System: Now, I will tell you the number of the ticket for your call. Your ticket number is 2609192. Have you noted it down?
User: Yes.

> System: Thank you! A technical expert from Help Desk will contact you as soon as possible. Have a nice day!

When the mixed-initiative strategy is used, both the user and the system can take the initiative in the dialog. The advantage is that the system can guide the user in the tasks that are to be performed, while the user can take the initiative, ask questions, introduce new topics, and provide overinformative responses. In the following example of a mixed-initiative dialog with the LUNA dialog system involving the provision of railway information related to timetables, fares, trip times, types of trains, and services the user responds to the system's question about the destination city by stating the day as well as the city (Griol et al. 2008).

> System: Welcome to the railway information system. How can I help you?
> User: I want to know timetables from Madrid.
> System: Do you want to leave from Madrid?
> User: Yes.
> System: Tell me the destination city.
> User: To Granada next Friday.
> System: Do you want timetables for next Friday?
> User: Yes
> System: I have the following information about trains from Madrid to Granada. There is only one train at 8.30. Anything else?
> User: No, thank you.
> System: Thanks for using this service. Have a good trip.

Although not shown in this dialog, an example of a mixed-initiative dialog in which the user takes the initiative instead of responding directly to the system's question is shown in the following example:

> System: Do you want timetables for next Friday?
> User: Are there trains before 6 in the morning?

The problem with mixed-initiative dialogs is that the user can potentially say anything and by introducing a different topic may cause the system to lose track of its agenda. Thus, mixed-initiative dialogs require advanced ASR and SLU capabilities as well as the ability to maintain and monitor the dialog history and the system's agenda.

10.2.2   Error Handling and Confirmation Strategies

Because of the limitations of current ASR and SLU technologies that are employed in conversational interfaces, it is necessary to assume that the information captured by these components (and possibly also by sensors) may be uncertain or ambiguous. Error propagation from the ASR and SLU modules is one of the main problems that the dialog manager must try to detect and correct.

To deal with the uncertainty problem, the ASR and SLU components typically employ confidence scores attached to the frame slots, for example, real numbers in the range between 0 and 1. A confidence score that falls below a threshold indicates that the data item in the slot must be either confirmed or rejected by the system.

Two types of confirmation strategy are often employed: explicit confirmation and implicit confirmation. With explicit confirmation, the system generates an additional dialog turn to confirm the data item obtained from the previous user turn, as in the following example:

> User: I want to know timetables from Madrid.
> System: Do you want to leave from Madrid?
> User: Yes.

The disadvantage of explicit confirmations is that the dialog tends to be lengthy due to these additional confirmation turns, and this makes the interaction less efficient and even excessively repetitive if all the data items provided by the user have to be confirmed.

The following is an example of an implicit confirmation:

> User: I want to know timetables from Madrid.
> System: What time do you want to leave from Madrid?

When the implicit confirmation strategy is used, the system includes some of the user's previous input in its next question. If the user answers the question directly, for example, in this case by stating a departure time, then it is assumed that the previous information about the destination is implicitly confirmed and no additional turns are required. However, it is the user's responsibility to make a correction if the system has misrecognized the information and this can lead to the user producing utterances that are beyond the scope of the ASR and SLU components, for example:

> User: I want to know timetables from Madrid.
> System: What time do you want to leave from Madrid?

> User: No, I just wanted to know about times from Madrid but I might be departing from somewhere else depending on whether I have the use of the car next Friday.

These confirmation strategies are useful for avoiding misunderstandings, for example, when the system has understood something from its interaction with the user but is uncertain about how accurate it is. One related, but different situation is non-understanding, which occurs when the system has not been able to collect any data from its interaction with the user. In this case, two typical strategies for handling the error are to ask the user to repeat the input, or to ask for it to be rephrased.

In the case of multimodal conversational interfaces, the input information can also be ambiguous. For example, input made with a pen on a touch-sensitive screen can have three different purposes: pointing (as a substitute for the mouse), hand-writing, and drawing. In order to address this problem, the system must employ some method to try to automatically decide the mode in which the pen is being used and/or employ an additional dialog turn to get a confirmation from the user about the intended mode.

A number of different approaches to DM have been developed within the research community and in industry (Lee et al. 2010; Wilks et al. 2011). These approaches can be classified into two main categories: handcrafted approaches using rules and statistical or data-driven approaches using machine learning methodologies. Hybrid approaches are also possible in which these two main approaches are combined. The following sections discuss approaches to DM.

10.3 Handcrafted Approaches to Dialog Management

One of the simplest DM strategies is finite state-based DM, in which a generic program implements the application with an interaction model based on finite state machines. This approach is usually confined to highly structured tasks in which system-directed initiative is used and the user's input is restricted to utterances within the scope of the ASR and SLU components (Barnard et al. 1999; Lee et al. 2010). This knowledge-based approach generally uses finite state automata with handcrafted rules. The user's actions determine the transitions between the system responses that constitute the nodes of the finite state automaton, and the user's responses to the system prompts are coded in recognition grammars.

Although this approach has been deployed in many practical applications because of its simplicity, these early applications only support a strict system-directed dialog interaction, in which at each turn the system directs the user by proposing a small number of choices for which there is a limited grammar or vocabulary to interpret the input. Directed dialog has been efficient in terms of accuracy and cost of development. However, although libraries and dialog modules have been created that can be reused and adapted to different applications, the weakest point of this approach is its lack of versatility and poor domain portability (Acomb et al. 2007; Pieraccini et al. 2009).

Unlike the finite state approach, frame-based dialog managers do not have a predefined dialog path but use a frame structure comprised of one slot for piece of information that the system has to gather from the user (McTear 2004). The advantage of this approach is that the system can capture several data at once and the information can be provided in any order (more than one slot can be filled per dialog turn and in any order). The form interpretation algorithm (FIA), the basis for the VoiceXML standard, is an example of a model of frame-based dialog management (see Chap. 11). Using frames, it is possible to specify the whole topic of a dialog. A study by Lemon et al. (2001) is an example of a frame-based system, as is the COMIC DM system (Catizone et al. 2003). The core idea is that humans communicate to achieve goals and during the interaction the mental state of the speakers may change. Thus, frame-based dialog managers model dialog as a cooperation between the user and the system to reach common goals. Utterances are not considered as text strings but as dialog acts in which the user communicates their intentions.

A more advanced approach is Information State Theory, also known as Information State Update (ISU), introduced in Chap. 4 (Traum and Larsson 2003). The information state of a dialog represents the information needed to uniquely distinguish it from all others. It comprises the accumulated user interventions and previous dialog actions on which the next system response can be based. The information state is also sometimes known as the *conversation store*, *discourse context*, or *mental state*. In the information state approach, the main tasks for the dialog manager are to update the information state based on the observed user actions and based on this update to select the next system action as specified in the update rules.

Plan-based approaches take the view that humans communicate to achieve goals, including changes to the mental state of the listener. Plan-based theories of communicative action and dialog (e.g., Allen and Perrault 1980; Appelt 1985; Cohen and Levesque 1990) claim that the speaker's speech act is part of a plan and that it is the listener's task to identify and respond appropriately to this plan (Wilks et al. 2011). Plan-based approaches attempt to model this claim and explicitly represent the (global) goals of the task.

Conversational games theory (Carletta et al. 1995; Kowtko et al. 1993) uses techniques from both discourse grammars and plan-based approaches by including a goal or plan-oriented level in its structural approach. It can be used to model conversations between a human and a computer in a task-oriented dialog (Williams 1996). This approach deals with discourse phenomena such as side sequences and clarifications by allowing games to have another game embedded within them (Wilks et al. 2011).

Additionally, when it is necessary to execute and monitor operations in a dynamically changing application domain, an agent-based approach can be employed. A modular agent-based approach to DM makes it possible to combine the benefits of different dialog control models, such as finite state-based dialog control and frame-based DM (Chu et al. 2005).

As previously discussed, in most settings, application developers, together with voice user interface (VUI) designers, typically handcraft DM strategies using rules and heuristics. As it is extremely challenging to anticipate every possible user input, handcrafting dialog management strategies is an error-prone process that needs to be iteratively refined and tuned, which requires considerable time and effort. The VoiceXML standard, which was introduced briefly in Chap. 4, is an example of the handcrafted approach that is used widely in industry to develop voice user interfaces. Chap. 11 provides an overview of VoiceXML along with exercises in how to build a simple dialog system using VoiceXML.

One of the main problems with handcrafted approaches to DM is that it is extremely challenging to anticipate every possible user input and design appropriate strategies to handle it. The process is error-prone and requires considerable time and effort to iteratively refine and tune the dialog strategies.

10.4 Statistical Approaches to Dialog Management

Machine learning approaches to DM try to reduce the effort and time required to handcraft DM strategies and, at the same time, facilitate the development of new dialog managers and their adaptation to deal with new domains. The application of machine learning approaches to DM strategy design is a rapidly growing research area. The main idea is to learn optimal strategies from corpora of real human–computer dialog data using automated "trial-and-error" methods instead of relying on empirical design principles (Young 2002).

Statistical approaches to DM present additional important advantages. Rather than maintaining a single hypothesis for the dialog state, they maintain a distribution over many hypotheses for the correct dialog state. In addition, statistical methodologies choose actions using an optimization process, in which a developer specifies high-level goals and the optimization works out the detailed dialog plan. Finally, statistical DM systems have shown, in research settings, more robustness to speech recognition errors, yielding shorter dialogs with higher task completion rates (Williams and Young 2007).

The main trend in this area is an increased use of data to improve the performance of the system. As described in Paek and Pieraccini (2008), there are three main aspects of spoken dialog interaction where the use of massive amounts of data can potentially improve the automation rate and ultimately the penetration and acceptance of speech interfaces in the wider consumer market. They are as follows:

- Task-independent behaviors (e.g., error correction and confirmation behavior).
- Task-specific behaviors (e.g., logic associated with certain customer care practices).
- Task interface behaviors (e.g., prompt selection).

Statistical models can be trained using corpora of human–computer dialogs with the goal of explicitly modeling the variability in user behavior that can be difficult to address by means of handwritten rules (Schatzmann et al. 2006). Additionally, it is possible to extend the strategy learned from the training corpus with handcrafted rules that include expert knowledge or specifications about the task (Suendermann and Pieraccini 2012; Laroche et al. 2008; Torres et al. 2008; Young et al. 2013).

The goal is to build systems that exhibit more robust performance, improved portability, better scalability, and easier adaptation to other tasks. However, model construction and parameterization are dependent on expert knowledge, and the success of statistical approaches is dependent on the quality and coverage of the models and data used for training (Schatzmann et al. 2006). Moreover, the training data must be correctly labeled for the learning process. The size of currently available annotated dialog corpora is usually too small to sufficiently explore the vast space of possible dialog states and strategies. Collecting a corpus with real users and annotating it requires considerable time and effort.

To address these problems, researchers have proposed alternative techniques that facilitate the acquisition and labeling of corpora, such as Wizard of Oz (Fraser and Gilbert 1991; Lane et al. 2004), bootstrapping (Fabbrizio et al. 2008; Abdennadher et al. 2007), active learning (Cohn et al. 1994; Venkataraman et al. 2005), automatic dialog act classification and labeling (O'Shea et al. 2012; Venkataraman et al. 2002), and user simulation (Schatzmann et al. 2006; Callejas et al. 2012).

Another relevant problem is how to deal with unseen situations, that is, situations that may occur during the dialog and that were not considered during training. To address this point, it is necessary to employ generalizable models in order to obtain appropriate system responses that enable the system to continue with the dialog in a satisfactory way.

Another difficulty is in the design of a good dialog strategy, which in many cases is far from being trivial. In fact, there is no clear definition of what constitutes a good dialog strategy (Schatzmann et al. 2006; Lemon and Pietquin 2012). Users are diverse, which makes it difficult to foresee which form of system behavior will lead to a quick and successful dialog completion, and speech recognition errors may introduce uncertainty about the user's intentions.

Statistical approaches to DM can be classified into three main categories: dialog modeling based on reinforcement learning (RL), corpus-based statistical dialog management, and example-based dialog management. Example-based approaches can be considered a specific case of corpus-based statistical dialog management, given that they usually perform dialog modeling by means of prepared dialog examples (Murao et al. 2003; Lee et al. 2009). These approaches assume that the next system action can be predicted when the dialog manager finds dialog examples that have a similar dialog state to the current dialog state (Lee et al. 2010). The best example is then selected from the candidate examples by calculating heuristic similarity measures between the current input and the example.

Hybrid approaches to DM combine statistical and rule-based approaches to try to reduce the amount of dialog data required for parameter estimation and to allow system designers to directly incorporate their expert domain knowledge into the dialog models (Lison 2015). In the following sections, we provide a detailed description of reinforcement learning and corpus-based approaches.

10.4.1 Reinforcement Learning

The most recent research advances in reinforcement learning (RL) for building spoken conversational interfaces have been reviewed and summarized in a survey paper by Frampton and Lemon (2009). An earlier survey can be found in Schatzmann et al. (2006). See also Rieser and Lemon (2011).

The most widespread methodology for machine learning of dialog strategies involves modeling human–computer interaction as an optimization problem using Markov decision processes (MDPs) and reinforcement learning methods (Levin and Pieraccini 1997; Levin et al. 2000; Singh et al. 1999). The main drawback of this approach is that the large state space required for representing all the possible dialog paths in practical spoken conversational interfaces makes its direct representation intractable. In addition, while exact solution algorithms do exist, they do not scale to problems with more than a few states/actions (Young et al. 2010, 2013).

Partially observable MDPs (POMDPs) outperform MDP-based dialog strategies since they provide an explicit representation of uncertainty (Roy et al. 2000). This enables the dialog manager to avoid and recover from recognition errors by sharing and shifting probability mass between multiple hypotheses of the current dialog state.

Another disadvantage of the POMDP methodology is that the optimization process is free to choose any action at any time. As a result, there is no obvious way to incorporate domain knowledge or constraints such as business rules. In addition, in the worst case, spurious actions might be taken with real users, an especially serious concern if POMDP-based systems are going to handle financial or medical transactions. POMDP-based systems have been limited to small-scale problems, since the state space would be huge and exact POMDP optimization is again intractable (Young et al. 2010).

Formally, a partially observable MDP is defined as a tuple $\{S, A, T, R, O, Z, \lambda, b_0\}$ where

- S is a set of the system states;
- A is a set of actions that the system may take;
- T defines a transition probability $P(s'|s, a)$;
- R defines the immediate reward obtained from taking a particular action in a particular state $r(s, a)$;
- O is a set of possible observations that the system can receive from the world;
- Z defines the probability of a particular observation given the state and machine action $P(o'|s'\ a)$;
- λ is a geometric discount factor $0 \leq \lambda \leq 1$; and
- b_0 is an initial belief state $b_0(s)$.

The operation of a POMDP is as follows. At each moment, the system is in an unobserved state s. The system selects an action a_m, receives a reward r, and transits to a state (unobserved) s', where s' only depends on s and a_m. The system receives an observation o', which depends on s' and a_m. Although the observation allows the system to have some evidences about the state s in which the system is now, s is not exactly known, and b(s) (belief state) is defined to indicate the probability of the system being in the state s.

Based on b, the machine selects an action a ∈ A, receives a reward r(s, a), and transitions to state s', which depends only on s and a. The machine then receives an observation o' ∈ O, which is dependent on s' and a. In each moment, the probability of the system being in a specific state is updated taking into account o' and a, as shown in Eq. 10.1.

$$
\begin{aligned}
b'(s') = P(s'|o', a, b) &= \frac{P(o'|s'_m, a_m, b)P(s'_m|a_m, b)}{P(o'|a_m, b)} \\
&= \frac{P(o'|s'_m, a_m, b)\sum_{s_m \in S_m} P(s'_m|a_m, b, s_m)P(s_m|a_m, b)}{P(o'|a_m, b)} \\
&= k \cdot P(o'|s', a) \sum_{s \in S} P(s'|a, s)b(s)
\end{aligned}
$$

$$(10.1)$$

where k = P (o'|a, b) is a normalization constant (Kaelbling et al. 1998). At each time t, the system receives a reward r(b_t, $a_{m,t}$), which depends on b_t and the selected action $a_{m,t}$. The reward accumulated during the dialog is called a *return* and can be calculated by means of Eq. 10.2.

$$
R = \sum_{t=0}^{\infty} \lambda^t R(b_t, a_{m,t}) = \sum_{t=0}^{\infty} \lambda^t \sum_{s \in S} b_t(s) r(s, a_{m,t})
$$

$$(10.2)$$

Each action $a_{m,t}$ is determined by the policy π(b_t), and the construction of the POMDP model implies to find the strategy π* which maximizes the return at every point b. Due to the vast space of possible belief states, however, the use of POMDPs for any practical system is far from straightforward. The optimal policy can be represented by a set of policy vectors where each vector v_i is associated with an action a(i) ∈ A_m and v_i(s) equals the expected value of taking action *a (i)* in state *s*. Given a complete set of policy vectors, the optimal value function and corresponding policy are computed as shown in Eq. 10.3.

$$
V^{\pi^*}(b) = \max_i \{v_i, b\}
$$

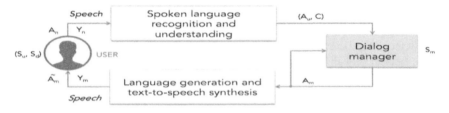

Fig. 10.1 Modeling a dialog system by means of POMDPs (Young et al. 2013)

and

$$V^{\pi^*}(b) = \max_i \{v_i, b\} \qquad (10.3)$$

The application of a POMDP to model a conversational interface is based on the classical architecture of these systems as shown in Fig. 10.1. As this figure shows, the user has an internal state S_u corresponding to a goal to be accomplished and the dialog state S_d represents the previous history of the dialog. Based on the user's goal prior to each turn, the user decides some communicative action (also called an intention) A_u, expressed in terms of dialog acts and corresponding to an audio signal Y_u.

Then, the speech recognition and language understanding modules take the audio signal Y_u and generate the pair $\left(\tilde{A}_u, C\right)$. This pair consists of an estimate of the user's action A_u and a confidence score that provides an indication of the reliability of the recognition and semantic interpretation results. This pair is then passed to the dialog model, which is in an internal state S_m and decides what action A_m the dialog system should take. This action is also passed back to the dialog manager so that S_m may track both user and machine actions. The language generator and the text-to-speech synthesizer take A_m and generate an audio response Y_m. The user listens to Y_m and attempts to recover A_m. As a result of this process, users update their goal state S_u and their interpretation of the dialog history S_d. These steps are then repeated until the end of the dialog.

One of the main challenges for conversational interfaces is that $\tilde{A}_u$ usually contains recognition errors (i.e., $\tilde{A}_u \neq A_u$). As a result, the user's action A_u, the user's state S_u, and the dialog history S_d are not directly observable and can never be known to the system with certainty. However, $\tilde{A}_u$ and the confidence scores C provide evidence from which A_u, S_u, and S_d can be inferred.

Therefore, when using POMDPs to model a conversational interface, the POMDP state S_m expresses the unobserved state of the world and can naturally be factored into three distinct components: the user's goal S_u, the user's action A_u, and the dialog history S_d. Hence, the factored POMDP state S is defined as $S_m = (S_u,$

a_u, s_d). The belief state b is then a distribution over these three components: $s_m = b_s = b(s_u, a_u, s_d)$. The observation o is the estimate of the user dialog act $\tilde{A}_u$. In the general case, this will be a set of N-best hypothesized user acts, each with an associated probability:

$$o = \left[(\tilde{a}_u^1, p_1), (\tilde{a}_u^2, p_2), \ldots, (\tilde{a}_u^N, p_N) \right] \tag{10.4}$$

where $p_n = P(\tilde{a}_u^N | o)$ for $n = 1 \ldots N$.

The transition function for an SDS-POMDP follows directly by substituting the factored state into the regular POMDP transition function and making independence assumptions:

$$\begin{aligned} P(s_m' | s_m, a_m) &= P(s_u', a_u', s_d' | s_u, a_u, s_d, a_m) \\ &= P(s_u' | s_u, a_m) P(a_u' | s_u', a_m) P(s_d' | s_u', a_u', s_d, a_m) \end{aligned} \tag{10.5}$$

The observation model is obtained by making similar reasonable independence assumptions regarding the observation function, giving

$$P(o' | s_m', a_m) = P(o' | s_u', a_u', s_d', a_m) = P(o' | a_u') \tag{10.6}$$

The above factoring simplifies the belief update equation as shown in Eq. 10.7.

$$\begin{aligned} b'(s_u', a_u', s_d') = k \times \underbrace{P(o' | a_u')}_{\text{Observation model}} \underbrace{P(a_u' | s_u', a_m)}_{\text{User action model}} \sum_{s_u} \underbrace{P(s_u' | s_u, a_m)}_{\text{User goal model}} \\ \times \sum_{s_d} \underbrace{P(s_d' | s_u', a_u', s_d, a_m)}_{\text{Dialog model}} b(s_u, s_d) \end{aligned} \tag{10.7}$$

As shown in the previous equation, the probability distribution for a_u' is called the user action model. It allows the observation probability to be scaled by the probability that the user would speak a_u' given the goal s_u' and the last system prompt a_m. The user goal model determines the probability of the user goal switching from s_u to s_u' following the system prompt a_m. Finally, the dialog model enables information relating to the dialog history to be maintained such as grounding and focus.

10.4.1.1 Reinforcement Learning: Some Problems and Some Solutions

Scaling the dialog model to handle real-world problems remains a significant challenge for RL-based systems, given that the complexity of a POMDP grows with the number of user goals, and optimization quickly becomes intractable. The summary POMDP method (Young et al. 2010) provides a way to scale up the POMDP model for so-called slot-filling spoken dialog systems. In this approach, the belief state and actions are mapped down to a summarized form where optimization becomes tractable.

The original belief space and actions are called *master space* and *master actions*, while the summarized versions are called *summary space* and *summary actions*. The updated belief state b is then mapped into a summary state $\tilde{b}$, where an optimized dialog policy is applied to compute a new summary machine action $\tilde{a}_m$. The summary machine action is then mapped back into the master space where it is converted to a specific machine dialog act a_m.

The optimization of the policy in these two spaces is usually carried out using techniques such as *point-based value iteration* or *Q-learning*, in combination with a user simulator. Q-learning is a technique for online learning where a sequence of sample dialogs is used to estimate the Q functions for each state and action. The optimal action for each point p is given by

$$\bar{a}_p = \underset{\bar{a}}{\operatorname{argmax}} \, \overline{Q}(a, p) \qquad (10.8)$$

Given that a good estimate of the true Q-value can be obtained if sufficient dialogs are completed, user simulation is usually introduced to reduce the time-consuming and expensive task of obtaining these dialogs with real users. Simulation is usually done at a semantic dialog act level to avoid having to reproduce the variety of user utterances at the word or acoustic levels.

Agenda-based state representations, like the one described in (Thomson et al. 2007), factor the user state into an agenda A and a goal G. The goal G consists of constraints C that specify the detailed goal of the dialog and requests R that specify the desired pieces of information.

The user agenda A is a stack-like structure containing the pending user dialog acts that are needed to elicit the information specified in the goal. At the beginning of each dialog, a new goal G is randomly selected. Then, the goal constraints C are converted into user and system inform acts (a_u and a_m acts) and the requests R into request acts. A bye act is added at the bottom of the agenda to close the dialog once the goal has been fulfilled. The agenda is ordered according to priority, with A[N] denoting the top item and A[1] denoting the bottom item. As the dialog progresses, the agenda and goal are dynamically updated and acts are selected from the top of the agenda to form user acts a_u.

Young et al. (2010) present an approach that scales the POMDP framework for the implementation of practical spoken conversational interfaces by defining two state spaces. Approximate algorithms have also been developed to overcome the intractability of exact algorithms, but even the most efficient of these techniques such as point-based value iteration (PBVI) cannot scale to the many thousands of states required by a statistical dialog manager (Williams et al. 2006).

Composite summary point-based value iteration (CSPBVI) has suggested the use of a small summary space for each slot where PBVI policy optimization can be applied. However, policy learning in this technique can only be performed off-line, i.e., at design time, because policy training requires an existing accurate model of user behavior. An alternative technique for online training based on Q-learning is presented in Thomson et al. (2007), which allows the system to adapt to real users as new dialogs are recorded. This technique does not require any model of user behavior, so user simulation techniques are proposed to iteratively learn the dialog model.

Other authors have combined conventional dialog managers with a fully observable Markov decision process (Singh et al. 2002; Heeman 2007), or proposed using multiple POMDPs and selecting actions using handcrafted rules (Williams et al. 2006). In Williams (2008), the robustness of the POMDP approach is combined with the developer control available in conventional approaches: The (conventional) dialog manager and POMDP run in parallel, but the dialog manager is augmented so that it outputs one or more allowed actions at each time step. The POMDP then chooses the best action from this limited set. Results from a real voice dialer application show that adding the POMDP machinery to a standard dialog system yields a significant improvement.

Crook et al. (2014) describe an evaluation of a POMDP-based spoken dialog system using crowd-sourced calls with real users. The evaluation compares a "hidden information state" POMDP system that uses a handcrafted compression of the belief space with the same system using instead an automatically computed belief space compression.

In Tetreault and Litman (2008), the authors aimed to evaluate the best state-space representations so that RL can be used to find an optimal dialog policy. The authors presented three metrics for the tutoring domain and ways to build confidence intervals for model switching. In the work reported in Gašić et al. (2011), online optimization of dialog policy was conducted in spoken dialog systems via live interaction with human subjects.

Jurčíček et al. (2012) presented two RL algorithms for learning the parameters of a dialog model. The Natural Belief Critic algorithm is designed to optimize the model parameters while the policy is kept fixed. The Natural Actor and Belief Critic algorithm jointly optimizes both the model and the policy parameters. The algorithms were evaluated on a statistical dialog system for the tourist information domain modeled as a POMDP. The experiments indicated that model parameters estimated to maximize the expected reward function provide improved performance compared to the baseline handcrafted parameters.

Thomson and Young (2010) used expectation–propagation (EP) to infer the unobserved dialog state together with the model parameters. The main advantage of this algorithm is that it is an off-line method and it does not rely on annotated data. However, it requires the model to be generative (i.e., the observations must be conditioned on the dialog state).

Wierstra et al. (2010) used recurrent neural networks (RNN) to approximate the policy. This method selects a new system action based on the accumulated information in the internal memory and the last observation. Png and Pineau (2011) presented a framework based on a Bayes-adaptive POMDP algorithm to learn an observation model. In this work, a dialog model was factored into a transition model between hidden dialog states and an observation model, and only learning of the observation model was considered.

Lopes et al. (2015) have very recently presented a data-driven approach to improve the performance of SDSs by automatically finding the most appropriate terms to be used in system prompts. Speakers use one another's terms (entrain) when trying to create common ground during a spoken dialog. Those terms are commonly called *primes*, since they influence the interlocutors' linguistic decision-making. The proposed approach emulates human interaction, with a system built to propose primes to the user and accept the primes that the user proposes. Live tests with this method show that the use of on-the-fly entrainment reduces out-of-vocabulary and word error rate and also increases the number of correctly transferred concepts.

Lison (2015) has also recently presented a modeling framework for DM based on the concept of probabilistic rules, which are defined as structured mappings between logical conditions and probabilistic effects. Probabilistic rules are able to encode the probability and utility models employed in DM in a compact and human-readable form. As a consequence, they can reduce the amount of dialog data required for parameter estimation and allow system designers to directly incorporate their expert domain knowledge into the dialog models.

Other interesting approaches for statistical DM are based on modeling the system by means of Hidden Markov models (HMMs) (Cuayáhuitl et al. 2005) or using Bayesian networks (Paek and Horvitz 2000; Meng et al. 2003).

10.4.2 Corpus-Based Approaches

Griol et al. (2014) describe a corpus-based approach to DM based on the estimation of a statistical model from the sequences of the system and user dialog acts obtained from a set of training data. The next system response is selected by means of a classification process that considers the complete history of the dialog.

Another main characteristic is the inclusion of a data structure that stores the information provided by the user. The main objective of this structure is to easily encode the complete information related to the task provided by the user during the dialog history and then to consider the specific semantics of the task and include this information in the proposed classification process.

In order to control the interactions with the user, the proposed dialog manager represents dialogs as a sequence of pairs (A_i, U_i), where A_i is the output of the dialog system (the system answer) at time i and U_i is the semantic representation of the user turn (the result of the understanding process of the user input) at time i; both expressed in terms of dialog acts (Griol et al. 2008). Each dialog is represented by:

$$(A_1, U_1), \ldots, (A_i, U_i), \ldots, (A_n, U_n)$$

where A_1 is the greeting turn of the system and U_n is the last user turn. We refer to a pair (A_i, U_i) as S_i, the dialog sequence at time i.

In this framework, we consider that, at time i, the objective of the dialog manager is to find the best system answer A_i. This selection is a local process for each time i that takes into account the previous history of the dialog, that is to say, the sequence of states of the dialog preceding time i:

$$\widehat{A}_i = \underset{A_i \in \mathcal{A}}{\operatorname{argmax}} P(A_i \mid S_1, \ldots, S_{i-1}) \tag{10.9}$$

where set A contains all the possible system answers.

Following Eq. 10.9, the dialog manager selects the next system prompt by taking into account the sequence of previous pairs (A_i, U_i). The main problem with resolving this equation is that the number of possible sequences of states is usually very large. To solve the problem, we define a data structure in order to establish a partition in this space, i.e., in the history of the dialog preceding time i. This data structure, which we call *Dialog Register* (DR), contains the information provided by the user throughout the previous history of the dialog.

After applying the above considerations and establishing the equivalence relation in the histories of the dialogs, the selection of the best A_i is given by:

$$\widehat{A}_i = \underset{A_i \in \mathcal{A}}{\operatorname{argmax}} P(A_i \mid DR_{i-1}, S_{i-1}) \tag{10.10}$$

Each user turn supplies the system with information about the task; i.e., the user asks for a specific concept and/or provides specific values for certain attributes. However, a user turn can also provide other kinds of information, such as task-independent information (for instance, Affirmation, Negation, and Not-Understood dialog acts). This kind of information implies some decisions that are different from simply updating the DR_{i-1}. Hence, for the selection of the best system response A_i, we take into account the DR that results from turn 1 to turn $i-1$, and we explicitly consider the last state S_{i-1}.

We propose solving Eq. 10.10 using a classification process, in which every dialog situation (i.e., each possible sequence of dialog acts) is classified taking into account a set of classes C, in which a class contains all the sequences that provide the same set of system actions (responses). The objective of the dialog manager at each moment is to select a class of this set $c \in C$, so that the system answer is the one associated with the selected class.

The classification function can be defined in several ways. Griol et al. (2014) propose the use of a multilayer perceptron (MLP) (Rumelhart et al. 1986), where the input layer holds the input pair (DR_{i-1}, S_{i-1}) corresponding to the Dialog Register and the state. The values of the output layer can be seen as an approximation of the a posteriori probability of the input belonging to the associated class $c \in C$.

As stated before, the DR contains information about concepts and values for the attributes provided by the user throughout the previous history of the dialog. For the dialog manager to determine the next answer, the exact values of the attributes are assumed to be not significant. They are important for accessing databases and for constructing the output sentences of the system. However, the only information necessary to predict the next action by the system is the presence or absence of concepts and attributes. Therefore, the codification proposed for each slot in the DR is in terms of three values, $\{0, 1, 2\}$, according to the following criteria:

- (0) The concept is unknown, or the value of the attribute is not given.
- (1) The concept or attribute is known with a confidence score that is higher than a given threshold.
- (2) The concept or attribute has a confidence score that is lower than the given threshold.

To decide whether the state of a certain value in the DR is 1 or 2, the system employs confidence measures provided by the ASR and SLU modules (Torres et al. 2005).

The previously described process allows every task to be modeled based only on the information provided by the user in the previous turns and its own model. In other dialog systems, the dialog manager generates the next system response taking also into account the information generated by the module that controls the application (that is denoted as the application manager (AM)). For example, the AM can validate restrictions, apply privacy policies, or carry out computations that define the next system response (for instance, selecting a different system action depending on the result of a query to the databases of the application). Thus, the output of this module has to be taken into account for the selection of the best system action.

For this reason, for this kind of task, two phases are proposed for the selection of the next system turn. In the first phase, the information contained in the DR and the last state S_{i-1} are considered to select the best request to be made to the AM:

$$\widehat{A}_i = \underset{A_{1_i} \in A_1}{\mathrm{argmax}}\, P(A_i \mid DR_{i-1}, S_{i-1}) \qquad (10.11)$$

where A_1 is the set of possible requests to the AM.

In the second phase, the system answer $\widetilde{A}_2$ is generated taking into account $\widetilde{A}_1$ and the information provided by the AM (AM_i):

$$\widehat{A}_{2_i} = \underset{A_{2_i} \in A_2}{\mathrm{argmax}}\, P(A_i | AM_i, A_{1_i}) \qquad (10.12)$$

where $\widetilde{A}_2$ is the set of possible system answers.

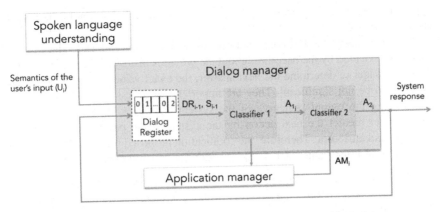

Fig. 10.2 Scheme of the architecture proposed in the corpus-based DM methodology

Figure 10.2 shows the scheme proposed for the development of the dialog manager for this kind of task, detailing the two phases described for the generation of the system response. The use of two MLPs is proposed to deal with the specific information defined for each phase.

The AM makes it possible to consider specific requisites (e.g., special requirements, policies, or specific routines) that endow conversational interfaces with a more sophisticated behavior that is different from only requiring information from the user and checking or updating a repository. This phase also makes it possible for systems to deal with specific cases for the different attributes, given that the exact values for each attribute are considered to access the data repositories. In addition, the statistical dialog model supports user adaptation, which makes it suitable for different application domains with varying degrees of complexity.

10.5 Summary

Given the current state of the dialog, the principal role of the dialog manager is to choose an action that will result in a change of dialog state. The strategy followed by the dialog manager, sometimes referred to as the policy, should be designed to enable successful, efficient, and natural conversations.

This is a challenging goal, and in most commercially deployed conversational interfaces, a human designer handcrafts the dialog manager. This handcrafted approach is limited for several reasons: it is not always easy to specify the optimal action at each state of the dialog; a dialog behavior that is generic and static is usually assumed for the entire user population; designing such strategies is labor-intensive, especially for large systems.

Machine learning approaches to DM try to reduce the effort and time required by handcrafted DM strategies; they isolate domain knowledge from the dialog strategy;

and they facilitate the development of new dialog managers and their adaptation to new domains.

DM is discussed further in Chap. 11 where we provide practical exercises related to the application of rule-based and statistical DM techniques for a specific task. Chap. 17 will discuss the most relevant approaches for the evaluation of DM.

Further Reading

Wilks et al. (2011) present a detailed survey of DM approaches and architectures, along with practical examples of dialog systems developed using these approaches and architectures. Lee et al. (2010) present a detailed survey covering design issues and approaches to DM and techniques for modeling. The paper also explains the use of user simulation techniques for the automatic evaluation of spoken conversational interfaces.

Some recent research advances in RL for building SDSs were reviewed and summarized in a survey paper by Frampton and Lemon (2009). An earlier survey can be found in Schatzmann et al. (2006). See also Rieser and Lemon (2011), Lemon and Pietquin (2012), and Thomson (2013). Young et al. (2013) provide an overview of the state of the art in the development of POMDP-based spoken dialog systems.

Meena et al. (2014) summarize the main approaches to turn taking in human conversations. They also explore a range of automatically extractable features for online use, covering prosody, lexicosyntax, and context, and different classes of learning algorithms for turn taking in human–machine conversations.

References

Abdennadher S, Aly M, Bühler D, Minker W, Pittermann J (2007) Becam tool—a semi-automatic tool for bootstrapping emotion corpus annotation and management. In: Proceedings of the international conference on spoken language processing (Interspeech'2007), Antwerp, Belgium, 27–31 Aug 2007, pp 946–949. http://met.guc.edu.eg/Repository/Faculty/Publications/69/Paper.pdf

Acomb K, Bloom J, Dayanidhi K, Hunter P, Krogh P, Levin E, Pieraccini R (2007) Technical support dialog systems: issues, problems, and solutions. In: Proceedings of the NAACL-HLT-Dialog'07 workshop on bridging the gap: Academic and Industrial Research in Dialog Technologies, Rochester, NY, USA, 26 Apr 2007, pp 25–31. http://dl.acm.org/citation.cfm?id=1556332&CFID=585421472&CFTOKEN=72903197

Allen JF, Perrault CR (1980) Analyzing intentions in dialogs. Artif Intell 15(3):143–178. doi:10.1016/0004-3702(80)90042-9

Appelt DE (1985) Planning English sentences. Cambridge University Press, Cambridge. doi:10.1017/CBO9780511624575

Barnard E, Halberstadt A, Kotelly C, Phillips M (1999) A consistent approach to designing spoken-dialog Systems. In: Proceedings of the IEEE workshop on automatic speech recognition and understanding (ASRU'99), Keystone, Colorado, USA, pp 1173–1176

Callejas Z, Griol D, Engelbrecht K, López-Cózar R (2012) A clustering approach to assess real user profiles in spoken dialogue systems. In: Mariani J, Rosset S, Garnier-Rizet M, Devilliers L (eds) Natural language interaction with robots: putting spoken dialog systems into practice. Springer, New York, pp 327–334. doi:10.1007/978-1-4614-8280-2_29

Carletta JC, Isard A, Isard S, Kowtko J, Doherty-Sneddon G, Anderson A (1995) The coding of dialog structure in a corpus. In: Andernach T, van de Burgt SP, van der Hoeven GF (eds) Proceedings of the Twente workshop on language technology: corpus-based approaches to dialogue modelling, University of Twente, Netherlands, June 1995

Catizone R, Setzer A, Wilks Y (2003) Multimodal dialogue management in the COMIC project. In: Jokinen K, Gamback B, Black W, Catizone R, Wilks Y (eds) Proceedings of the 2003 EACL workshop on dialogue systems: interaction, adaptation and styles of management, Budapest, Hungary, 13–14 Apr 2003. http://aclweb.org/anthology/W/W03/W03-2705.pdf

Chu S, O'Neill I, Hanna P, McTear M (2005) An approach to multistrategy dialog management. In: Proceedings of the 9th international conference on spoken language processing (Interspeech2005), Lisbon, Portugal, pp 865–868. http://www.isca-speech.org/archive/archive_papers/interspeech_2005/i05_0865.pdf

Cohen P, Levesque H (1990) Rational interaction as the basis for communication. In: Cohen P, Morgan J, Pollack M (eds) Intentions in communication. MIT Press, Cambridge, MA, pp 221–256. https://www.sri.com/work/publications/rational-interaction-basis-communication. Accessed 20 Jan 2016

Cohn DA, Atlas L, Ladner R (1994) Improving generalization with active learning. Mach Learn 15 (2):201–221. doi:10.1007/BF00993277

Crook PA, Keizer S, Wang Z, Tang W, Lemon O (2014) Real user evaluation of a POMDP spoken dialog system using automatic belief compression. Comput Speech Lang 28(4):873–887. doi:10.1016/j.csl.2013.12.002

Cuayáhuitl H, Renals S, Lemon O, Shimodaira H (2005) Human-computer dialogue simulation using Hidden Markov models. In: Proceedings of the IEEE workshop on automatic speech recognition and understanding (ASRU2005), San Juan, Puerto Rico, 27 Nov 2005, pp 290–295. doi:10.1109/ASRU.2005.1566485

Fabbrizio GD, Tur G, Hakkani-Tür D, Gilbert M, Renger B, Gibbon D, Liu Z, Shahraray B (2008) Bootstrapping spoken dialogue systems by exploiting reusable libraries. Nat Lang Eng 14 (3):313–335. doi:10.1017/S1351324907004561

Frampton M, Lemon O (2009) Recent research advances in reinforcement learning in spoken dialog systems. Knowl Eng Rev 24(4):375–408. doi:10.1017/S0269888909990166

Fraser M, Gilbert G (1991) Simulating speech systems. Comput Speech Lang 5(1):81–99. doi:10.1016/0885-2308(91)90019-M

Gašić M, Jurčíček F, Thomson B, Yu K, Young S (2011) On-line policy optimisation of spoken dialog systems via live interaction with human subjects. In: Proceedings of IEEE workshop on automatic speech recognition and understanding (ASRU), Waikoloa, Hawaii, 11–15 Dec 2011, pp 312–317. doi:10.1109/ASRU.2011.6163950

Griol D, Hurtado LF, Segarra E, Sanchis E (2008) A statistical approach to spoken dialog systems design and evaluation. Speech Commun 50(8–9):666–682. doi:10.1016/j.specom.2008.04.001

Griol D, Callejas Z, López-Cózar R, Riccardi G (2014) A domain-independent statistical methodology for dialog management in spoken dialog systems. Comput Speech Lang 28 (3):743–768. doi:10.1016/j.csl.2013.09.002

Heeman, P (2007) Combining reinforcement learning with information-state update rules. In: Proceedings of the 8th annual conference of the North American chapter of the Association for Computational Linguistics (HLT-NAACL2007), Rochester, New York, USA, 22–27 Apr 2007. http://aclweb.org/anthology/N07-1034

Jurčíček F, Thomson B, Young S (2012) Reinforcement learning for parameter estimation in statistical spoken dialog systems. Comput Speech Lang 26(3):168–192. doi:10.1016/j.csl.2011.09.004

Kaelbling LP, Littman ML, Cassandra AR (1998) Planning and acting in partially observable stochastic domains. Artif Intell 101(1–2):99–134. doi:10.1016/s0004-3702(98)00023-x

Kowtko JC, Isard SD, Doherty, GM (1993) Conversational games within dialogue. Human Communication Research Centre, University of Edinburgh, (HCRC/RP-31). doi:10.1.1.52.5350

Lane I, Ueno S, Kawahara T (2004) Cooperative dialogue planning with user and situation models via example-based training. In: Proceedings of workshop on man-machine symbiotic systems, Kyoto, Japan, 23–24 Nov 2004, pp 93–102

Laroche R, Putois G, Bretier P, Young S, Lemon O (2008) Requirements analysis and theory for statistical learning approaches in automaton-based dialogue management. CLASSiC Project Deliverable 1.1.1. Edinburgh University, Edinburgh, UK. http://www.classic-project.org/deliverables/d1.1.1.pdf

Lee C, Jung S, Kim S, Lee G (2009) Example-based dialog modeling for practical multi-domain dialog system. Speech Commun 51(5):466–484. doi:10.1016/j.specom.2009.01.008

Lee CJ, Jung SK, Kim KD, Lee DH, Lee GG (2010) Recent approaches to dialog management for spoken dialog systems. J Comput Sci Eng 4(1):1–22. doi:10.5626/JCSE.2010.4.1.001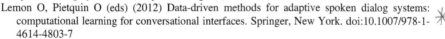

Lemon O, Pietquin O (eds) (2012) Data-driven methods for adaptive spoken dialog systems: computational learning for conversational interfaces. Springer, New York. doi:10.1007/978-1-4614-4803-7

Lemon O, Bracy A, Gruenstein A, Peters S (2001) The Witas multimodal dialog system I. In: Proceedings of the 7th Eurospeech conference on speech communication and technology (INTERSPEECH'01), Aalborg, Denmark, 3–7 Sept 2001, pp 1559–1562. http://www.isca-speech.org/archive/eurospeech_2001/e01_1559.html

Levin E, Pieraccini R (1997) A stochastic model of human-machine interaction for learning dialog strategies. In: Proceedings of the 5th European conference on speech communications and technology (Eurospeech1997), Rhodes, Greece, pp 1883–1886. http://www.isca-speech.org/archive/eurospeech_1997/e97_1883.html

Levin E, Pieraccini R, Eckert W (2000) A stochastic model of human-machine interaction for learning dialog strategies. IEEE T on Speech Audi P 8(1):11–23. doi:10.1109/89.817450

Lison P (2015) A hybrid approach to dialogue management based on probabilistic rules. Comput Speech Lang 34(1):232–255. doi:10.1016/j.csl.2015.01.001

Lopes J, Eskenazi M, Trancoso I (2015) From rule-based to data-driven lexical entrainment models in spoken dialog systems. Comput Speech Lang 31(1):87–112. doi:10.1016/j.csl.2014.11.007

McTear M (2004) Spoken dialogue technology: toward the conversational user interface. Springer, New York. doi:10.1007/978-0-85729-414-2

Meena R, Skantze G, Gustafson J (2014) Data-driven models for timing feedback responses in a pap task dialogue system. Comput Speech Lang 28(4):903–922. doi:10.1016/j.csl.2014.02.002

Meng HH, Wai C, Pieraccini R (2003) The use of belief networks for mixed-initiative dialog modeling. IEEE Trans Speech Audio Process 11(6):757–773. doi:10.1109/TSA.2003.814380

Murao HK, Kawaguchi N, Matsubara S, Ymaguchi Y, Inagaki Y (2003) Example-based spoken dialogue system using WOZ system sog. In: Proceedings of the 4th SIGDIAL workshop on discourse and dialogue, Sapporo, Japan, 5–6 July 2003, pp 140–148. http://www.aclweb.org/anthology/W/W03/W03-2112.pdf

O'Shea J, Bandar Z, Crockett K (2012) A multi-classifier approach to dialog act classification using function words. In: Nguyen NT (ed) Transactions on computational collective intelligence VII. Lecture notes in computer science, vol 7270, pp 119–143. doi:10.1007/978-3-642-32066-8_6

Paek T, Horvitz E (2000) Conversation as action under uncertainty. In: Proceedings of the 16th conference on uncertainty in artificial intelligence, Stanford, CA, USA, pp 455–464. http://arxiv.org/pdf/1301.3883.pdf

Paek T, Pieraccini R (2008) Automating spoken dialog management design using machine learning: an industry perspective. Speech Commun 50:716–729. doi:10.1016/j.specom.2008.03.010

Pieraccini R, Suendermann D, Dayanidhi K, Liscombe J (2009) Are we there yet? Research in commercial spoken dialog systems. In: Matoušek V, Mautner P (eds) Text, speech and dialogue: 12th international conference, TSD 2009, Pilsen, Czech Republic, 13–17 Sept 2009, pp 3–13. doi:10.1007/978-3-642-04208-9_3

Png S, Pineau J (2011) Bayesian reinforcement learning for POMDP-based dialogue systems. In: Proceedings of international conference on acoustics, speech and signal processing (ICASSP2011), Prague, Czech Republic, 22–27 May 2011, pp 2156–2159. doi:10.1109/ICASSP.2011.5946754

Rieser V, Lemon O (2011) Reinforcement learning for adaptive dialogue systems: a data-driven methodology for dialogue management and natural language generation. Springer, New York. doi:10.1007/978-3-642-24942-6

Roy N, Pineau J, Thrun S (2000) Spoken dialogue management using probabilistic reasoning. In: Proceedings of the 38th annual meeting of the Association for Computational Linguistics (ACL2000), Hong Kong, China, 1–8 Oct 2000. https://aclweb.org/anthology/P/P00/P00-1013.pdf

Rumelhart DE, Hinton GE, Williams RJ (1986) Learning internal representations by error propagation. In: Rumerhart DE, McClelland JL (eds) Parallel distributed processing: explorations in the microstructure of cognition, vol 1. MIT Press, Cambridge, pp 318–362. http://dl.acm.org/citation.cfm?id=104293

Schatzmann J, Weilhammer K, Stuttle M, Young S (2006) A survey of statistical user simulation techniques for reinforcement-learning of dialogue management strategies. Knowl Eng Rev 21 (2):97–126. doi:10.1017/s0269888906000944

Singh S, Kearns M, Litman D, Walker M (1999) Reinforcement learning for spoken dialog systems. In: Proceedings of neural information processing systems (NIPS 1999), Denver, USA, pp 956–962. http://papers.nips.cc/paper/1775-reinforcement-learning-for-spoken-dialogue-systems.pdf

Singh S, Litman D, Kearns M, Walker M (2002) Optimizing dialogue management with reinforcement learning: experiments with the NJFun system. J Artif Intell Res 16:105–133. doi:10.1613/jair.859

Suendermann D, Pieraccini R (2012) One year of contender: what have we learned about assessing and tuning industrial spoken dialog systems? In: Proceedings of the NAACL-HLT workshop on future directions and needs in the spoken dialog community: tools and data (SDCTD 2012), Montreal, Canada, 7 June 2012, pp 45–48. http://www.aclweb.org/anthology/W12-1818 Accessed 20 Jan 2016

Tetreault JR, Litman D (2008) A reinforcement learning approach to evaluating state representations in spoken dialogue systems. Speech Commun 50(8–9):683–696. doi:10.1016/j.specom.2008.05.002

Thomson B (2013) Statistical methods for spoken dialog management. Springer theses. Springer, New York. doi:10.1007/978-1-4471-4923-1

Thomson B, Young S (2010) Bayesian update of dialog state: a POMDP framework for spoken dialogue systems. Comput Speech Lang 24(4):562–588. doi:10.1016/j.csl.2009.07.003

Thomson B, Schatzmann J, Weilhammer K, Ye H, Young S (2007) Training a real-world POMDP-based dialogue system. In: Proceedings of NAACL-HLT-Dialog'07 workshop on bridging the gap: academic and industrial research in dialog technologies, Rochester, NY, USA, pp 9–16. http://dl.acm.org/citation.cfm?doid=1556328.1556330

Torres F, Hurtado LF, García F, Sanchis E, Segarra E (2005) Error handling in a stochastic dialog system through confidence measures. Speech Commun 45:211–229. doi:10.1016/j.specom.2004.10.014

Torres F, Sanchis E, Segarra E (2008) User simulation in a stochastic dialog system. Comput Speech Lang 22(3):230–255. doi:10.1016/j.csl.2007.09.002

Traum DR, Larsson S (2003) The information state approach to dialog management. In: Smith R, Kuppevelt J (eds) Current and new directions in discourse and dialog. Kluwer Academic Publishers, Dordrecht, pp 325–353. doi:10.1007/978-94-010-0019-2_15

Venkataraman A, Stolcke A, Shriberg E (2002) Automatic dialog act labeling with minimal supervision. In: Proceedings of the 9th australian international conference on speech science and technology, Melbourne, Australia, 2–5 Dec 2002. https://www.sri.com/sites/default/files/publications/automatic_dialog_act_labeling_with_minimal.pdf

Venkataraman A, Liu Y, Shriberg E, Stolcke A (2005) Does active learning help automatic dialog act tagging in meeting data. In: Proceedings of interspeech-2005, Lisbon, Portugal, 4–8 Sept 2005, pp 2777–2780. http://www.isca-speech.org/archive/interspeech_2005/i05_2777.html

Wierstra D, Förster A, Peters J, Schmidhuber J (2010) Recurrent policy gradients. Logic J IGPL 18 (5):620–634. doi:10.1093/jigpal/jzp049

Wilks Y, Catizone R, Worgan S, Turunen M (2011) Some background on dialogue management and conversational speech for dialogue systems. Comput Speech Lang 25(2):128–139. doi:10. 1016/j.csl.2010.03.001

Williams S (1996) Dialogue management in mixed-initiative, cooperative, spoken language system. Proceedings of 11th twente workshop on language technology (TWLT11) dialogue management in natural language systems, Enschade, Netherlands. doi: http://users.mct.open.ac. uk/sw6629/Publications/twlt96.pdf

Williams, JD (2008) The best of both worlds: Unifying conventional dialog systems and POMDPs. In: Proceedings of the international conference on spoken language processing (InterSpeech-2008), Brisbane, Australia, 22–16 Sept 2016, pp 1173–1176. http://www.isca-speech.org/archive/interspeech_2008/i08_1173.html

Williams JD, Young S (2007) Partially observable Markov decision processes for spoken dialog systems. Comput Speech Lang 21(2):393–422. doi:10.1016/j.csl.2006.06.008

Williams JD, Poupart P, Young S (2006) Partially observable Markov decision processes with continuous observations for dialog management. In: Dybkær L, Minker W (eds) Recent trends in discourse and dialogue. Springer, New York, PP 191–217. doi: 10.1007/978-1-4020-6821-8_8

Young S (2002) Talking to machines (statistically speaking). In: Proceedings of the 7th international conference on spoken language processing, Denver, Colorado, USA, 16–20 sept 2002, pp 9–16. http://www.isca-speech.org/archive/archive_papers/icslp_2002/i02_0009.pdf

Young S, Gašić M, Keizer S, Mairesse F, Schatzmann J, Thomson B, Yu K (2010) The Hidden Information State model: a practical framework for POMDP-based spoken dialogue management. Comp Speech Lang 24(2):150–174. doi:10.1016/j.csl.2009.04.001

Young S, Gašić M, Thomson B, Williams J (2013) POMDP-based statistical spoken dialog systems: a review. In: Proceedings of the IEEE 101(5), Montreal, Canada, pp 1160–1179. doi:10.1109/JPROC.2012.2225812

Chapter 11
Implementing Dialog Management

Abstract There is a wide range of tools that support the generation of rule-based dialog managers for conversational interfaces. However, it is not as easy to find toolkits to develop statistical dialog managers based on reinforcement learning and/or corpus-based techniques. In this chapter, we have selected the VoiceXML standard to put into practice the handcrafted approach, given that this standard is used widely in industry to develop voice user interfaces. The second part of the chapter describes the use of a statistical dialog management technique to show the application of this kind of methodology for the development of practical conversational interfaces.

11.1 Introduction

As we saw in Chap. 10, a number of different approaches to dialog management (DM) have been developed within the research community and in industry. These approaches can be classified into two main categories: handcrafted approaches using rules and statistical or data-driven approaches using machine learning methodologies. Hybrid approaches are also possible in which these two main approaches are combined.

The Form Interpretation Algorithm (FIA), the basis for the Voice Extensible Markup Language (VoiceXML) standard, is an example of a model of handcrafted DM. Section 11.2 provides a description of the main features of the World Wide Web Consortium (W3C) VoiceXML standard along with an exercise consisting of several steps to iteratively develop a VoiceXML-based conversational interface acting as a pizzeria service.

Statistical approaches to DM can be classified into three main categories: DM based on reinforcement learning (RL), corpus-based statistical DM, and example-based DM. Example-based approaches can be considered a specific case of corpus-based statistical DM, given that they also make use of transcribed examples of dialogs. These approaches assume that the next system action can be predicted when the dialog manager finds dialog examples that have a similar dialog state to the current dialog state. Section 11.3 describes the use of a corpus-based

© Springer International Publishing Switzerland 2016
M. McTear et al., *The Conversational Interface*,
DOI 10.1007/978-3-319-32967-3_11

DM technique, as described in Chap. 10, to develop a conversational interface for the same practical application domain.

11.2 Development of a Conversational Interface Using a Rule-Based Dialog Management Technique

VoiceXML was introduced briefly in Chaps. 4 and 10. In this section, we provide an overview of the main elements of VoiceXML that will be used in the exercises, in which we create a VoiceXML-based conversational interface. For more detail, see the W3C VoiceXML specification.[1]

VoiceXML supports the development of interactive spoken conversational interfaces that include the recognition of spoken and DTMF input, text-to-speech synthesis, dialog management, audio playback, recording of spoken input, telephony, and mixed-initiative conversations.

Figure 11.1 shows a comparative architecture of a VoiceXML system and an Internet application. Although the figure clearly shows the similarity between the two approaches, the VoiceXML application has some additional complexity when compared to the Internet application. Internet users enter a URL to access an application, while VoiceXML users dial a telephone number. Once connected, the public switched telephone network (PSTN) or mobile network communicates with the voice gateway. The gateway then forwards the request over hypertext transfer protocol (HTTP) to a Web server that can service the request. On the server, standard server-side technologies such as PHP, JSP, ASP, or CGI can be used to dynamically generate the VoiceXML content and grammars that are then returned to the voice gateway. On the gateway, a voice browser interprets the VoiceXML code using a voice browser. The content is then spoken to the user over the telephone using prerecorded audio files or synthesized speech. If user input is required at any point during the application cycle, it can be entered via either speech or DTMF. This process is repeated several times during the use of a typical VoiceXML-based application.

As shown in Fig. 11.1, the voice gateway incorporates many important voice technologies, including ASR, telephony dialog control, DTMF, TTS, and prerecorded audio playback. According to the VoiceXML specification, a VoiceXML platform must support document acquisition (acquire VoiceXML documents for use within the voice browser), audio output (TTS and prerecorded audio files), audio input (spoken and DTMF inputs), and transfer (make connections to a third party through a communications network such as the telephone network).

The top-level element of a VoiceXML file is <vxml>, which is mainly a container for dialogs. There are two types of dialog constructs: <form> and <menu>. Forms present information and gather input, while menus offer choices of what to do next. Code 11.1 shows an example of a VoiceXML form.

---

[1]http://www.w3.org/TR/voicexml20/. Accessed February 26, 2016.

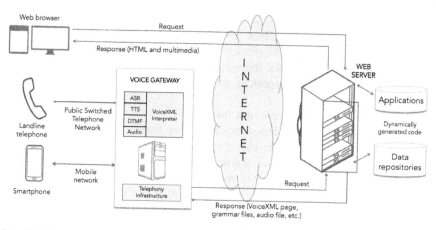

Fig. 11.1 Comparative architecture of a VoiceXML system and an Internet application

```
<?xml version="1.0"?>

<vxml version="2.0" xmlns="http://www.w3.org/2001/vxml">

<form>

<field name="survey">
<prompt>Do you prefer watching films at the cinema or at
home?</prompt>

<grammar src="survey.grxml"  type="application/srgs+xml"/>

</field>

<filled>

<if cond="survey == 'cinema'">

<prompt> I have understood that you prefer going to the cin-
ema.

</prompt>

<else/>

<submit next="vxml.com/next_question.php"/>

</if>

</form>

</vxml>
```

Code 11.1 A VoiceXML form

In this form, there is a field called "survey" in which there is a prompt that asks the user a question and a grammar providing the recognition vocabulary for the user's response. The <filled> element is executed when the user has supplied a value for the field. In this case, there is a condition stating that if the response is "cinema," the system will speak a prompt, otherwise it transfers to another VoiceXML document.

Code 11.2 shows an example of a menu.

In this menu, the system offers a choice between watching films at the cinema or at home and, depending on the user's response, transfers to another VoiceXML document. Menus can be seen as shorthand for forms, containing a single anonymous field that prompts the user to make a choice and transitioning to different places based on that choice. Like a regular form, a menu can have its grammar scoped so that it is active when the user is executing another dialog.

Forms are the key component of VoiceXML documents. A form contains the following:

- A set of form items—input items that can be completed by user input and control items that cannot, such as <block>.
- Declarations of non-form item variables.
- Event handlers.
- Actions within the <filled> element—procedural logic that is executed when certain combinations of input item variables are assigned.

Forms are interpreted by the Form Interpretation Algorithm (FIA). The FIA has a main loop that repeatedly selects a form item and then visits it. The selected form

```
<?xml version="1.0"?>

<vxml version="2.0" xmlns="http://www.w3.org/2001/vxml">
<menu>

<prompt>

Welcome. Say one of: <enumerate/>

</prompt>

<choice next="vxml.com/cinema.vxml">
Watch films at the cinema</choice>

<choice next="vxml.com/home.vxml ">

Watch films at home.</choice>

</menu>

</vxml>
```

Code 11.2 A VoiceXML menu

item is the first in document order whose guard condition is not satisfied. The default guard condition verifies if the field's form item variable has a value i.e., it is not undefined, so that if a simple form contains only fields, the user will be prompted for each field in turn. This way, the default DM process that can be specified in VoiceXML consists of asking for the values of each of the input fields in the same order as they were specified by the developer of the application (system-directed dialog applications).

Each input item specifies an input item variable to gather from the user. Input items have prompts to tell the user what to say or key in, grammars that define the allowed inputs, and event handlers that process any resulting events (e.g., help required, no input provided, or no match with the inputs specified in the grammar).

The <prompt> element controls the output of synthesized speech and prerecorded audio. Attributes of <prompt> control whether a user can interrupt the prompt, define conditions that must evaluate to true in order for the prompt to be played, emit different prompts if the user is doing something repeatedly, or specify speech markup elements using Speech Synthesis Markup Language (SSML), for example, emphasis, prosody, and pauses.

The <grammar> element specifies the utterances that a user may speak to perform an action or supply information and, for a matching utterance, returns a corresponding semantic interpretation. Formats supported by VoiceXML platforms include the W3C Speech Recognition Grammar Specification (SRGS),[2] the Augmented BNF (ABNF),[3] and the Java Speech Grammar Format (JSGF).[4]

Grammars are specified using the <grammar> element within a <field>. Grammars can be inline, i.e., they are specified within the VoiceXML document, or they can be external, located using a Uniform Resource Identifier (URI) that can be absolute or relative. The <option> element can be used when a simple set of alternatives is all that is needed to specify the valid input values for a field. Each <option> element contains PCDATA that is used to generate a speech grammar according to the same grammar generation method used by the <choice> element (see the example of a menu above).

Code 11.3 is an example of an inline grammar using the ABNF grammar format. In this example, the user can say "blue," "green," or "red" in response to the system's prompt.

Code 11.4 shows a grammar specified using the <option> element. Here, the user can either say the words specified in the options or use the keypad to input DTMF.

[2]Speech Recognition Grammar Specifications. Available at: http://www.w3.org/TR/speech-grammar/. Accessed February 26, 2016.

[3]Augmented BNF for Syntax Specifications. Available at: https://cafe.bevocal.com/docs/grammar/abnf.html. Accessed February 26, 2016.

[4]JSpeech Grammar Format Specifications. Available at: https://www.w3.org/TR/jsgf/. Accessed February 26, 2016.

```
<field name="color">

<prompt>Which is your favorite color?</prompt>

<help>Say one of blue, green or red</help>

<grammar mode="voice" type="application/srgs">

#ABNF 1.0;

$options = blue | green | red

</grammar>

</field>
```

Code 11.3 Inline grammar using the ABNF grammar format

```
<field name="color">

<prompt> Please select a color. </prompt>

<option dtmf="1" value="blue"> blue </option>

<option dtmf="2" value="pink"> pink </option>

<option dtmf="3" value="red">  red </option>

</field>
```

Code 11.4 Grammar using the <option> element

```
<field name="command">

<prompt>Select a figure and a color.</prompt>

<grammar src="FiguresAndColors.grxml"/>

</field>
```

Code 11.5 Reference to an external grammar

Code 11.5 shows how an external grammar is referenced using the src attribute within the <grammar> element, while Code 11.6 is a grammar in grxml format that allows the user to state a color and a figure within the same utterance.

Input item grammars are always scoped to the containing input item. Link grammars are given the scope of the element that contains the link. Form grammars are by default given dialog scope, so that they are active only when the user is in the

```
<grammar xmlns="http://www.w3.org/2001/06/grammar"
xml:lang="en-US" tag-format="semantics/1.0" root = "ColorsFigures"
<rule id="ColorsFigures" scope="public">
<ruleref   uri="determiner_rule"/>   <ruleref   uri="colors_rule"/>   <ruleref
uri="figures_rule"/>
</rule>

<rule id="determiner_rule" scope="public">
<one-of>
<item> a </item>
<item> the </item>
</one-of>
</rule>

<rule id="colors_rule" scope="public">
<one-of>
<item> blue </item>
<item> pink </item>
<item> red </item>
</one-of>
</rule>

<rule id="figures_rule" scope="public">
<one-of>
<item> triangle </item>
<item> square </item>
<item> circle </item>
</one-of>
</rule>
</grammar>
```

Code 11.6 The `FiguresAndColors.grxml` external grammar

form. Finally, menu grammars are also by default given dialog scope and are active only when the user is in the menu.

Event handlers in VoiceXML define what to do when the user asks for help (`<help>` event), the user has asked to exit (`<exit>` event), the user has not responded within the timeout interval (`<noinput>` event), and the user's input was not recognized (`<nomatch>` event). They also deal with several error events (e.g., when a fetch of a document has failed, a requested platform resource was not available during execution, or an operation is not authorized by the platform). The `<throw>` element throws an event, for which the corresponding actions can be defined by means of the `<catch>` element.

The FIA can be customized in several ways to develop mixed-initiative dialog applications. The first possibility is to assign a value to a form item variable so that its form item will not be selected. A second possibility is to use the `<clear>` element to set a form item variable to undefined, which forces the FIA to revisit the form item again.

Input fields can also have guard conditions that activate only when other fields have been filled or when more advanced conditions hold true. These advanced conditions can, for instance, be defined using the shadow variables of the `<field>` element. These variables include the raw string of the words that were recognized (name$.utterance), the mode in which user input was provided: DTMF or voice (name$.inputmode), script variables containing the semantic interpretation of the field (name$.interpretation), and the confidence level for the field in a range from 0.0 to 1.0 (name$.confidence).

Another method is to explicitly specify the next form item to visit using <goto nextitem>. This forces an immediate transition to that form item even if any cond attribute that is present evaluates to "false." No variables, conditions, or counters in the targeted form item will be reset. The form item's <prompt> will be played even if it has already been visited. If the <goto nextitem> occurs in a <filled> action, the rest of the <filled> action and any pending <filled> actions will be skipped.

A mixed-initiative dialog may be completed in several ways. One common authoring style combines an <initial> element that prompts for a general response with <field> elements that prompt for specific information. More complex techniques, such as using the cond attribute on <field> and <filled> elements and the previously described shadow variables, may achieve a similar effect.

The <initial> element is visited when the user is initially being prompted for form-wide information, and the FIA has not yet entered into the directed mode where each field is visited individually. Like <field> items, the <initial> element has prompts, catches, and event counters. The <filled> element performs actions that are executed when a combination of one or more input items in the form is filled. This element usually contains conditional logic specified by means of the <if>, <else>, and <elseif> elements, for example, to verify that an origin city is not the same as the city provided for the destination. If a form has form-level grammars, its input items can be filled in any order, and more than one input item can be filled as a result of a single user utterance. Form grammars can be active when the user is in other dialogs. The user can speak to any active grammar and can have input items set and actions taken in response.

11.2.1 Practical Exercises Using VoiceXML

The main objective of this exercise is to develop a spoken conversational interface acting as a pizzeria service. Different platforms and interpreters are available to test conversational interfaces developed using the VoiceXML standard.[5] These platforms usually provide developers with certified interpreters for specific versions of the VoiceXML standard, the ASR and TTS interfaces, and the VoIP and telephony technologies.

To carry out the exercise that is described in this section, we propose the use of the Voxeo Evolution IVR Developer Portal.[6] With this platform, you can develop

---

[5]A detailed list of VoiceXML platforms can be found at: http://www.voicexml.org/solutions/category/voicexml-platforms/. Accessed February 27, 2016.

[6]https://voxeo.com/developers/the-evolution-ivr-voip-developer-portal/. Accessed February 27, 2016.

and test VoiceXML-based applications for free. The platform also offers discussion forums and extensive tutorials, sample applications, sample grammars, and comprehensive reference guides. The Web interface can be accessed here.[7] There is detailed guide explaining how to create an account to join the Voxeo Community, how to upload files to the IVR server by means of the `Files, Logs, & Reports` functionality, and how to use the `Application Manager` to map the application to a telephone number.[8]

You can download the code for the exercises from GitHub, in the ConversationalInterface repository,[9] in the folder called *chapter11/PizzaVXML*.

Exercise 11.1
Develop a VoiceXML-based pizzeria service (`pizza1.vxml`) including:

- A welcome message.
- Two fields: type and size. Include in each field:

  - A system query to ask for information (prompt).
  - A grammar with some keywords for recognition (`type.grxml` and `size.grxml` grammars).

The following instructions explain how to create an application and associate it with a starting VoiceXML file.

1. Click on the Account tab and in Files, Logs, Reports, create a new directory called `pizzeria` in the folder `root/www/`.
2. In the folder `pizzeria`, create a new file called `pizza1.vxml`, using the code listed in Code 11.7.
3. In Application Manager, create a new application called PizzaRules and select the radio button voice phone calls. Under `Voice Application Type` for `Region` select USA, for `App Type` VoiceXML, and for `ASR/TTS` Nuance 9.
4. To associate your application with a VoiceXML file, at `Voice URL`, click on `file manager`, and then in the pop-up window, click on the folder where your file was created (`pizzeria`) and on the file `pizza.vxml`. Note that you will have to change this mapping when you want to test later versions, e.g., `pizza2.vxml`. Note also that it may take a short time before the application is updated on the server and available to test.
5. Click on `Create Application` and your application will be created.
6. You will see some telephone numbers (Skype and inum) on the right-hand side of the page that you can use to call and test your application.

Solution to Exercise 11.1: Code 11.7 shows the VoiceXML code for `pizza1.vxml`. Codes 11.8 and 11.9 show the grammars. The grammar files should be placed in the grammar folder.

---

[7]http://evolution.voxeo.com/. Accessed February 26, 2016.
[8]http://help.voxeo.com/go/help/. Accessed February 26, 2016.
[9]http://zoraidacallejas.github.io/ConversationalInterface/. Accessed March 2, 2016.

```
<?xml version="1.0" encoding="UTF-8"?>

<vxml version="2.1" >

<form>

<block>

<prompt bargein="false"> Welcome to the Pizzeria. </prompt>

</block>

<field name="type">

<prompt bargein="false"> What kind of pizza do you want?

</prompt>

<grammar src="../grammar/type.grxml"/>

</field>

<field name="size">

<prompt bargein="false"> Tell me the size. </prompt>

<grammar src="../grammar/size.grxml"/>

</field>

</form>

</vxml>
```

Code 11.7 pizza1.vxml

```
<?xml version= "1.0"?>

<grammar xmlns="http://www.w3.org/2001/06/grammar"

xml:lang="en-US" tag-format="semantics/1.0" root = "type"

<rule id = "type" scope = "public">

<one-of>

<item> marinara</item>

<item> margherita</item>

<item> barbecue</item>

<item> Sicilian</item>

</one-of>

</rule>

</grammar>
```

Code 11.8 type.grxml

```
<?xml version= "1.0"?>

<grammar xmlns="http://www.w3.org/2001/06/grammar"

xml:lang="en-US" tag-format="semantics/1.0" root = "size"

<rule id = "size" scope = "public">

<one-of>

<item> small</item>

<item> large</item>

<item> extra large</item>

</one-of>

</rule>

</grammar>
```

Code 11.9 `size.grxml`

Run your application using a service such as Skype and test it with the inputs specified in the grammars. You can add more items to the grammars to allow a wider range of inputs.

Exercise 11.2
Extend `pizza1.vxml` to create `pizza2.vxml`:

- Ask the user to repeat what they said if they are wrong, and provide all possible options if they are wrong a second time. (Use of `<nomatch>` with `count`).
- When all fields are filled, synthesize a message indicating that the order will be ready soon and play an audio track. (Use of the `<filled>` element at form level and the `audio` label).

Solution to Exercise 11.2: Code 11.10 shows the VoiceXML code for `pizza2.vxml`. Use the grammars that were used in Exercise 11.1 (Codes 11.8 and 11.9).

Test the application by saying words that are not in the vocabulary in order to throw a `<nomatch>` event—for example, "vegetarian" for `type` and "medium" for `size`.

```
<?xml version="1.0" encoding="UTF-8"?>

<vxml version="2.1" >

<form>

<block>

<prompt bargein="false"> Welcome to the Pizzeria. </prompt>

</block>

<field name="type">

<prompt bargein="false"> What kind of pizza do you want?

</prompt>

<grammar src="../grammar/type.grxml"/>

<nomatch>

<prompt bargein="false"> Please say the name of a pizza, for
example Sicilian. </prompt>

</nomatch>

<nomatch count ="2">

<prompt bargein="false"> Please say the name of a pizza. The
four options are marinara, margherita, barbecue and Sicilian.
</prompt>

</nomatch>

</field>

<field name="size">

<prompt bargein="false"> Tell me the size. </prompt>

<grammar src="../grammar/size.grxml"/>

<nomatch>

 <prompt bargein="false"> Please say the size of the pizza,
for example large </prompt>

</nomatch>

<nomatch count ="2">

 <prompt bargein="false"> Please say the size of the pizza.
The options are small, large, and extra large.
```

Code 11.10 pizza2.vxml

```
</prompt>

</nomatch>

</field>

<filled>

<prompt bargein="false"> Your pizza will be ready soon. Thank
you for using the pizzeria service.

<audio src="./audio/pizza.wav"/></prompt>

</filled>

</form>

</vxml>
```

Code 11.10 (continued)

Exercise 11.3
Extend pizza2.vxml as follows to create pizza3.vxml:

- After filling all the fields, confirm if the order is correct and if it is not, reorder all the data pieces.

  - Incorporate a confirmation field with yes/no options.
  - If the user says "no," empty the corresponding field value (<clear> tag).

Solution to Exercise 11.3: Code 11.11 shows the VoiceXML code for pizza3. vxml. Use the grammars that were used in Exercise 11.1 (Codes 11.8 and 11.9) as well as the grammar in Code 11.12.

Exercise 11.4
Extend pizza3.vxml as follows to create pizza4.vxml:

- Allow mixed-initiative dialogs in which the second time that the system does not understand the long request, ask for each data piece step by step.

Solution to Exercise 11.4: Code 11.13 shows the VoiceXML code for pizza4. vxml. Use the grammars that were used in Exercise 11.3 (Codes 11.8, 11.9, and 11.12) as well as the grammar in Code 11.14.

1. Test the application with inputs such as "a large Sicilian pizza," "a small margherita pizza," and various other combinations of the sizes and types specified in the grammar.
2. Try inputs that do not match the order rule in complete.grxml (e.g., "large" or "Sicilian" to see how the FIA moves into directed dialog mode.

3. You can add further variation to the inputs that can be recognized by using the repeat attribute (repeat = "0-1"). For help on how to do this, consult the VoiceXML documentation[10] or the Voxeo tutorials.[11]

Exercise 11.5

Extend pizza4.vxml as follows to create pizza5.vxml:

- Allow users to ask for help at any moment of the dialog.

Solution to Exercise 11.5: Code 11.15 shows the VoiceXML code for pizza5.vxml. Use the grammars that were used in Exercise 11.4 (Codes 11.8, 11.9, 11.12, and 11.14). Test the application by saying "help" at different times during the interaction.

```
<?xml version="1.0" encoding="UTF-8"?>

<vxml version="2.1">

<form>

<block>

<prompt bargein="false">Welcome to the Pizzeria. </prompt>

</block>

<field name="type">

<prompt bargein="false">What kind of pizza do you want?

</prompt>

<grammar src="../grammar/type.grxml"/>

<nomatch>

 <prompt bargein="false"> Please say the name of a pizza, for
example Sicilian </prompt>

</nomatch>

<nomatch count ="2">

<prompt bargein="false">Please say the name of a pizza. The
four options are marinara, margherita, barbecue and Sicilian.
</prompt>
```

Code 11.11 pizza3.vxml

---

[10]https://www.w3.org/TR/voicexml20/. Accessed February 27, 2016.
[11]http://help.voxeo.com/go/help/xml.vxml. Accessed February 27, 2016.

```
</nomatch>

</field>

<field name="size">

<prompt bargein="false">Tell me the size</prompt>

<grammar src="../grammar/size.grxml"/>

<nomatch>

<prompt bargein="false">Please say the size of the pizza, for
example large </prompt>

</nomatch>

<nomatch count ="2">

<prompt bargein="false"> Please say the size of the pizza.
The options are small, large, and extra large.

</prompt>

</nomatch>

</field>

<field name="confirmation">

<prompt bargein="false"> You have ordered a <value
expr="size"/><value expr="type"/> pizza Is this correct?
</prompt>

<grammar src="../grammar/confirmation.grxml"/>

<filled>

<if cond="confirmation =='no' ">

<clear namelist="type"/>

<clear namelist="size"/>

<clear namelist="confirmation"/>

</if>

</filled>

</field>
```

Code 11.11 (continued)

```
<filled>

<prompt bargein="false"> Your pizza will be ready soon. Thank
you for using the pizzeria service.

<audio src="../audio/pizza.wav"/>

</prompt>

</filled>

</form>

</vxml>
```

Code 11.11 (continued)

```
<grammar xmlns="http://www.w3.org/2001/06/grammar"

xml:lang="en-US" tag-format="semantics/1.0" root = "yesno">

<rule id = "yesno" scope = "public">

<one-of>

<item> yes</item>

<item> no</item>

</one-of>

</rule>

</grammar>
```

Code 11.12 confirmation.grxml

```xml
<?xml version="1.0" encoding="UTF-8"?>

<vxml version="2.1">

<form>

<grammar src = "../grammar/complete.grxml"/>

<initial name="complete">

<prompt bargein="false"> Welcome to the Pizzeria. Please tell
me your order. </prompt>

<catch event ="nomatch noinput">

<prompt bargein="false"> Please say the size and type of
pizza you want. For example, a large Sicilian pizza</prompt>

</catch>

<catch event ="nomatch noinput" count ="2">

<prompt bargein="false"> I am sorry, I have not understood
your order. Let's go step by step.

</prompt>

<assign name="complete" expr="true"/>

</catch>

</initial>

<field name="type">

<prompt bargein="false">What kind of pizza do you want?

</prompt>

<grammar src="../grammar/type.grxml"/>

<nomatch>

<prompt bargein="false"> Please say the name of a pizza, for
example Sicilian. </prompt>

</nomatch>

<nomatch count ="2">

<prompt bargein="false"> Please say the name of a pizza. The
four options are marinara, margherita, barbecue and Sicilian.
</prompt>
```

Code 11.13 pizza4.vxml

```
</nomatch>

</field>

<field name="size">

<prompt bargein="false">Tell me the size</prompt>

<grammar src="../grammar/size.grxml"/>

<nomatch>

<prompt bargein="false"> Please say the size of the pizza,
for example large. </prompt>

</nomatch>

<nomatch count ="2">

<prompt bargein="false"> Please say the size of the pizza.
The options are small, large, and extra large.

</prompt>

</nomatch>

</field>

<field name="confirmation">

<prompt bargein="false"> You have ordered a <value
expr="size"/> <value expr="type"/> pizza

Is this correct? </prompt>

<grammar src="../grammar/confirmation.grxml"/>

<filled>

<if cond="confirmation =='no' ">

<clear namelist="type"/>

<clear namelist="size"/>

<clear namelist="confirmation"/>

</if>

</filled>

</field>

<filled>
```

Code 11.13 (continued)

```
<prompt bargein="false"> Your pizza will be ready soon. Thank
you for using the pizzeria service.

<audio src="./audio/pizza.wav"/></prompt>

</filled>

</form>

</vxml>
```

Code 11.13 (continued)

```
<?xml version= "1.0"?>

<grammar xmlns="http://www.w3.org/2001/06/grammar"

xml:lang="en-US" tag-format="semantics/1.0" root = "order">

<rule id="order" scope="public">

a

<item>

<item> <ruleref uri = "size_rule"/>

<tag>out.size=rules.size_rule.size</tag>

</item>

<item> <ruleref uri = "type_rule"/>

<tag>out.type=rules.type_rule.type</tag>

</item>

</item>

pizza

</rule>

<rule id = "size_rule" scope = "public">

<one-of>

<item> small <tag>out.size ="small" </tag> </item>

<item> large <tag>out.size ="large" </tag> </item>

<item> extra <tag>out.size ="extra large" </tag> </item>
```

Code 11.14 complete.grxml

```
</one-of>

</rule>

<rule id = "type_rule" scope = "public">

<one-of>

<item> marinara <tag>out.type ="marinara" </tag> </item>

<item> margherita <tag>out.type =" margherita " </tag>
</item>

<item> barbecue <tag>out.type =" barbecue " </tag> </item>

<item> Sicilian <tag>out.type =" Sicilian " </tag> </item>

</one-of>

</rule>

</grammar>
```

Code 11.14 (continued)

```xml
<?xml version="1.0" encoding="UTF-8"?>

<vxml version="2.1">

<form>

<!-- enable the 'help' universal grammar -->

<property name="universals" value="help" />

<grammar src = "../grammar/complete.grxml" />

<initial name="complete">

<prompt bargein="false"> Welcome to the Pizzeria. Please tell
me your order.</prompt>

<catch event ="nomatch noinput">

<prompt bargein="false"> Please say the size and type of
pizza you want. For example, a large Sicilian pizza.

</prompt>

</catch>

<help>

<prompt bargein="false"> The four types are marinara,  mar-
gherita, barbecue and Sicilian. The sizes are small, large,
and extra large.</prompt>

</help>

<catch event ="nomatch noinput" count ="2">

<prompt bargein="false"> I am sorry, I have not understood
your order. Let's go step by step.</prompt>

<assign name="complete" expr="true"/>

</catch>

</initial>

<field name="type">

<prompt bargein="false">What kind of pizza do you want?

</prompt>

<grammar src="../grammar/type.grxml"/>

<nomatch>
```

Code 11.15 pizza5.vxml

```
<prompt bargein="false"> Please say the name of a pizza, for
example Sicilian. </prompt>

</nomatch>

<nomatch count ="2">

<prompt bargein="false"> Please say the name of a pizza. The
four options are marinara, margherita, barbecue and Sicilian.
</prompt>

</nomatch>

<help>

<prompt bargein="false"> The four options are marinara, mar-
gherita, barbecue and Sicilian. </prompt>

</help>

</field>

<field name="size">

<prompt bargein="false">Tell me the size</prompt>

<grammar src="../grammar/size.grxml"/>

<nomatch>

<prompt bargein="false"> Please say the size of the pizza,
for example large </prompt>

</nomatch>

<nomatch count ="2">

<prompt bargein="false"> Please say the size of the pizza.
The options are small, large, and extra large.

</prompt>

</nomatch>

<help>

<prompt bargein="false"> The sizes are small, large, and ex-
tra large. </prompt>

</help>

</field>

<field name="confirmation">
```

Code 11.15 (continued)

```
<prompt bargein="false"> You have ordered a pizza <value
expr="type"/> <value expr="size"/> Is this   correct?
</prompt>

<grammar src="../grammar/confirmation.grxml"/>

<filled>

<if cond="confirmation =='no'">

<clear namelist="type"/>

<clear namelist="size"/>

<clear namelist="confirmation"/>

</if>

</filled>

</field>

<filled>

<prompt bargein="false"> Your pizza will be ready soon. Thank
you for using the pizzeria service. <audio
src="./audio/pizza.wav"/>

</prompt>

</filled>

</form>

</vxml>
```

Code 11.15 (continued)

11.3 Development of a Conversational Interface Using a Statistical Dialog Management Technique

In this section, we will generate a statistical dialog manager to solve the same problem: a pizza ordering service. In this case, instead of designing the dialog flow, we will use a corpus of dialogs in a pizza service to automatically learn the best dialog strategy.

The statistical methodology for dialog management that is proposed to solve this exercise was described in Chap. 10 (Sect. 10.4.2). This corpus-based approach to dialog management is based on the estimation of a statistical model from the sequences of system and user dialog acts obtained from a set of training data. The next system response is selected by means of a classification process that considers the complete history of the dialog.

Another main characteristic is the inclusion of a data structure that stores the information provided by the user. The main objective of this structure, called the *Dialog Register* (*DR*), is to easily encode the complete information related to the task provided by the user during the dialog history, then to consider the specific semantics of the task, and to include this information in the proposed classification process. See Griol et al. (2014) for a complete description of this statistical methodology for DM.

We have extended the definition of the pizza ordering service after Exercise 11.5 to allow users to provide the following task-dependent information items:

- Type of order (delivery or pick up),
- Number of pizzas,
- Types of pizzas,
- Sizes of pizzas,
- Types of pizza dough, and
- Drinks (optional field).

Users can also provide the following task-independent information items:

- Acceptance,
- Rejection, and
- Not-understood.

The *Dialog Register* defined for the task consists of 6 fields corresponding to the previously described task-dependent items of information:

Type_Order	Number_Pizzas	Types_Pizzas	Sizes_Pizzas	Types_Doughs	Drinks

The set of system actions (responses) defined for the task includes:

1. Welcome (*Opening*).
2. Ask the type of order (*Ask_Type_Order*).
3. Ask the number of pizzas (*Ask_Number_Pizzas*).
4. Ask the types of pizzas (*Ask_Types_Pizzas*).
5. Ask the sizes of pizzas (*Ask_Sizes_Pizzas*).
6. Ask the types of dough (*Ask_Types_Doughs*).
7. Ask the drinks (*Ask_Drinks*).
8. Confirm the type of order (*Confirm_Type_Order*).
9. Confirm the number of pizzas (*Confirm _Number_Pizzas*).
10. Confirm the types of pizza (*Confirm _Types_Pizzas*).
11. Confirm the sizes of pizzas (*Confirm _Sizes_Pizzas*).
12. Confirm the types of dough (*Confirm _Types_Doughs*).
13. Confirm the drinks (*Confirm _Drinks*).
14. Closing (*Closing*).

A set of 100 dialogs was automatically generated by means of a dialog simulation technique, which allows to acquire a dialog corpus annotated in terms of the user and system dialog acts defined for the task (Griol et al. 2013). These dialogs allow users to provide and/or confirm one or more task-independent items of information in a single user turn and also perform these actions without following a strict order.

As described in Chap. 10, for the dialog manager to determine the next system answer the exact values of the attributes in the *DR* are assumed to be not significant. They are important for accessing databases and for constructing the output sentences of the system. However, the only information necessary to predict the next action by the system is the presence or absence of concepts and attributes. Therefore, the codification proposed for each slot in the *DR* is in terms of three values, {0, 1, 2}, according to the following criteria:

- (0) The value for the slot is unknown or not given.
- (1) The value of the slot is known with a confidence score that is higher than a given threshold.
- (2) The value of the slot has a confidence score that is lower than the given threshold.

To decide whether the state of a certain value in the *DR* is 1 or 2, the system employs confidence measures provided by the ASR and SLU modules.

Using this codification for the *DR*, when a dialog starts (in the greeting turn) all the values in the Dialog Register are initialized to 0. The information provided by the users at each dialog turn is used to update the previous *DR* and obtain the current one, as shown in Fig. 11.2.

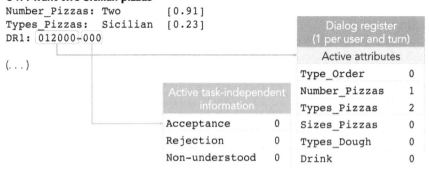

System1: Welcome to the Pizzeria. How can I help you?
```
A1: (Opening)
DR0: 000000-000
```

U1: I want two Sicilian pizzas
```
Number_Pizzas: Two       [0.91]
Types_Pizzas:  Sicilian  [0.23]
DR1: 012000-000
```

	Dialog register (1 per user and turn)	
	Active attributes	
Type_Order	0
Number_Pizzas	1
Types_Pizzas	2
Sizes_Pizzas	0
Types_Dough	0
Drink	0

Active task-independent information
Acceptance	0
Rejection	0
Non-understood	0

Fig. 11.2 Excerpt of a simulated dialog with its corresponding Dialog Register and active task-independent information for one of the turns

This figure shows the semantic interpretation and confidence scores (in brackets) for a user's utterance provided by the SLU module. In this case, the confidence score assigned to the slot *Types_Pizzas* is very low. Thus, a value of 2 is added in the corresponding position of DR_1. The slot *Number_Pizzas* is recognized with a high confidence score, adding a value of 1 in the corresponding position of DR_1.

The updated DR, the active task-independent information (*Acceptance, Rejection*, and *Not-Understood* dialog acts), and the codification of the labeling of the last system turn (A_1) are considered as an input to a classifier that provides the probabilities of selecting each next system response according to the current state of the dialog represented by means of the described input. This process is repeated to predict the next system response after each user turn.

You can find the set of simulated dialogs in the ConversationalInterface repository,[12] in the folder called *chapter11/PizzaStat*.

Each dialog sample in the corpus consists of 11 elements delimited by a comma:

- Last system response: The last system response is coded using a natural number between 1 and 14, which represents each one of the system responses in the same order that they were previously presented in the chapter.
- Current Dialog Register (e.g., 012000): The Dialog Register consists of the 6 fields representing task-dependent dialog acts, which are coded as previously described.
- Task-independent information provided in the current user turn (e.g., 000): This information corresponds to the *Acceptance, Rejection,* and *Not-Understood* dialog acts. These dialogs are represented using the same codification defined for the *DR*.
- Next system response (natural number between 1 and 14).

To train a classifier using the set of provided samples, we propose the use of the Waikato Environment for Knowledge Analysis (Weka) Software.[13] Detailed documentation about this software can be found at here.[14] Different classifier functions can be selected using the software (artificial neural networks, decision trees, Naïve Bayes, etc.). You can test each one of these classifiers with the Weka ARFF file that has been generated with the samples in the corpus using the file `Dialog_Pizza.arff` that you can download from the ConversationalInterface repository,[15] in the folder called *chapter11/PizzaStat*.

Open this file with Weka, select a specific classifier at the `Classify` tab of the Weka Explorer, and specify 5-fold cross-validation with the 80 % of the corpus for training and the remaining 20 % of the corpus as a test partition. Repeat the Weka prediction with different classifiers and study which one provides better results

---

[12]http://zoraidacallejas.github.io/ConversationalInterface/. Accessed March 2, 2016.

[13]http://www.cs.waikato.ac.nz/ml/weka/. Accessed February 26, 2016.

[14]http://www.cs.waikato.ac.nz/ml/weka/documentation.html. Accessed February 26, 2016.

[15]http://zoraidacallejas.github.io/ConversationalInterface/. Accessed March 2, 2016.

Table 11.1 Confusion matrix (Weka)

a	b	c	d	e	f	g	h	i	j	k	l	m	n	<!-- classified as
0	0	0	0	0	0	0	0	0	0	0	0	0	0	a = 1
0	60	0	0	0	0	0	0	0	0	0	0	0	0	b = 2
0	0	35	0	0	0	0	0	0	0	0	0	0	0	c = 3
0	0	0	89	0	0	0	0	0	0	0	0	0	0	d = 4
0	0	0	0	110	0	0	0	0	0	0	0	0	0	e = 5
0	0	0	0	0	156	0	0	0	0	0	0	0	0	f = 6
0	0	0	0	0	0	166	0	0	0	0	0	0	0	g = 7
0	0	0	0	0	0	0	24	0	0	0	0	0	0	h = 8
0	0	0	0	0	0	0	0	24	0	0	0	0	0	i = 9
0	0	0	0	0	0	0	0	0	28	0	0	0	0	j = 10
0	0	0	0	0	0	0	0	0	0	21	0	0	0	k = 11
0	0	0	0	0	0	0	0	0	0	0	10	0	0	l = 12
0	0	0	0	0	0	0	0	0	0	0	0	58	0	m = 13
0	0	0	0	0	0	0	0	0	0	0	0	0	174	n = 14

regarding the number of system responses selected by the classifier that are correctly predicted.

As an example, the results that are provided following this process when the multilayer perceptron (MLP) is selected as the classifier functions are as follows:

```
Correctly Classified Instances: 955 (98.657 %)
Incorrectly Classified Instances: 13 (1.343 %)
Kappa statistic: 0.9847
Mean absolute error: 0.0063
Root mean squared error: 0.0441
Relative absolute error: 5.0366 %
Root relative squared error: 17.6224 %
Total Number of Instances: 968
```

As can be observed, the statistical dialog manager successfully predicts the next system response at a rate of 98.657 %, where this percentage denotes the cases in which the system response that is selected by the MLP is the same as the reference response annotated in the corresponding dialog sample. The confusion matrix that is provided by Weka is shown in Table 11.1.

You can also try to design an ensemble of classifiers to try to improve the results that are obtained.

11.4 Summary

In this chapter, we have provided exercises showing how to develop practical dialog managers by means of rule-based and statistical techniques. The VoiceXML language has been selected given that it is an approach used widely in industry to develop handcrafted dialog managers for spoken conversational interfaces. We have selected the same practical application domain to implement a corpus-based statistical dialog manager.

The next stage for the conversational interface is to take the action decided by the dialog manager and determine the content of the response and how best to express it. Chapter 12 describes the fundamental aspects and main approaches for response generation.

Further Reading
A detailed explanation of the use of the VoiceXML standard can be found in the references provided in the chapter. Detailed tutorials and examples of VoiceXML applications are also available at the Voxeo Web site.[16]

---

[16]http://help.voxeo.com/go/help/xml.vxml.voicexml. Accessed 17 April 2016.

Further practical applications of the proposed statistical methodology for dialog management are described in Griol et al. (2008, 2009), while additional tutorials about the Weka software and the ARFF data format can be found here[17,18].

References

Griol D, Hurtado LF, Segarra E, Sanchis E (2008) A dialog management methodology based on neural networks and its application to different domains. Lect Notes Comput Sci 5197:643–650. doi:10.1007/978-3-540-85920-8_78

Griol D, Riccardi G, Sanchis E (2009) A statistical dialog manager for the LUNA project. In: Proceedings of 10th annual conference of the international speech communication association (INTERSPEECH), Brighton, UK, pp 272–275. 6–10 Sept 2009. http://disi.unitn.it/~riccardi/papers/Interspeech09-DM.pdf

Griol D, Carbo J, Molina JM (2013) An automatic dialog simulation technique to develop and evaluate interactive conversational agents. Appl Artifi Intell 27(9):759–780. doi:10.1080/08839514.2013.835230

Griol D, Callejas Z, López-Cózar R, Riccardi G (2014) A domain-independent statistical methodology for dialog management in spoken dialog systems. Comput Speech Lang 28 (3):743–768. doi:10.1016/j.csl.2013.09.002

---

[17]https://sourceforge.net/projects/weka/files/documentation/Initial%20upload%20and%20presentations/Weka_a_tool_for_exploratory_data_mining.ppt/download?use_mirror=netix. Accessed 17 April 2016.

[18]http://weka.wikispaces.com/ARFF. Accessed 17 April 2016.

Chapter 12
Response Generation

Abstract Once the dialog manager has interpreted the user's input and decided how to respond, the next step for the conversational interface is to determine the content of the response and how best to express it. This stage is known as response generation (RG). The system's verbal output is generated as a stretch of text and passed to the text-to-speech component to be rendered as speech. In this chapter, we provide an overview of the technology of RG and discuss tools and other resources.

12.1 Introduction

Once the dialog manager has interpreted the user's input and decided how to respond, the next step is to determine the content of the response (content determination) and how best to express it (content realization). Compared with the other components of a conversational interface, there are no readily available tools for response generation (RG).

There are two main approaches to RG. In many spoken dialog systems and especially in commercial voice user interfaces, a simple approach is adopted in which the system outputs predetermined responses, using either canned text or templates in which the values of variables can be inserted at runtime. This approach works well in fairly restricted interactions. However, a more elaborate approach is required where the content to be output cannot be determined in advance and where it needs to be edited and cast in a form suitable for spoken output. In this case, techniques from natural language generation (NLG) are required.

12.2 Using Canned Text and Templates

The following are examples of the types of output provided by a typical spoken dialog system or voice user interface when engaged in a fairly tightly constrained task such as flight booking:

© Springer International Publishing Switzerland 2016
M. McTear et al., *The Conversational Interface*,
DOI 10.1007/978-3-319-32967-3_12

- Prompts to elicit information from the user, e.g., "What time do you want to leave?"
- Messages indicating some problem, e.g., "Sorry, I didn't get that."
- Requests for confirmation, e.g., "So you want to fly to London?"
- Retrieved information from a knowledge source, e.g., "There is a flight from Belfast to London on Friday morning departing at 7.05 a.m."

These outputs can be handled using simple methods such as canned text and template filling.

Canned text can be used in interactions where the system has to elicit a predetermined set of values from the user—such as departure time, destination, and airline. The prompts can be designed in advance along with messages indicating problems and errors and can be executed at the appropriate places in the dialog. Templates provide some degree of flexibility by allowing information to be inserted into the prompt or message. For example, to confirm that the system has understood, the following prompt might be used:

So you want to go to $Destination on $Day?

Here, $Destination and $Day are filled by values elicited in the preceding dialog. Templates can become quite complex and can involve some form of computation using rules to construct the message, as in the responses provided by chatbots using markup languages such as AIML (see Chap. 7).

There is an extensive literature on the design of effective prompts, based on best practice guidelines as well as studies of usability (Balentine and Morgan 2001; Cohen et al. 2004; Hura 2008; Lewis 2011).

Prompts can be directive or non-directive. Directive prompts state explicitly what the user should say, for example, "Select savings account or current account," while non-directive prompts are more open-ended, for example, "How may I help you?" Usability studies of directive versus non-directive prompts have found that directive prompts are more effective as they make users more confident in what they are required to say (Balentine and Morgan 2001). One way to improve the usability of non-directive prompts is to include an example prompt, for example, "You can say transfer money, pay a bill, or hear last 5 transactions" (Balentine and Morgan 2001). Lewis (2011: 222–229) provides a detailed overview of experimental studies of non-directive prompts, summarizing key recommendations for design.

Prompts that present menu choices are another design challenge. Given a large number of menu choices, the Voice User Interface (VUI) designer has to choose whether to present more options in each menu, leading to fewer menus (i.e., a broader menu design), or whether to divide the choices into a menu hierarchy with more menus but fewer options in each menu (i.e., a deeper design). One consideration that has guided menu design is the limits of human working memory —for example, if a large number of options are provided for each menu. Various experimental studies are summarized in Lewis (2011: 231–239).

Designing reprompts is another consideration. If a prompt has to be repeated, either because the user has not responded at all or because the user has responded incorrectly, it is preferable not to simply repeat the prompt but rather to change it depending on the circumstances. For example, if the original prompt was unsuccessful in eliciting more than one item of information, the reprompt can be shortened (or tapered) to ask for less information:

> System: Please tell me your home address, including postal code and city name
> User: (answers, but system fails to understand)
> System: Sorry I didn't get that, please repeat your home address

Another situation is where it appears that the user does not know what to do or say, in which case an incremental prompt can be used that provides more detailed instructions:

> System: how many would you like?
> User: what?
> System: how many shares do you want to buy? For example, one hundred
> User: a thousand
> System: I'm sorry, I still didn't get that. Please state the number of shares you would like to buy or enter the number using your keypad

Finally, a leading prompt indicates which words the user should provide in their response:

> System: welcome to Automated Banking Services. To transfer money, say transfer, to pay a bill, say pay a bill, …

The main problem with canned text and templates is a lack of flexibility. Designers have to anticipate all the different circumstances that might occur in a dialog and design rules to appropriately adapt the system output.

12.3   Using Natural Language Generation Technology

In some cases, the methods of canned text and templates are not sufficient for generating an appropriate response. For example, the content to be output may need to be structured in such a way that it is easy for the user to understand. There may be a large number of flights that match the user's requirements, so that the RG component has to decide how many to present at a time. Or there may be different ways of presenting the information, for example, comparing different attributes of the flights—price, departure times, and stopovers—or recommending a particular flight based on known user preferences. Another issue is that when the user has requested some information, as in a Web search, the retrieved content may need to be summarized and structured in a form suitable for spoken output by the TTS component and not simply read out in its entirety.

NLG technology can handle information that cannot be predicted at design time, such as the results of a Web search that need to be processed and cast into a form suitable for spoken output. Applications that use NLG technology include weather forecasts generated from weather prediction data, textual summaries of electronic medical records, financial and business reports based on the analysis of retail sales data, and flight information obtained from airline schedules.

NLG has been viewed as consisting of several stages of processing that are ordered in a pipeline model (Fig. 12.1).

When the input to the pipeline takes the form of data, such as the numerical data in weather predictions, there are two preliminary stages (Reiter 2007):

1. Signal analysis—the analysis of numerical and other data, involving looking for patterns and trends using numerical pattern recognition algorithms.
2. Data interpretation—identification of complex messages, in which there are causal and other relations between the messages. Generally, symbolic reasoning approaches based on domain knowledge have been used for this stage.

Reiter (2007) discusses an example from the BABYTALK project[1] in which textual summaries of medical data about babies in a neonatal intensive care unit are generated (Portet et al. 2009). The data include numerical sensor data, such as heart rate, blood pressure, temperature, and medical records of medications administered and blood test results. Data interpretation involves creating the text of the messages by deciding how important particular events are and detecting relationships between events.

The next stages consist of the stages in the original pipeline model presented in Reiter and Dale (2000): *document planning*, *microplanning*, and *realization*. These stages are applied to textual content that has been retrieved from sources such as knowledge bases and databases.

---

[1]http://www.abdn.ac.uk/ncs/departments/computing-science/babytalk-308.php.   Accessed February 21, 2016.

Fig. 12.1 The NLG pipeline

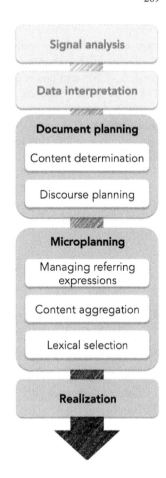

12.3.1   Document Planning

Document planning consists of two substages: *content determination* and *discourse planning*. Content determination involves deciding what information to communicate. Not all of the information that has been retrieved may need to be spoken to the user, or it may need to be delivered in stages. For example, it is not appropriate to use speech to convey long lists of information to a user. Similarly, the information to be conveyed might need to be adapted to the needs of different users. Content determination involves filtering the information and summarizing it, where required. For example, in the Mercury Flight information system described in Chap. 4, there were rules specifying how the output should be expressed depending on the number of flights retrieved (Seneff 2002).

Once the content has been determined, the next stage—discourse planning— involves structuring and organizing the content so that it is easy for the user to process. In some cases, a particular message type will have a regular structure. For example, a set of instructions typically describes the required actions in the order of their execution, using connecting words such as "first," "next," and "finally." Well-structured texts can be modeled using schemas that set out the main components of the text and reflect how the text is organized sequentially (McKeown 1985). A more elaborate approach, derived from rhetorical structure theory, describes the relations between the elements of a text, for example, indicating that a second sentence is an *elaboration* of a preceding sentence or that it expresses a *contrast* (Mann and Thompson 1988). The following is an example of a contrast, indicated by the cue word "however":

> There is an early morning flight to London. However, it has been delayed by 90 minutes.

The output of the document planning stage is a document plan.

12.3.2 Microplanning

Microplanning, also known as sentence planning, is concerned with creating the sentences to convey the information. There are three main tasks in microplanning:

1. Referring expressions.
2. Aggregation.
3. Lexical selection.

Dealing with referring expressions involves determining how to refer to an entity in a text within a particular context. For example, if some entity that has already been mentioned is to be referred to again, it should be possible to refer to that entity using a pronoun, as in the following example:

> There is a flight to London at 3 p.m. It arrives at 4.05.

Aggregation is the issue of combining content into meaningful portions, for example, by using conjunctions, ellipsis, or combinations of these, as in the following examples:

> The flight departs at 9. It arrives at 10 (no aggregation).
> The flight departs at 9, and it arrives at 10 (aggregation with conjunction).
> The flight departs at 9 and [ ] arrives at 10 (aggregation with conjunction and ellipsis).

Lexical selection involves choosing the appropriate words to express the content. In most simple systems, a single lexical item is associated with each entity in the database. However, using different expressions provides more variation in the text, as in the following examples:

> The first flight departs at 9. The second flight departs at 10. The third flight departs at 11 (no variation).
> The first flight departs at 9. The departure times of the next flights are 10 and 11.

Variation in lexical selection may be useful when outputting text to different users with different levels of expertise—for example, more technical terms for experienced users of some technology and more general descriptive terms to novice users.

12.3.3   Realization

Realization is the stage involving linguistic analysis in which rules of grammar are applied, for example, to enforce subject–verb agreement, insert function words, choose the correct inflections of content words, order the words within the sentence, and apply rules of spelling. A grammar is used that provides a set of choices for linguistic realization, for example, between active and passive sentences, as in the following examples:

> Bad weather has delayed the flight.
> The flight has been delayed by bad weather.

Where the output takes the form of speech, as in most of the output from a conversational interface, marking the text for prosody is also important (Theune 2002). The use of prosody makes spoken output more natural and easier to process. Prosody covers variations in pitch, loudness, tempo, and rhythm within an utterance. For example, some of the words in an utterance are more accented (or emphasized) than others, particularly if they express information that is assumed to be new to the hearer. Also, the utterance may be divided into intonational phrases, sometimes with a brief pause between the phrases. Failure to consider prosody results in the robotic speech typical of early TTS systems. In cases where the output can be predicted in advance, it can be marked up by hand using markup languages such as the W3C's Speech Synthesis Markup Language (SSML) that includes markup tags such as <emphasis>, <break>, and <prosody>.[2] A compatible TTS engine would be required that could interpret and render the tags. In cases where the output text cannot be determined in advance, as in arbitrary text retrieved from the Web, prosodic markup has to be done automatically at runtime.

12.4 Statistical Approaches to Natural Language Generation

NLG technology has been applied mainly to texts such as monologues or written texts where the content is known in advance. However, in spoken dialog, the content emerges dynamically as the dialog proceeds so that a more adaptive approach is required. The pipeline architecture used in NLG is also problematic, as the processing at earlier stages such as document planning cannot take account of issues that might arise at later stages in the pipeline. For example, there might be problems with realizing the content at the syntactic, lexical, or phonetic levels that should have been taken into account at the earlier content determination stage. Recent work using statistical machine learning models has begun to address these issues. The following example is taken from a research project in which the system had to present information about restaurants to a user (Lemon 2011; Lemon et al. 2010; Rieser et al. 2010, 2014).

Information presentation requires the system to maintain a balance between presenting enough information to respond to the user's requirements while avoiding overloading the user with too much information. Three strategies were identified for the presentation of information, given a set of records returned from a database of restaurants:

- *summary* (e.g., "I found twenty-six restaurants … eleven are in the expensive price range …").
- *recommend* (e.g., "The restaurant XX has the best overall quality …").
- *compare* (e.g., "The restaurant XX is in the cheap price range … and the restaurant YY is in the moderate price range …").

These could also be combined, so that, for example, a summary could be followed by a recommend, or by a compare followed by a recommend. Furthermore, decisions were required to determine how many attributes of a restaurant to present and in what order, and at the realization stage how many sentences to use and what lexical choices to make.

The decisions involved in RG were modeled using a Markov decision process (MDP) and reinforcement learning (RL), in which each state in the MDP represented a dialog context and system actions represented decisions that received a reward on transition to the next state (see Chap. 10 for a more detailed discussion

---

on RL). The aim was to maximize the long-term expected reward of taking decisions that would result in an optimal policy for generating the content. Moreover,
by treating information presentation as a joint optimization problem combining
content structuring and selection of attributes, it was possible to consider
lower-level actions at the same time as higher-level actions, which would not be
possible using the traditional pipelined architecture for NLG. The following is an
example of a strategy learnt by the RL policy:

```
If the initial number of items returned from the database is
high, start with a summary.
```

```
If the user selects an item, stop generating.
Otherwise, continue with a recommend.
```

```
If the number of database items is low, start with a compare
and then continue with a recommend, unless the user selects an
item.
```

An important issue is that the strategies are context dependent as the system
bases its decisions for generation on features in the state space at each generation
step. It would be a difficult task to hand-code all the rules required to handle such a
large number of permutations of contextual features and provide similar adaptivity.

12.5   Response Generation for Conversational Queries

Finding answers to questions posed to spoken dialog systems and voice user
interfaces has traditionally involved sending a query to a structured information
source such as a database—for example, to get flight information, stock quotes, and
weather reports. The system first engages with the user in a slot-filling dialog to
elicit the values of a number of parameters and then queries the information source
with these values to obtain the required information. This information is translated
into natural language text using the NLG techniques described earlier.

However, conversational interfaces also support general knowledge questions
such as the following:

Who is the president of France?
When was Google founded?
Where is the Eiffel Tower?

Although it might be possible to find answers to questions like these in a
database, more typically the answers are to be found in digitally stored documents
on the World Wide Web and in other online sources such as books and journals.
Because these sources consist of text and do not conform to a predefined data

model, they are viewed as being *unstructured*. Modern search engines can handle queries to such unstructured sources with amazing efficiency and speed, although generally the response to a search query takes the form of a list of Web pages where the required information has to be retrieved by consulting one or more of the links.

For a conversational interface, a different type of response is required in the form of a short stretch of text that can be spoken to the user. One approach uses the technology of question answering (QA) to find answers in large collections of unstructured text on the World Wide Web. Alternatively structured knowledge sources, often referred to as semantic repositories, have been created in which knowledge is structured and linked.[3] The following section provides an overview of these technologies followed by a brief discussion of how text summarization can also be used to create a response that can be spoken by a conversational interface.

12.5.1   Question Answering

QA is one of the earliest applications of natural language processing, going back to systems developed in the 1960s. Typically, most systems deal with factoid questions, for which answers can usually be found in short stretches of text that have been retrieved using information retrieval techniques from a collection of documents. Finding an answer may involve identifying named entities in the question, such as a person, time, or location, as in the examples above, and retrieving a ranked set of passages of text containing matching named entities from a large collection of documents.

QA has three main stages: question processing, passage retrieval and ranking, and answer processing (see Jurafsky and Martin (2009, Chap. 23) for a detailed account of QA). As far as the RG component of a conversational interface is concerned, the main task is to take the output of the QA system and generate a stretch of text that can be spoken as an answer to the user's question. In many cases, the output may be suitable without further editing, while in other cases some editing may be required, using the NLG techniques discussed earlier.

There are many systems that provide QA technology, including IBM's Watson and the Evi system. Watson[4] is a powerful question answering system that was originally developed using IBM's DeepQA architecture to answer questions on the Jeopardy quiz show, where it won first prize in 2011.[5] Watson uses a combination of NLP, hypothesis generation and evaluation, and dynamic machine learning to extract knowledge from millions of documents containing structured as well as unstructured contents. Watson is now being used for QA in domains such as health

---

[3]http://linkeddata.org/. Accessed February 21, 2016.

[4]http://www.ibm.com/smarterplanet/us/en/ibmwatson/. Accessed February 21, 2016.

[5]http://ibmresearchnews.blogspot.co.uk/2011/12/dave-ferrucci-at-computer-history.html. Accessed February 21, 2016.

care to provide clinical decision support. Ferrucci et al. (2010) provide an overview of the DeepQA project and the implementation of Watson. There is also a description of how to build a simple demo application that allows users to ask questions about films and filmmaking.[6]

Evi[7] runs on Android and iPhone smartphones and Kindle Fire tablets. Users can ask questions using speech. Evi examines all the possible meanings of the words in the user's question to find the most likely meaning of the question and draws on a database of nearly a billion facts to find an answer, using a series of logical deductions. Evi was developed by a technology company based in Cambridge, England (formerly known as True Knowledge). Evi was acquired by Amazon in 2012 and is now part of the Amazon group.

12.5.2 Structured Resources to Support Conversational Question Answering

In order to be able to find answers to a user's query, a conversational interface has to be able to access knowledge bases that contain the relevant information. Increasingly, there has been a trend toward developing large semantic repositories in which this knowledge is stored in a machine-readable and accessible form (de Melo and Hose 2013). We briefly describe three of these repositories: WolframAlpha, Google's Knowledge Graph, and DBpedia.

WolframAlpha is a system for dynamically computing answers to free-form natural language questions by drawing on a vast store of curated, structured data.[8] WolframAlpha provides several of its own personal assistant apps in areas such as personal fitness, travel, and personal finance and is used as a knowledge source by a number of virtual personal assistants to generate answers to user's questions.

Knowledge Graph was introduced by Google in 2012 as a knowledge base that is constructed mainly from crowd-sourced and manually curated data. Knowledge Graph provides structured information about a topic as well as a short summary that may be used as a spoken answer in queries to Google Now.[9] In a new venture, Google introduced Knowledge Vault in 2014, which includes information from structured sources as well as unstructured sources across the entire World Wide Web.

DBpedia extracts structured information from Wikipedia and makes it available on the Web.[10] The English version of DBpedia includes 4.58 million things, most

---

[6]http://www.ibm.com/developerworks/cloud/library/cl-watson-films-bluemix-app/. Accessed February 21, 2016.

[7]https://www.evi.com/. Accessed February 21, 2016.

[8]http://www.wolframalpha.com/tour/what-is-wolframalpha.html. Accessed February 21, 2016.

[9]https://googleblog.blogspot.co.uk/2012/05/introducing-knowledge-graph-things-not.html. Accessed February 21, 2016.

[10]http://wiki.dbpedia.org/about. Accessed February 21, 2016.

of which are classified in an ontology, including persons, creative works, music albums, films, and video games, and there are also versions in 125 languages. The DBpedia dataset can be accessed online via a SPARQL query endpoint and as linked data.[11] See Lehmann et al. (2015) for a recent overview paper about the DBpedia project.

12.5.3 Text Summarization

In some cases, an answer to a question might take the form of a summary of the contents of one or more documents. A summary is a reduction of the original text that preserves its most important points. The snippets produced by Web search engines are examples of such summaries. In the simplest kind of summary, phrases and sentences from the document(s) are selected and combined to form the summary, known as an *extract*. This is the most common form of summary using current technology. A different approach produces an *abstract*, in which a semantic representation is built of the contents of the document(s) and used to create a summary involving similar processes to those used in NLG, i.e., content determination, discourse planning (structuring), and microplanning (sentence realization) (see Jurafsky and Martin 2009, Chap. 23 for a detailed account). There may also be a process of sentence simplification to produce a more suitable form of output as a spoken message.

12.6 Summary

RG involves determining the content of the response to the user's query and deciding how best to express it. The simplest approach is to use predetermined responses, either in the form of canned text or in the form of templates in which the variables can be inserted at runtime. Using NLG technology provides greater flexibility as the text to be output can be tailored to the needs of the user and adapted or summarized to meet the requirements of a spoken response. NLG has traditionally been viewed as consisting of several stages of processing that are ordered in a pipeline model. Recently, statistical models have been applied to RG. In these models, there is greater flexibility as the content can emerge dynamically over the course of the dialog, whereas in the traditional pipeline approach processing at earlier stages cannot take account of issues that might arise at later stages in the pipeline.

RG for spoken dialog systems and voice user interfaces normally involves translating structured information retrieved from a database into a form suitable for a spoken response. A different approach is required for queries that access unstructured

---

[11]http://dbpedia.org/OnlineAccess. Accessed February 21, 2016.

content on the World Wide Web. Here, techniques drawn from QA, text summarization, and sentence simplification are proving useful.

So far, we have focused on the spoken language aspects of conversational interfaces, where the interaction has involved queries and commands to virtual personal assistants on smartphones. In the next part, we look first in Chap. 13 at other smart devices that require a conversational interface, such as smart watches, wearables, and social robots. Then, in Chaps. 14 and 15, we show how conversational interfaces can be made more believable and acceptable by incorporating the ability to recognize and produce emotions, affect, and personality.

Further Reading

Reiter and Dale (2000) is a collection of chapters dealing with standard text on NLG. Here, the main elements of NLG are described on which most NLG research has been based. Stent and Bangalore (2014) is a recent collection of chapters dealing with NLG for interactive systems, including approaches using RL and data-driven methods. See also the collection edited by Krahmer and Theune (2010). Reiter (2010) reviews recent developments in NLG, while Lemon et al. (2010) present several case studies illustrating statistical approaches. See also Janarthanam and Lemon (2014) and Mairesse and Young (2014).

Information about NLG can be found at the Web page of the ACL Special Interest Group on Natural Language Generation (SIGGEN).[12] There is also an NLG Systems Wiki that lists NLG systems, including links to home pages and key references.[13] See also the Web page for the NLG Group at the University of Aberdeen.[14] SimpleNLG is a Java API that was developed at the University of Aberdeen to facilitate the generation of natural language text.[15] A list of tools for implementing NLG can be found at the ACL Wiki page.[16]

Until recently, NLG has not been used widely in real-world applications, but since 2009 several companies have been launched that generate text from data using NLG techniques. These include ARRIA,[17] Automated Insights,[18] NarrativeScience,[19] and Yseop Smart NLG.[20]

De Melo and Hose (2013) provide a comprehensive tutorial on searching the Web of data. For a collection of readings on QA, see Maybury (2004) and Strzalkowski

---

[12]http://www.siggen.org/. Accessed February 21, 2016.

[13]http://www.nlg-wiki.org/systems/. Accessed February 21, 2016.

[14]http://www.abdn.ac.uk/ncs/departments/computing-science/natural-language-generation-187. php. Accessed February 21, 2016.

[15]https://github.com/simplenlg/simplenlg. Accessed February 21,2016.

[16]http://aclweb.org/aclwiki/index.php?title=Downloadable_NLG_systems. Accessed February 21, 2016.

[17]http://www.arria.com/. Accessed February 21, 2016.

[18]http://automatedinsights.com/. Accessed February 21, 2016.

[19]http://www.narrativescience.com/. Accessed February 21, 2016.

[20]http://www.slideshare.net/gdm3003/searching-the-web-of-data-tutorial. Accessed February 21, 2016.

and Harabagiu (2008). For a fairly non-technical coverage of the technologies behind IBM's Watson, see Baker (2012). The standard textbook on text summarization is Mani (2001). See also Mani and Maybury (1999) and Jones (2007).

References

Baker S (2012) Final Jeopardy: the story of Watson, the computer that will transform our world. Houghton Mifflin Harcourt, New York

Balentine B, Morgan DP (2001) How to build a speech recognition application: a style guide for telephony dialogs, 2nd edn. EIG Press, San Ramon

Cohen MH, Giangola JP, Balogh J (2004) Voice user interface design. Addison Wesley, New York

de Melo G, Hose K (2013) Searching the web of data. In: Serdyukov P et al (ed) Advances in information retrieval: Proceedings of 35th European conference on IR research, ECIR 2013, Moscow, Russia, 24–27 Mar 2013. Lecture Notes in Computer Science, vol 7814. Springer Publishing Company, pp 869–873. doi:10.1007/978-3-642-36973-5_105

Ferrucci D, Brown E, Chu-Carroll J, Fan J, Gondek D, Kalyanpur AA, Lally A, Murdock JW, Nyberg E, Prager J, Schlaefer N, Welty C (2010) Building Watson: an overview of the DeepQA project. AI Mag 31(3):59–79. http://dx.doi.org/10.1609/aimag.v31i3.2303. Accessed 20 Jan 2016

Hura S (2008) Voice user interfaces. In: Kortum P (ed) HCI beyond the GUI: design for haptic, speech, olfactory, and other non-traditional interfaces. Morgan Kaufmann, Burlington, pp 197–227. doi:10.1016/b978-0-12-374017-5.00006-7

Janarthanam S, Lemon O (2014) Adaptive generation in dialogue systems using dynamic user modeling. Comp Linguist 40(4):883–920. doi:10.1162/coli%5Fa%5F00203

Jones KS (2007) Automatic summarizing: the state of the art. Inf Proc Manage 43(6):1449–1481. doi:10.1016/j.ipm.2007.03.009

Jurafsky D, Martin JH (2009) Speech and language processing: an introduction to natural language processing, computational linguistics, and speech recognition, 2nd edn. Prentice Hall, Upper Saddle River

Krahmer E, Theune M (eds) (2010) Empirical methods in natural language generation: data-oriented methods and empirical methods. Springer, New York. doi:10.1007/978-3-642-15573-4

Lehmann J, Isele R, Jakob M, Jentzsch A, Kontokostas D, Mendes PN, Hellmann S, Morsey M, van Kleef P, Auer S, Bizer C (2015) DBpedia—a large-scale, multilingual knowledge base extracted from Wikipedia. Semant Web J 6(2):167–195. doi:10.3233/SW-140134

Lemon O (2011) Learning what to say and how to say it: joint optimisation of spoken dialogue management and natural language generation. Comp Speech Lang 25(2):210–221. doi:10.1016/j.csl.2010.04.005

Lemon O, Janarthanam S, Rieser V (2010) Statistical approaches to adaptive natural language generation. In: Lemon O, Pietquin O (eds) Data-driven methods for adaptive spoken dialogue systems: computational learning for conversational interfaces. Springer, New York. doi:10.1007/978-1-4614-4803-7_6

Lewis JR (2011) Practical speech user interface design. CRC Press, Boca Raton. doi:10.1201/b10461

Mani I (2001) Automatic summarization. John Benjamins, Amsterdam. doi:10.1075/nlp.3

Mani I, Maybury M (eds) (1999) Advances in automatic text summarization. MIT Press, Cambridge

Mann WC, Thompson SA (1988) Rhetorical structure theory: toward a functional theory of text organization. Text 8(3):243–281. doi:10.1515/text.1.1988.8.3.243

Mairesse F, Young S (2014) Stochastic language generation in dialogue using factored language models. Comp Linguist 40(4):763–799. doi:10.1162/coli%5Fa%5F00199

Maybury M (ed) (2004) New directions in question answering. AAAI/MIT Press, Menlo Park and Cambridge

McKeown KR (1985) Text generation: using discourse strategies and focus constraints to generate natural language text. Cambridge University Press, Cambridge. doi:10.1017/cbo9780511620751

Portet F, Reiter E, Gatt A, Hunter J, Sripada S, Freer Y, Sykes C (2009) Automatic generation of textual summaries from neonatal intensive care data. Artif Intell 173:789–816. doi:10.1016/j.artint.2008.12.002

Reiter E (2007) An architecture for data-to-text systems. In: Proceedings of ENLG-2007 11th European workshop on natural language generation. Schloss Dagstuhl, Germany, pp 97–104. doi:10.3115/1610163.1610180

Reiter E (2010) Natural language generation. In: Clark A, Fox C, Lappin S (eds) Handbook of computational linguistics and natural language processing. Wiley, Chichester, pp 574–598. doi:10.1002/9781444324044.ch20

Reiter E, Dale R (2000) Building natural language generation systems. Cambridge University Press, Cambridge. doi:10.1017/cbo9780511519857

Rieser V, Lemon O, Liu X (2010) Optimising information presentation for spoken dialog systems. In: Proceedings of the 48th annual meeting of the association for computational linguistics, Uppsala, Sweden, 11–16 July 2010, pp 1009–1018. http://www.aclweb.org/anthology/P10-1103. Accessed 20 Jan 2016

Rieser V, Lemon O, Keizer S (2014) Natural language generation as incremental planning under uncertainty: adaptive information presentation for statistical dialogue systems. IEEE/ACM Trans Audio Speech Lang Proc 22(5):979–994. doi:10.1109/tasl.2014.2315271

Seneff S (2002) Response planning and generation in the Mercury flight reservation system. Comp Speech Lang 16:283–312. doi:10.1016/s0885-2308(02)00011-6

Stent A, Bangalore S (2014) Natural language generation in interactive systems. Cambridge University Press, Cambridge. doi:10.1017/cbo9780511844492

Strzalkowski T, Harabagiu S (eds) (2008) Advances in open domain question answering. Springer, New York. doi:10.1007/978-1-4020-4746-6

Theune M (2002) Contrast in concept-to-speech generation. Comp Speech Lang 16:491–531. doi:10.1016/s0885-2308(02)00010-4

Part III
Conversational Interfaces and Devices

Chapter 13
Conversational Interfaces: Devices, Wearables, Virtual Agents, and Robots

Abstract We are surrounded by a plethora of smart objects such as devices, wearables, virtual agents, and social robots that should help to make our life easier in many different ways by fulfilling various needs and requirements. A conversational interface is the best way to communicate with this wide range of smart objects. In this chapter, we cover the special requirements of conversational interaction with smart objects, describing the main development platforms, the possibilities offered by different types of device, and the relevant issues that need to be considered in interaction design.

13.1 Introduction

So far we have discussed conversational interfaces on smartphones. In this chapter, we turn to other smart objects that also require a conversational interface, such as various types of wearable device, virtual agents, and social robots.

Smartphones and wearable devices have built-in sensors and actuators that gather data about the user and the environment, including location, motion, orientation, and biosignals such as heart rate. The interpretation of the data from the sensors is sometimes performed in a small built-in processor, but it is usually performed outside the wearable in another device with higher computational power such as a smartphone, usually through Bluetooth or Wi-fi communication. As discussed in Sect. 13.2, this is one of the reasons why wearables are not as widespread as other technologies, as in many cases they are used just as another interface to the smartphone.

Currently, wearables can obtain data from users that until recently was not accessible on regular consumer gadgets at affordable prices. This opens a new world of possibilities for developers wishing to exploit this data and to create exciting applications. For example, the "quantified self" movement[1] aims to exploit

---

[1]http://quantifiedself.com. Accessed February 22, 2016.

© Springer International Publishing Switzerland 2016
M. McTear et al., *The Conversational Interface*,
DOI 10.1007/978-3-319-32967-3_13

this technology by allowing users to quantify their daily activities, mainly in terms of physical and physiological data (e.g., heart rate, sleeping hours, etc.), so that they can monitor their activity and gain a better understanding of themselves (Chan et al. 2012). Many applications are being developed to foster health, self-knowledge, motivation, and active and healthy living. Calvo and Peters (2014) have called applications such as these "positive computing".

Designing conversational interfaces is even more critical in the case of robots. As robots move from industrial applications to other domains in which a relationship with the user is crucial, e.g., companionship, health care, education, and entertainment, there has been an increasing focus on making robots more human-like in their appearance and, more importantly, in their communicative capabilities.

In the following sections, we describe the issues involved in designing and implementing conversational interfaces for these wearables, virtual agents, and social robots.

13.2   Wearables

Wearable computing devices (wearables) have expanded rapidly in recent years as a result of factors such as the availability of wireless access and acceptance by the public of wearable designs (Baker et al. 2015). Initially, wearables were seen as the next stage in a movement in personal computing from fixed desktop PCs to portable devices such as laptops, then to smaller devices such as smartphones and tablets, and finally to wearables. Wearables are small computing systems that the user can carry comfortably, just like an item of clothing or jewelry. However, it soon became apparent that, in addition to being portable, having the devices near to the user's body could also provide additional sources of valuable information.

13.2.1   Smartwatches and Wristbands

Smartwatches and wristbands are the most common wearable technologies. They can be used as an interface to a smartphone so that the user can receive notifications and messages without having to take the phone out of a bag or pocket. Users can specify that only urgent notifications should appear on their smartwatches so that they are only interrupted when something important happens (e.g., calls from certain contacts, or messages with a certain content). However, some users like to stay constantly connected and do not want to miss a single thing, so that the wearable provides a stimulus that is nearer to them, such as a vibration on the wrist as opposed to the vibration of the mobile phone inside a purse, or a stimulus that might otherwise be missed, for example, when exercising.

However, some wearables such as smartwatches can also run apps that are developed specifically for the device. Many apps for mobile phones also have smartwatch versions that have been developed using special APIs. Some smartwatches can be used with different wearable vendors, such as Android Wear,[2] and others are vendor specific, such as Pebble Developer,[3] or WatchKit for Apple Watch.[4] Chapter 16 presents a laboratory on how to develop multimodal applications with Android Wear.

Smartphones and wristbands can also make use of sensors to measure pulse, body temperature, galvanic skin response (GSR), heart rate, and skin temperature. In some cases, the devices have displays that provide feedback to the user, while in other cases the information gathered by the sensors is sent to a smartphone where different apps can display the interpreted data. For example, heart rate data is typically shown on the screen of the device, while sleep-tracking data acquired during the night is usually shown as a graphic on a smartphone.

Usually, smartphone apps that allow users to monitor their data are vendor specific; that is, the company that sells the wearable device provides the app. Apps may contain historical data, for example, by establishing and tracking goals, such as the number of steps to walk during the week or the number of hours of sleep, and by linking with a community of users and providing challenges, for example, to see who exercises more during a weekend. This is the case with apps provided by companies such as Adidas, Fitbit, Garmin, Jawbone, MisFit, Nike, and Polar. Many of these companies also provide developer APIs and SDKs, for example, Fitbit, Garmin, and Polar. There are also solutions for developers who want to integrate training data into their applications, for example, Google Fit[5] and Apple HealthKit.[6] With these APIs, health and fitness data is stored in secure locations and can be easily discovered and tracked.

13.2.2 Armbands and Gloves

Armbands and gloves are used mainly for gesture control. Their positioning allows muscle sensors to detect changes in movements of the arm and gestures of the hands (see Fig. 13.1).[7] This allows them to capture movements to control devices remotely, for example, by defining a gesture to play music, or making a robot reproduce hand movements.

---

[2]http://developer.android.com/wear/index.html. Accessed February 22, 2016.
[3]http://developer.getpebble.com/. Accessed February 22, 2016.
[4]https://developer.apple.com/watchkit/. Accessed February 22, 2016.
[5]https://developers.google.com/fit/. Accessed February 22, 2016.
[6]https://developer.apple.com/healthkit/. Accessed February 22, 2016.
[7]https://www.myo.com/. Accessed February 22, 2016.

Fig. 13.1 The Myo gesture
control armband, made by
Thalmic Labs (reproduced
with permission from Thalmic
Labs)

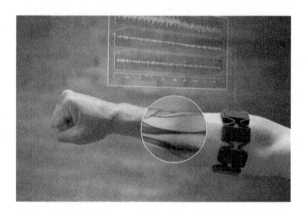

13.2.3   Smart Glasses

Glasses with mountable microphones or video cameras can function as augmented
reality glasses for navigation. Using a wireless connection, they can provide virtual
information to the user that is superposed on what they are looking at (Fig. 13.2).

Despite the huge enthusiasm that greeted the appearance of Google glasses, their
development is still in its infancy. In fact, Google stopped their glasses beta program in January 2015, although the project has not been officially canceled. There
are several glasses in the market, but most of them are beta versions, for example,
the Sony SmartEyeglass, for which there is a developer version.[8]

Glasses can also incorporate holographic technology, as in Microsoft's
HoloLens.[9] A Developer Edition was made available in 2016. Interestingly,
Microsoft has paid special attention to ways of interacting with the glasses, focusing
primarily on spoken communication, as this enhances the feeling of immersion
created by the combination of augmented and virtual reality.

Glasses should be light to wear, and the superposed information should not be
disruptive for the user. Currently, smart glasses are still quite large and heavy
compared with normal glasses, and they may result in some discomfort for users.
Sony recommend in their terms and conditions that the use of their glasses should
be limited to 2 hours a day to reduce discomfort, eye strain, fatigue, and dizziness.
Smart glasses can also help users who regularly use normal glasses by monitoring
their sight problems. For example, Shima glasses[10] offer developers and beta testers
the possibility to have their prescription embedded within the device.

There are some issues with smart glasses that still need to be resolved. One of
these is privacy, since users can record video and audio with the glasses and this
could infringe on the privacy of other people. Another issue is safety, as a user may

---

[8]http://developer.sonymobile.com/products/smarteyeglass/. Accessed February 22, 2016.

[9]https://www.microsoft.com/microsoft-hololens/en-us/development-edition. Accessed 17 April 2016.

[10]http://www.laforgeoptical.com/. Accessed February 22, 2016.

Fig. 13.2 Scenarios for smart glasses

be reading the information displayed in the glasses while driving or doing other critical and potentially harmful activities.

13.2.4   Smart Jewelry

Smart jewelry is a more fashionable alternative to smartwatches and fitness trackers. Different start-ups are creating smart jewelry. For example, Vinaya presents Bluetooth-connected smart pendants that connect to the iPhone and vibrate to provide notifications of important events. Indeed, their Web page,[11] in which they show their ring sketched like a *pret-a-porter* dress design, looks more like the Web page of a fashion magazine than a technology company. Ringly[12] displays rings with different colors and materials that notify text messages, e-mail, WhatsApp messages, phone calls, social networks, etc., and MEMI[13] presents bracelets with similar functionalities.

However, though these examples of smart jewelry look similar to a normal piece of jewelry, their capacity is limited to notifications and they do not have sensing

---

[11]http://www.vinaya.com/. Accessed February 22, 2016.

[12]https://ringly.com/. Accessed February 22, 2016.

[13]http://www.memijewellery.com/. Accessed February 22, 2016.

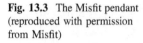

Fig. 13.3 The Misfit pendant (reproduced with permission from Misfit)

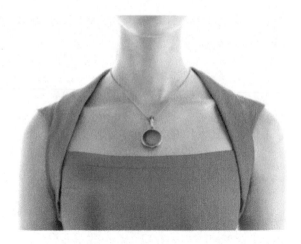

capabilities. Integrating sensors would require a larger piece of jewelry, as is the case with the Bellabeat LEAF.[14] LEAF can be used as a pendant or clip, making it less like the jewelry of Vinaya, although it still has a very aesthetic design. The functionalities of LEAF are similar to those of fitness trackers, for example, activity tracking and sleep monitoring. This is the same for the Misfit pendant[15] (Fig. 13.3). Other companies plan to offer sensing functionalities on devices that look like real jewelry, for example, EarO-Smart.[16] These devices are usually targeted at female customers and sometimes include applications designed for women, for example, to track sleep patterns during menstrual cycles.

13.2.5 Smart Clothing

Clothes with embedded sensors are a relatively new technology that has been emerging recently. This technology is being embraced mainly in the health and fitness domains, as many of the available products monitor vital signs and biomechanics. Athletes can wear smart clothing that allows coaches to monitor them and to spot who is under pressure, how to avoid and control injuries, as well as enabling them to compare players and to compare the same player in different positions.[17]

Some items of smart clothing such as shirts or body suits can collect data such as heart rate, breathing rate, or the intensity of a workout and can provide feedback on

---

[14]https://www.bellabeat.com/. Accessed February 22, 2016.

[15]https://store.misfit.com/. Accessed February 22, 2016.

[16]http://earosmart.com/. Accessed February 22, 2016.

[17]http://www.catapultsports.com/uk/. Accessed February 22, 2016.

which parts of the body are under too much pressure.[18,19,20] There are also socks that analyze and improve running form by tracking the position of the feet and the foot-landing technique, helping to prevent injuries while also tracking data such as the number of steps taken, the distance covered, calories, and cadence.[21] There are also belts that adjust automatically when you eat too much. Generally, these items of smart clothing are connected to specific apps that can be used on a mobile phone to monitor the information coming from the shirt sensors.

In addition to applications for athletes, smart clothing can help with health monitoring by keeping track of cardiac, respiratory, and activity data for people suffering from diverse conditions. Another application is monitoring the sleep of babies.

13.3   Multimodal Conversational Interfaces for Smart Devices and Wearables

Smart devices and wearables have introduced new interaction scenarios that have different implications for interface development. With smaller and simpler wearables such as fitness bands, communication between the system and the user can be restricted to small buttons for user–system interaction and light-emitting diodes (LEDs) and vibration for system–user interaction. With more sophisticated devices such as smartwatches, spoken interaction is augmented with visual responses based on cards.

The principles of conversation described in Chap. 3 apply also to wearables and smartphones with respect to prompt and menu design, relevant dialog act selection, the design of confirmations, turn taking, and grounding strategies. However, there are some additional aspects that must be considered. For example, developers must take into account that users have preconceived ideas about how to operate a particular device. Currently, the spoken interaction paradigm for wearables and smartphones is more command-like than conversational; thus, designers who intend to build a conversational interface for these devices must be aware that users may not address the device in a conversational way unless they are instructed on how to do so by the system.

Another relevant aspect is an Internet connection. Many systems still perform speech recognition online and thus require an active Internet connection while the user is speaking to the device. Thus, developers must consider whether voice is the best alternative depending on whether the device is likely to be connected to the Internet, and even when the device is always likely to be connected, they must

---

[18]http://www.hexoskin.com/. Accessed February 22, 2016.

[19]http://omsignal.com/pages/omsignal-bra. Accessed February 22, 2016.

[20]http://www.heddoko.com/. Accessed February 22, 2016.

[21]http://www.sensoriafitness.com/. Accessed February 22, 2016.

predict mechanisms to maintain communication with users if the connection is temporarily lost. In this situation, the solution is usually to balance the different modalities that are available on the device to obtain the best combination of oral, visual, and gestural communication. Unfortunately, guidelines for developers using Android[22] and iOS[23] focus mainly on how to program the interfaces rather than on design issues, though Microsoft provides some speech design guidelines for Windows Phones.[24]

With respect to visual interfaces, cards are becoming a useful design pattern since they can be placed beneath or beside one another and stacked on top of each other so that they can be swiped and easily navigated. The content of Web pages and apps is increasingly becoming an aggregation of many individual pieces of content from heterogeneous services on to cards, and interaction with our smartphones and devices is more and more a flow of notifications from a wide range of different apps.[25] Many companies now use cards, from social networks such as Twitter (Twitter Cards for multimedia) and Facebook (each input in the wall is shown as a card in the history of the user), blogs (e.g., Pinterest was one of the first to move the blog concept from posts to visual cards), all sorts of apps (e.g., Apple Passbook), and even operating systems (e.g., cards on Windows 8) and Web applications (e.g., Google Now uses a wide range of cards[26]).

Chris Tse, from cardstack.io, discusses patterns of card UI design and good design practice.[27] He places the types of card in a continuum from long-lived to short-lived cards. At the long-lived end of the spectrum, cards function as records, for example, Apple Passbook, while at the medium end they function as teasers and at the short-lived end they function as alerts.

The anatomy of a card is usually context, lens, and triggers (Fig. 13.4). For example, in the figure, we can see that cards present small pieces of information in a highly browsable way that some people might even find addictive.[28]

Hierarchy is not relevant with cards. Card collections display cards that are at the same level of importance, even if they have varied layouts (see Fig. 13.5) or are related to different issues. The focus is on the ability of the user to scan through them. Card collections usually scroll vertically, though there are many different

---

[22]http://developer.android.com/intl/es/training/wearables/apps/voice.html. Accessed February 22, 2016.

[23]https://developer.apple.com/library/ios/documentation/AVFoundation/Reference/AVSpeech Synthesizer_Ref. Accessed February 22, 2016.

[24]https://msdn.microsoft.com/en-us/library/windows/apps/jj720572%28v=vs.105%29.aspx. Accessed February 22, 2016.

[25]https://blog.intercom.io/why-cards-are-the-future-of-the-web/. Accessed February 22, 2016.

[26]https://www.google.com/landing/now/#cards. Accessed February 22, 2016.

[27]https://speakerdeck.com/christse/patterns-of-card-ui-design. Accessed February 22, 2016.

[28]More in: https://www.google.com/design/spec/components/cards.html#cards-actions. Accessed February 22, 2016.

Fig. 13.4 A sample card

Fig. 13.5 Sample card collection showing cards with different layouts (https://www.google.com/design/spec/components/cards.html#cards-content. Accessed February 24, 2016). Google and the Google logo are registered trademarks of Google Inc., used with permission

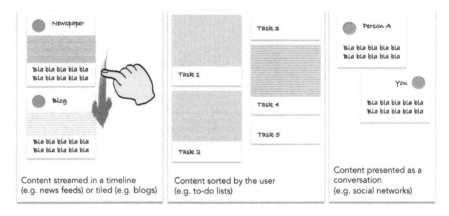

Fig. 13.6 Different types of card container

containers that can be used[29] that allow a seamless experience with many different screen sizes and devices (see Fig. 13.6). However, card-centric user interfaces may not be suitable for all contexts.[30]

Speech and card interfaces may be complemented by augmented reality applications. Molineux and Cheverst (2012) outline very interesting case studies of museum guidance supported by on-device object recognition, phone–cam interactions for large public displays, way finding for individuals with cognitive impairments, and hand-gesture manipulation of projected content.

Smartphones and wearables allow gestural interaction. This can be done in general-purpose devices such as smartphones thanks to sensors like the accelerometer and in specific wearables such as the armbands shown in Fig. 13.1. Dunne et al. (2014) present a study of the effect of gestural interaction on wearability. The authors distinguish two types of gestural interactions: passive and active. In passive interactions, the device listens for movements that trigger certain actions. In active interactions, the user consciously performs movements to provide instructions to the device.

For active input, designers must find a trade-off between clarity and visual distinction of the input. That is, if a gesture is remarkably different from everyday movements, it will be easily recognizable by the device, but also by other people (it has a "social weight"). On the other hand, if the gesture is more natural, it is less noticed as it is more likely that everyday movements are interpreted as an input gesture by the device.

Currently, there is no standard vocabulary for gestures, which makes it difficult to generate interfaces that are usable. In fact, we have learnt from visual languages

---

[29]http://thenextweb.com/dd/2015/06/16/how-cards-are-taking-over-web-design/.   Accessed February 22, 2016.

[30]http://ux.stackexchange.com/questions/60495/what-are-the-advantages-of-a-card-centric-user-interface. Accessed February 22, 2016.

(like sign language) that visual expressions are inherently ambiguous and that general-purpose visual languages often fail. Instead, experts recommend focusing design on specific domains and contexts (Ardito et al. 2014).

Other authors are working on silent speech interfaces (Bedri et al. 2015), where the user "talks silently," moving the mouth and/or tongue as if to pronounce a phrase that is not vocalized. These interfaces are usually meant for people who have lost their capacity to produce intelligible speech because of neurological or motor injuries but who are still able to articulate mouth and tongue. To build these interfaces, different sensors can be placed in the mouth and in earplugs to recognize tongue and jaw movements. However, it is difficult to distinguish silent speech from other actions such as eating. Similarly, Jeon and Lee (2013) have studied the use of non-speech sounds on mobile devices.

As can be observed, wearable devices and smartphones have opened many new possibilities for multimodal interfaces that must be addressed from a multidisciplinary perspective, bringing together interaction designers, usability researchers, and general human–computer interaction (HCI) practitioners to analyze the opportunities and directions to take in designing more natural interactions based on spoken language. This has been a topic for recent technical and scientific workshops, some of which have the aim of gaining more widespread acceptance of speech and natural language interaction (Munteanu et al. 2014).

13.4 Virtual Agents

Virtual characters that are able to display multimodal behaviors are being used nowadays for a variety of purposes, from unidirectional communication in which they take the role of a presenter and the user simply listens as if they are watching a TV show, to conversational partners from a wide spectrum of more directed information-providing tasks, to open tasks such as artificial companions.

These characters have been endowed with different visual appearances. Some early characters were cartoon like. For example, Smartakus, an animated character with the shape of an "i," was used in the SmartKom Project to present information (Wahlster et al. 2001). Then, more anthropomorphic agents appeared. For example, the August talking head had lip-synchronized speech synthesis, nonverbal behavior, and approach and gaze behavior to show awareness of the user's actions (Gustafson et al. 1999), while the REA agent used eye gaze, body posture, hand gestures, and facial displays to contribute to the conversation and organize her own interventions (Cassell et al. 2000).

The focus in current systems is on developing agents with a full body. Humans depend to a great extent on embodied behaviors to make sense and engage in face-to-face conversations. The same happens with machines: embodied agents help to leverage naturalness and users judge the system's understanding to be worse when it does not have a body (Cassell 2001). According to Cassell et al. (2000), the body is the best way to alternate multiple representations in order to convey multimodal information and to regulate conversation. Embodied conversational agents (ECAs) exhibit multimodal communicative capabilities comprising voice, gestures, facial expressions, gaze, and body posture and may play different roles of varying complexity, for example, as companions for the elderly, as toys, virtual trainers, intelligent tutors, or as Web/sales agents.

However, embodiment plays a central role for the system's output, enabling the agent to produce gestures and behaviors that enhance the image its projects, and also for its perceptual functions. Advances in the understanding of human cognition have demonstrated that our minds are not reasoning devices that can be isolated from our bodies. Rather, they are tied to the physical world to the extent that we understand concepts as relations between our bodies and the world. Early agents had limited perceptual abilities and the knowledge they had about the environment and the user was limited. According to André and Pelachaud (2010), for an ECA to be believable, it must be equipped with a sensory mechanism that makes it possible to render sophisticated attending behaviors.

13.5   Multimodal Conversations with Virtual Agents

ECAs should be endowed with refined communicative, emotional, and social capabilities. This means that apart from task-oriented functions, they should also integrate interpersonal goals. Many studies have demonstrated that there is a significant improvement in engagement and likeability when interacting with agents that display believable nonverbal behaviors. For example, Bickmore and Cassell (2005) show the importance of small talk to build rapport and trust with the REA agent, an ECA that acted as a real estate agent. Interactional functions helped create and maintain an open channel of communication between the user and the agent.

André and Pelachaud (2010) provide a concise but comprehensive overview of the design of ECAs. According to their description, many ECAs rely on Information State dialog managers like TRINDI (Traum and Larsson 2003) (see Chaps. 4 and 10). Also, their multimodal behavior is learnt from human–human conversations from which models are extracted and refined. Data-driven approaches are still not fully adopted, and so a vast amount of data must be gathered and annotated to observe the wide range of gestures and expressions that occur in face-to-face communication. In addition, not only the gestures themselves must be simulated, but also special attention must be paid to their temporal dynamics, co-occurrence, and the different meanings that may be conveyed when merging several gestures.

The design and development of the multimodal behaviors of ECAs has focused on issues such as the reusability of the components and the separation of behavior planning from behavior generation and realization. Different standards are being defined to establish common architectures and languages, such as the Situation, Agent, Intention, Behavior, Animation (SAIBA) framework, the Behavior Markup Language (BML), and the Functional Markup Language (FML) (described in

Chap. 15). These elements are able to encode affect, coping strategies, emphasis, turn management strategies, as well as head, face, gaze, body movements, gestures, lip movements, and speech. Other languages, e.g., Multimodal Utterance Representation Markup Language (MURML), focus on coupling verbal utterances with gestures that are associated with linguistic elements (Kopp and Wachsmuth 2004).

A lot of effort has also been put on building emotional models for ECAs, as will be described in Chaps. 14 and 15. For example, ECAs may be built to be artificial companions, and in that case, the objective of the system may be more related to emotion (e.g., making someone happy or confident) than accomplishing a certain task. As stated by Cowie, "companionship is an emotional business," and this encompasses several social, psychological, and ethical issues that are described in detail in Wilks (2010).

13.6 Examples of Tools for Creating Virtual Agents

Greta.[31] Greta is a real-time three-dimensional ECA developed by the Greta Team at Telecom ParisTech. Greta is based on a 3D model of a woman compliant with the Moving Picture Experts Group (MPEG-4) animation standard and is able to communicate using a rich palette of verbal and nonverbal behaviors in standard languages. Greta can talk and simultaneously show facial expressions, gestures, gaze, and head movements.

The Virtual Human Toolkit.[32] The Institute for Creative Technologies (ICT) Virtual Human Toolkit is a collection of modules, tools, and libraries designed to aid and support researchers and developers with the creation of ECAs. It provides modules for multimodal sensing, character editing and animation, and nonverbal behavior generation.

SmartBody.[33] SmartBody is a character animation platform developed originally at the University of Southern California that is included in the Virtual Human Toolkit but can also be used separately. It provides locomotion, steering, object manipulation, lip-syncing, gazing, and other nonverbal behaviors. The software is provided free and open source under the GNU Lesser General Public License (LGPL) and is multiplatform (it works on Windows, Linux, OSx, Android, and iOS).

MAX and the Articulated Communicator Engine (ACE).[34] ACE is a toolkit for building ECAs with a kinematic body model and multimodal utterance generation based on MURML. MAX is an ECA developed for cooperative construction tasks that has been under development at the University of Bielefeld for more than a decade.

---

[31] http://perso.telecom-paristech.fr/ ∼ pelachau/Greta/. Accessed February 24, 2016.
[32] https://vhtoolkit.ict.usc.edu/. Accessed February 24, 2016.
[33] http://smartbody.ict.usc.edu. Accessed February 24, 2016.
[34] http://www.techfak.uni-bielefeld.de/ ∼ skopp/max.html. Accessed February 24, 2016.

13.7 Social Robots

Robots are moving out of factories and increasingly entering our homes. This has provoked a paradigm shift: in this new scenario, users are not trained to operate the robots; instead, the users are naïve and untrained and so the robots must be able to communicate with them in an intuitive and natural fashion (Mathur and Reichling 2016). This can be achieved by endowing robots with the ability to hold conversations with their users. The complexity of these interactions may vary depending on the type and function of the robot.

On the one hand, robots may be understood as tools that can be used to access functionality and request information using a command-like interface. On the other hand, robots may be conceptualized as "hedonic" systems with which humans can engage in more complex relationships using a conversational interface. Robots such as these are known as *social robots*. With social robots, humans apply the social interaction models that they would employ with other humans, since they perceive the robots as social agents with which humans can engage in stronger and more lasting relationships. Social robots can also provide entertainment, sociability, credibility, trust, and engagement (de Graaf et al. 2015).

In the literature, there are many examples demonstrating that human beings attribute social identity to robots, even when the robots are seen as tools. Sung et al. (2007) show how some users attribute personalities to their cleaning robots. Hoenen et al. (2016) discuss how robots (in particular, the non-anthropomorphic ones) can be considered as social entities and how the social identity of a robot can be established through social interaction. Peca et al. (2015) show that interactivity between humans and objects can be a key factor in whether infants perceive a robot as a social agent. In this study, infants aged 9–17 months witnessed an interaction between an adult and a robot and they made inferences regarding its social agency based on the responsiveness of the robot.

Children often address robots as social agents. For example, Kahn et al. (2012) show how children believed that the robot used in experiments had feelings and was a social being that they considered as a friend with whom they could entrust secrets. Given findings such as these, one of the most promising application domains for social robots is to build robots for children, for entertainment and pedagogic purposes, and also to provide help for children suffering from conditions such as autism. However, currently, social interactions with most commercial robots are usually very predictable, so the robot loses its magic with continued use and children eventually lose interest.

Robots are also considered as an aid for aging populations by improving their quality of life and helping them to stay fit and healthy, supporting autonomy, and mitigating loneliness. To obtain these benefits, adults must accept robots as part of their home environment, find them easy to operate, and perceive them as social counterparts. Different studies have shown that people's acceptance of robots depends on a variety of social factors including safety, fun, social presence, and perceived sociability (Heerink et al. 2010). Also it is important that the robots adhere to human social rules including friendliness, speech styles, and ways of addressing the user. Other studies highlight barriers to the acceptance of robots by older adults, including older adults' uneasiness with technology, a feeling of stigmatization, and ethical/societal issues associated with robot use (Wu et al. 2014).

Some authors have warned about particular undesired effects of social robots. For example, Turkle (2012) discusses the negative impact that robots may have on our ability to build human relationships and deal with complexities and problems when we have robots as companions that can cater for every need:

> Our population is aging; there will be robots to take care of us. Our children are neglected; robots will tend to them. We are too exhausted to deal with each other in adversity; robots will have the energy. Robots won't be judgmental.

However, other authors even find it plausible that robots may be used in the future to influence their users to become more ethical (Borenstein and Arkin 2016).

13.8 Conversational Interfaces for Robots

Interacting with social robots puts several unique requirements on the conversational interface (Cuayáhuitl et al. 2015). As far as spoken language interaction is concerned, a robot has to be able to predict the direction of the arrival of speech within a wide area and be able to distinguish voice from noise, whereas with other devices, speech is directed toward a microphone that is usually held close to the user's mouth. This is known as *speech localization*. Other aspects of speech localization include exhibiting social interaction cues such as approaching the speaker or looking at them and also managing turn taking in single and multiparty conversations. As far as language understanding is concerned, robots need to be able to understand and use language beyond a restricted set of commands. Given that they operate in a situated environment, robots have the advantage that they can learn language by extracting representations of the meanings of natural language expressions that are tied to perception and actuation in the physical world (Matuszek et al. 2012). Flexible and optimized dialog management is also crucial. Robots should be able to engage in mixed initiative dialog and perform affective interaction (Mavridis 2015). They should also be able to recognize and produce multimodal behaviors that accompany speech (Lourens et al. 2010). See Chap. 15 for a more detailed discussion of expressive behaviors.

All these challenges must be addressed in order to develop social robots. According to Looije et al. (2010), social behaviors such as turn taking and emotional expressions are essential for a robot to be perceived as trustworthy. Social robots must also be compliant with social norms. Awaad et al. (2015) maintain that this involves a mixture of knowledge about procedures (knowing how to accomplish tasks) and functional affordances of objects (knowing what objects are used for). For example, they argue that robots should know that if no glasses are available when serving water, then a mug is a valid substitution, and that such a substitution is socially acceptable. Also, there may be aspects of human-human interactions that users may not wish to see in robots, such as social control or criticism (Baron 2015). Breazeal (2003, 2004) argues for taking the robot's perspective when tackling the relevant design issues, including naturalness, user expectation, user–robot relationship, and teamwork.

There are various requirements that need to be considered in the design of a social robot if it is to act as a companion. Luh et al. (2015) developed a scale of "companionship" for virtual pets based on the companionship features of real pets. The most important factors were enjoyment, psychological satisfaction, autonomy, responsibility, and interactive feedback. Benyon and Mival (2007) describe personification technologies in term of utility, form, emotion, personality, trust, and social attitudes, all of which should be considered during design. Pearson and Borenstein (2014) emphasize ethical aspects of creating companions for children, while Leite et al. (2013) present a detailed survey of studies of long-term human–robot interactions and provide directions for future research, including the need for continuity and incremental novel behaviors, affective interactions, empathy, and adaptation.

Looking at negative attitudes toward robots, the Negative Attitudes toward Robots Scale (NARS) and Robot Anxiety Scale (RAS) study negative attitudes and anxiety toward robots that may lead to users adopting a strategy of avoiding communication with robots (Nomura et al. 2006; Kanda and Ishiguro 2012). Other authors have related these factors to their perceived ease of use, which is directly related to the interface and how it influences social presence and perceived enjoyment (Heerink et al. 2010).

In summary, the integration of social robots into everyday life depends to a great extent on their ability to communicate with users in a satisfying way, for which multimodal conversation is of paramount importance (Fortunati et al. 2015). The effects of expressive multimodal behaviors and the display of emotions and personality are discussed in Chap. 15.

13.9 Examples of Social Robots and Tools for Creating Robots

13.9.1 Aldebaran Robots

The Aldebaran robots are the most widespread robots within the scientific community. Their family of robots includes NAO,[35] Pepper, and Romeo (Fig. 13.7). NAO is a small robot that has been used extensively for research and educational purposes. Pepper and Romeo are more recent. The former was created for SoftBank

---

[35]https://www.aldebaran.com/en/humanoid-robot/nao-robot. Accessed February 22, 2016.

Fig. 13.7 Aldebaran's NAO, Romeo, and Pepper robots (reproduced with permission from Aldebaran)

Mobile (an important mobile phone operator in Japan)[36] and has been endowed with emotion recognition capabilities, and the latter is a robot intended for research purposes.[37] All of the robots include sensors and actuators and incorporate a microphone and speakers to allow conversational interaction.[38]

13.9.2   Jibo

Jibo[39] is a social robot that was not designed with humanoid characteristics but more like a Disney or Pixar character with a single eye and a moving head and body that are used to give him a personality and promote social engagement (Fig. 13.8). Jibo can track faces and capture photographs, process speech, and respond using natural social and emotive cues. Developers can add skills and content to Jibo by using the Jibo SDK that provides animation tools for movements and displays, timeline tools for sequencing, behavior tools for engagement, and a visual simulator. You can see a video of Jibo here.[40]

---

[36]https://www.aldebaran.com/en/a-robots/who-is-pepper. Accessed February 22, 2016.

[37]http://projetromeo.com/en/. Accessed February 22, 2016.

[38]http://www.theverge.com/2016/1/6/10726082/softbank-pepper-ibm-watson-collaboration. Accessed February 22, 2016.

[39]https://www.jibo.com/. Accessed February 22, 2016.

[40]https://www.youtube.com/watch?v=3N1Q8oFpX1Y. Accessed February 22, 2016.

Fig. 13.8 Jibo, the social
robot (reproduced with
permission from Jibo)

13.9.3 FurHat

The FurHat platform was developed by human–computer interaction experts at
Furhat Robotics with a strong background in dialog systems.[41] FurHat (Fig. 13.9) is
a robotic head based on a projection system that renders facial expressions, with
motors to move the neck and head. Developers can use an open-source SDK, and
there are libraries for speech recognition and synthesis as well as face recognition
and tracking (Al Moubayed et al. 2012). You can see a video of FurHat here.[42]

13.9.4 Aisoy

Aisoy[43] is a programmable robot to encourage creative thinking in children and
improve their ability to solve challenging problems. Aisoy can be programmed by
children (with Scratch or Blocky), but it also has an SDK for programming in
higher-order languages. It is based on the Raspberry Pi and can be used for conversational applications as it incorporates a microphone and speakers (Fig. 13.10).

---

[41]http://www.furhatrobotics.com/. Accessed February 22, 2016.

[42]https://www.youtube.com/watch?v=v84e6HMFbyc. Accessed February 22, 2016.

[43]http://www.aisoy.com/.

Fig. 13.9 Furhat (reproduced with permission from Furhat Robotics)

Fig. 13.10 The Aisoy robot (reproduced with permission from Aisoy Robotics)

13.9.5 Amazon Echo

Amazon Echo[44] is similar to social robots such as Jibo and Pepper except that it does not provide an anthropomorphic physical embodiment. Instead, it has the form of a cylinder about 9 inches tall containing a microphone array and speakers. Echo is connected to Alexa, a cloud-based voice service that provides a range of capabilities known as *skills*, including information, creating shopping lists, providing news and traffic information, streaming music, and also some home control functions such as controlling lights. The use of far-field speech recognition and beam-forming technology means that Echo can hear from any direction and cope with ambient noise such as music playing in the background.

13.9.6 Hello Robo

The idea behind Hello Robo is that personal robotics should be more accessible and affordable to everyone.[45] Open-source robots have been developed that can be replicated using a desktop 3D printer. Examples are maki and poly.[46]

13.9.7 The Open Robot Hardware Initiative

Open robot hardware was created to provide resources and open-source hardware for developers of robotics applications. The Web site has information about different projects and provides tutorials on topics including robotic arms and hands, humanoid robots, vehicles and drones, legged robots, swarm robots, actuators and sensors, and modules for specific application domains such as social, health, and educational robotics.[47]

13.9.8 iCub.org: Open-Source Cognitive Humanoid Robotic Platform

The EU project RobotCub generated the iCub humanoid robot (Fig. 13.11) that is currently used worldwide and can be obtained from the Italian Institute of Technology for a fee. It has 53 motors that move the head, arms and hands, waist,

[44]http://www.amazon.com/echo. Accessed February 22, 2016.

[45]http://www.hello-robo.com/. Accessed March 1, 2016.

[46]http://inmoov.fr. Accessed March 1, 2016.

[47]http://www.openrobothardware.org/linkedprojects. Accessed February 22, 2016.

Fig. 13.11 The iCub robot
(reproduced with permission)

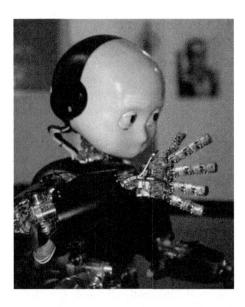

and legs. It can see and hear, and it has a sense of proprioception (body configuration) and movement (using accelerometers and gyroscopes). It is open source, and its code and even the production drawing describing its mechanical and electronic parts are available on the web page.[48]

13.9.9 SPEAKY for Robots

SPEAKY for Robots[49] (Bastianelli et al. 2015) aims to foster the definition and deployment of voice user interfaces (VUIs) in robotic applications where human–robot interaction is required. The goal is to develop a Robotic Voice Development Kit (RVDK).

13.9.10 The Robot Operating System (ROS)

ROS is an open-source project that aims to develop a platform for writing robot software and sharing code solutions and algorithms.[50] It is particularly interesting for students as it breaks the expert-only barrier.

---

[48]http://www.icub.org/. Accessed February 22, 2016.

[49]http://www.dis.uniroma1.it/ ~ labrococo/?q=node/373. Accessed February 22, 2016.

[50]http://www.ros.org/. Accessed February 22, 2016.

13.10 Summary

A variety of smart devices, wearables, virtual agents, and social robots are being developed that provide new ways to interact with Web services and with our environment. However, the potential for these devices still has to be realized as often their interface does not go beyond the command-and-control metaphor. In this chapter, we have addressed the possibilities and challenges for designers and developers of multimodal conversational interfaces to smart devices, wearables, virtual agents, and robots.

Further Reading

Trappl (2013) is a book of readings about agents and robots as butlers and companions. The chapters cover psychological and social considerations, experiences with and prerequisites for virtual or robotic companions, acceptability, trustworthiness, social impact, and usage scenarios involving spoken communication. Nishida et al. (2014) cover various aspects of conversational artifacts with a special emphasis on conversational agents. Markowitz (2014) is a comprehensive examination of conversational robots from technical, functional, and social perspectives, including aspects such as how to endow robots with conversational capabilities and how they can autonomously learn language. Also covered are the social aspects of spoken interaction with robots and how they will shape the future. There is also a special issue of the Information Society Journal about social robots and how robots are moving from the industrial to the domestic sphere.[51] Roberto Pieraccini, Director of Advanced Conversational Technologies at Jibo, Inc., reviews the challenges that social robots bring to voice interaction and how the technologies for interacting with social robots differ from those for telephone applications and personal assistants.[52]

The Mobile Voice Conference is a forum for industrial perspectives and new advances in speech interfaces for mobile devices.[53] The Conversational Interaction Technology Web site is an excellent source of information about recent innovations in speech technology, including wearables, devices, and robots.[54] Trends in wearables can be found here.[55,56]

Hexoskin has created a community of researchers that use their smart clothes for remote monitoring and provide software for data analysis and a list of scientific projects and papers.[57] Alpha2 is a programmable robot with a built-in speech

---

[51]http://www.tandfonline.com/toc/utis20/31/3. Accessed February 22, 2016.

[52]http://robohub.org/the-next-era-of-conversational-technology/. Accessed February 22, 2016.

[53]http://mobilevoiceconference.com/. Accessed February 22, 2016.

[54]http://citia.lt-innovate.eu/. Accessed February 22, 2016.

[55]https://www.wearable-technologies.com. Accessed February 22, 2016.

[56]http://urbanwearables.technology/. Accessed February 22, 2016.

[57]http://www.hexoskin.com/pages/health-research. Accessed February 22, 2016.

system that incorporates voice search as well as giving verbal reminders and receiving speech commands.[58]

There are numerous conferences on the challenges of social robots, for example, the International Conference on Social Robotics (ICSR).[59] There is also an International Journal on Social Robotics that covers the latest developments in all aspects of social robotics.[60] Royakkers and vanEst (2015) present a literature review of some relevant questions raised by the new robotics, including ethical issues.

RoboHelper is a human–human dialog corpus between an elderly person and a human helper that is being used as a baseline for training robotic companions (Chen et al. 2015). The corpus contains the transcribed dialogs that have been annotated using the Anvil tool.[61]

Exercises

1. Visit the Web pages of companies specializing in smart jewelry or smart clothes and examine what sorts of conversational interface are provided in the products.
2. Consider some new forms of conversational interface. There is a series of demos from the Interaction and Communication Design Lab at Toyohashi University of Technology in Japan of interactions with objects in the environment such as a Sociable Trash Box and a Sociable Dining Table.[62] Look at these demos and consider the usefulness of conversational interfaces for these sorts of objects.

References

Al Moubayed S., Beskow J, Skantze G, Granström B (2012) Furhat: A Back-projected human-like robot head for multiparty human-machine interaction. In: Esposito A, Esposito AM, Vinciarelli A, Hoffmann R, Müller VC (eds) Cognitive Behavioural Systems. Lecture Notes in Computer Science Vol. 7403, Springer Verlag, Berlin:114–130. doi:10.1007/978-3-642-34584-5_9

André E, Pelachaud C (2010) Interacting with embodied conversational agents. In: Chen F, Jokinen K (eds) Speech technology: theory and applications. Springer, New York, pp 122–149. doi:10.1007/978-0-387-73819-2_8

Ardito C, Costabile MF, Jetter H-C (2014) Gestures that people can understand and use. J Vis Lang Comput 25:572–576. doi:10.1016/j.jvlc.2014.07.002

Awaad I, Kraetzschmar GK, Hertzberg J (2015) The role of functional affordances in socializing robots. Int J Soc Robot 7:421–438. doi:10.1007/s12369-015-0281-3

---

[58]http://www.ubtrobot.com/en/html/archive/2015092816.html. Accessed February 22, 2016.

[59]http://www.icsoro.org/. Accessed February 22, 2016.

[60]http://link.springer.com/journal/12369. Accessed February 22, 2016.

[61]http://www.anvil-software.org/.

[62]http://www.icd.cs.tut.ac.jp/en/project.html. Accessed 12 April 2016.

Baker PMA, Gandy M, Zeagler C (2015) Innovation and wearable computing: a proposed collaborative policy design framework. IEEE Internet Comput 19:18–25. doi:10.1109/MIC. 2015.74

Baron NS (2015) Shall we talk? Conversing with humans and robots. Inf Soc 31:257–264. doi:10. 1080/01972243.2015.1020211

Bastianelli E, Nardi D, Aiello LC et al (2015) Speaky for robots: the development of vocal interfaces for robotic applications. Appl Intell 1–24. doi:10.1007/s10489-015-0695-5

Bedri A, Sahni H, Thukral P et al (2015) Toward silent-speech control of consumer wearables. Computer 48:54–62. doi:10.1109/MC.2015.310

Benyon D, Mival O (2007) Introducing the Companions project: Intelligent, persistent, personalised interfaces to the Internet. In: Proceedings of the 21st British HCI group annual conference on people and computers: HCI...but not as we know it (BCS-HCI'07), pp 193–194. http://dl.acm.org/citation.cfm?id=1531462&dl=ACM&coll=DL&CFID= 566912806&CFTOKEN=59937217

Bickmore T, Cassell J (2005) Social dialogue with embodied conversational agents. In: Kuppevelt J, Dy L, Bernsen NO (eds) Advances in natural multimodal dialogue systems. Springer, Netherlands, pp 23–54. doi:10.1007/1-4020-3933-6_2

Borenstein J, Arkin R (2016) Robotic nudges: the ethics of engineering a more socially just human being. Sci Eng Ethics 22:31–46. doi:10.1007/s11948-015-9636-2

Breazeal C (2003) Emotion and sociable humanoid robots. Int J Hum-Comput Stud 59:119–155. doi:10.1016/S1071-5819(03)00018-1

Breazeal C (2004) Social interactions in HRI: the robot view. IEEE Trans Syst Man Cybern Part C Appl Rev 34:181–186. doi:10.1109/TSMCC.2004.826268

Calvo RA, Peters D (2014) Positive computing: technology for wellbeing and human potential. The MIT Press, Cambridge, MA

Cassell J (2001) Embodied conversational agents. representation and intelligence in user interfaces. In: Proceedings of the American Association for the advancement of artificial intelligence (AAAI'01), pp 67–83. http://dx.doi.org/10.1609/aimag.v22i4.1593

Cassell J, Sullivan J, Prevost S, Churchill EF (eds) (2000) Embodied conversational agents. MIT Press, Cambridge

Chan M, Estève D, Fourniols J-Y, Escriba C, Campo E (2012) Smart wearable systems: current status and future challenges. Artif Intell Med 56:137–156. doi:10.1016/j.artmed.2012.09.003

Chen L, Javaid M, Di Eugenio B, Zefran M (2015) The roles and recognition of haptic-ostensive actions in collaborative multimodal human-human dialogues. Comp Speech Lang 34(1):201–231. doi:10.1016/j.csl.2015.03.010

Cuayáhuitl H, Komatani K, Skantze G (2015) Introduction for speech and language for interactive robots. Comput Speech Lang 34:83–86. doi:10.1016/j.csl.2015.05.006

De Graaf MMA, Allouch SB, Klamer T (2015) Sharing a life with Harvey: exploring the acceptance of and relationship-building with a social robot. Comput Hum Behav 43:1–14. doi:10.1016/j.chb.2014.10.030

Dunne LE, Profita H, Zeagler C et al (2014) The social comfort of wearable technology and gestural interaction. In: Engineering in Medicine and Biology Society (EMBC), 2014 36th annual international conference of the IEEE, Chicago, IL, pp:4159–4162, 26–30 Aug 2014. doi:10.1109/EMBC.2014.6944540

Fortunati L, Esposito A, Lugano G (2015) Introduction to the special issue "Beyond industrial robotics: social robots entering public and domestic spheres". Inf Soc 31:229–236. doi:10. 1080/01972243.2015.1020195

Gustafson J, Lindberg N, Lundeberg M (1999) The August spoken dialogue system. Proceedings of the 6th European conference on speech and communication technology (EUROSPEECH'99), Budapest, Hungary, pp 1151–1154, 5–9 Sept 1999. http://www.isca-speech.org/archive/eurospeech_1999/e99_1151.html

Heerink M, Kröse B, Evers V, Wielinga B (2010) Assessing acceptance of assistive social agent technology by older adults: the almere model. Int J Soc Robot 2:361–375. doi:10.1007/s12369-010-0068-5

Hoenen M, Lübke KT, Pause BM (2016) Non-anthropomorphic robots as social entities on a neurophysiological level. Comput Hum Behav 57:182–186. doi:10.1016/j.chb.2015.12.034

Jeon M, Lee J-H (2013) The ecological AUI (Auditory User Interface) design and evaluation of user acceptance for various tasks on smartphones. In: Kurosa M (ed) Human-computer interaction modalities and techniques: 15th international conference, HCI International 2013, pp 49-58, Las Vegas, USA, 21–26 July. doi:10.1007/978-3-642-39330-3_6

Kahn PH, Kanda T, Ishiguro H, Freier NG, Severson RL, Gill BT, Ruckert JH, Shen S (2012) "Robovie, you'll have to go into the closet now": children's social and moral relationships with a humanoid robot. Dev Psychol 48(2):303–314. doi:10.1037/a0027033

Kanda T, Ishiguro H (2012) Human-robot interaction in social robotics. Edición: New. CRC Press, Boca Raton. doi: http://www.crcnetbase.com/doi/book/10.1201/b13004

Kopp S, Wachsmuth I (2004) Synthesizing multimodal utterances for conversational agents. Comput Anim Virtual Worlds 15(1):39–52. doi:10.1002/cav.6

Leite I, Martinho C, Paiva A (2013) Social robots for long-term interaction: a Survey. Int J Soc Robot 5:291–308. doi:10.1007/s12369-013-0178-y

Looije R, Neerincx MA, Cnossen F (2010) Persuasive robotic assistant for health self-management of older adults: design and evaluation of social behaviors. Int J Hum Comput Stud 68:386–397. doi:10.1016/j.ijhcs.2009.08.007

Lourens T, van Berkel R, Barakova E (2010) Communicating emotions and mental states to robots in a real time parallel framework using Laban movement analysis. Robot Auton Syst 58:1256–1265. doi:10.1016/j.robot.2010.08.006

Luh D-B, Li EC, Kao Y-J (2015) The development of a companionship scale for artificial pets. Interact Comput 27:189–201. doi:10.1093/iwc/iwt055

Markowitz JA (ed) (2014) Robots that talk and listen: technology and social impact. Walter de Gruyter GmbH & Co. KG, Berlin; Boston. doi:http://dx.doi.org/10.1515/9781614514404

Mathur MB, Reichling DB (2016) Navigating a social world with robot partners: a quantitative cartography of the Uncanny Valley. Cognition 146:22–32. doi:10.1016/j.cognition.2015.09.008

Matuszek C, FitzGerald N, Zettlemoyer L, Bo L, Fox D (2012) A joint model of language and perception for grounded attribute learning. In: Proceedings of the 29th international conference on machine learning (ICML'12), Edinburgh, Scotland, pp 1671–1678. https://homes.cs.washington.edu/~lsz/papers/mfzbf-icml12.pdf

Mavridis N (2015) A review of verbal and non-verbal human–robot interactive communication. 63. Robot Auton Syst 63:22–35. doi:10.1016/j.robot.2014.09.031

Molineux A, Cheverst K (2012) A survey of mobile vision recognition applications. In: Tiwary US, Siddiqui TJ (eds) Speech, image and language processing for human computer interaction. IGI Global, New York. doi:10.4018/978-1-4666-0954-9.ch014

Munteanu C, Jones M, Whittaker S, Oviatt S, Aylett M, Penn G, Brewster S, d'Alessandro N (2014) Designing speech and language interactions. In: CHI '14 extended abstracts on human factors in computing systems (CHI EA '14). ACM, New York, USA, pp 75–78. doi:10.1145/2559206.2559228

Nishida T, Nakazawa A, Ohmoto Y (eds) (2014) Conversational informatics: a data-intensive approach with emphasis on nonverbal communication. Springer, New York. doi:10.1007/978-4-431-55040-2

Nomura T, Suzuki T, Kanda T, Kato K (2006) Measurement of negative attitudes toward robots. Interact Stud 7:437–454. doi:10.1075/is.7.3.14nom

Pearson Y, Borenstein J (2014) Creating "companions" for children: the ethics of designing esthetic features for robots. AI Soc 29:23–31. doi:10.1007/s00146-012-0431-1

Peca A, Simut R, Cao H-L, Vanderborght B (2015) Do infants perceive the social robot Keepon as a communicative partner? Infant Behav Dev. doi:10.1016/j.infbeh.2015.10.005

Royakkers L, van Est R (2015) A literature review on new robotics: automation from love to war. Int J Soc Robot 7:549–570. doi:10.1007/s12369-015-0295-x

Sung J-Y, Guo L, Grinter RE, Christensen HI (2007) "My Roomba is Rambo": intimate home appliances. In: Krumm J, Abowd GD, Seneviratne A, Strang T (eds) UbiComp 2007: ubiquitous computing. Springer, Berlin, pp 14–162. doi:10.1007/978-3-540-74853-3_9

Trappl R (ed) (2013) Your virtual butler: the making of. Springer, Berlin. doi:10.1007/978-3-642-37346-6

Traum DR, Larsson S (2003) The information state approach to dialog management. In: Smith R, Kuppevelt J (eds) Current and new directions in discourse and dialog. Kluwer Academic Publishers, Dordrecht, pp 325–353. doi:10.1007/978-94-010-0019-2_15

Turkle S (2012) Alone together: why we expect more from technology and less from each other. Basic Books, New York

Wahlster W, Reithinger N, Blocher A (2001) Smartkom: Multimodal communication with a life-like character. In: Proceedings of the 7th European conference on speech communication and technology (Eurospeech 2001), Aalborg, Denmark, pp 1547–1550, 3–7 Sept 2001. http://www.isca-speech.org/archive/eurospeech_2001/e01_1547.html

Wilks Y (ed) (2010) Close engagements with artificial companions. Key social, psychological, ethical and design issues. John Benjamins Publishing Company, Amsterdam. doi:10.1075/nlp.8

Wu Y-H, Wrobel J, Cornuet M, Kerhervé H, Damnée S, Rigaud A-S (2014) Acceptance of an assistive robot in older adults: a mixed-method study of human-robot interaction over a 1-month period in the Living Lab setting. Clin Interv Aging 9:801–811. doi:10.2147/CIA. S56435

Chapter 14
Emotion, Affect, and Personality

Abstract Affect is a key factor in human conversation. It allows us to fully understand each other, be socially competent, and show that we care. As such, in order to build conversational interfaces that display credible and expressive behaviors, we should endow them with the capability to recognize, adapt to, and render emotion. In this chapter, we explain the background to how emotional aspects and personality are conceptualized in artificial systems and outline the benefits of endowing the conversational interface with the ability to recognize and display emotions and personality.

14.1 Introduction

Emotion plays a key role in human interaction. For this reason, communication between a conversational system and humans should be more effective if the system can process and understand the human's emotions as well as displaying its own emotions. Picard (1997) coined the term *affective computing* at a time when emotion was not considered a relevant aspect of the design of artificial systems. Now there is very active research community working on affective computing, with several international conferences and journals. This work has demonstrated the many benefits that emotion can bring to artificial systems, including increased usability, efficiency, trust, rapport, and improved communication. Affective computing is also used in commercial applications such as sentiment analysis and opinion mining for marketing and branding.

Affective computing is cross-disciplinary, involving research in computer science and engineering, cognitive science, neuroscience, and psychology. Each of these fields has its own terminology and sometimes within the same field there is a lack of consensus on the meaning of some concepts and how they can be represented and formalized. For example, terms such as emotion, feeling, affect, affective states, mood, and social and interpersonal stances are used. For a review of different research traditions and theories of emotions, see Johnson (2009).

© Springer International Publishing Switzerland 2016
M. McTear et al., *The Conversational Interface*,
DOI 10.1007/978-3-319-32967-3_14

14.2   Computational Models of Emotion

Computational models of emotion are based partly on theoretical models from psychology, neuroscience, sociology, and linguistics. In the following sections, we provide an overview of three approaches that are used widely in computational models: the dimensional, discrete, and appraisal approaches.

14.2.1   The Dimensional Approach

In the dimensional approach, emotion is represented as a point in a continuous dimensional space. Dimensional theories understand emotion as a label or artificial psychological concept that is assigned to positions in this space.

Many systems focus on the single dimension of valence, i.e., whether the emotion is positive or negative (some authors use the terms "pleasant" or "unpleasant" instead). Though the full spectrum in the valence axis could be considered (i.e., a range of different values from totally positive to totally negative), usually this approach is used for polarity analyses consisting only of a positive versus a negative classification.

A second dimension is arousal, representing how energetic the emotion is (also known as activation/deactivation). Some authors use an intensity (or strength) dimension instead. Although arousal and intensity are often used interchangeably, there is a slight difference in meaning. For example, depression has a very low rate of arousal but can still be a very intense emotion.

To place categories (emotion adjectives) in the valence–arousal space, a circular representation has been adopted and different authors provide catalogs of emotion words placed as points in this space (see an example in Fig. 14.1).

This model allows emotion to be represented in a way that is easily computable and also easy for humans to interpret. For example, it was used by the FEELTRACE tool (Cowie et al. 2000) to represent how emotion develops over time. Figure 14.2 shows an example where the user being monitored was initially afraid (a negative and active emotion) and progressively calmed down to finish up being serene (a positive and passive emotion).

Although two dimensions usually suffice for most systems, there are also models with more dimensions. The best known is the pleasure, arousal, and dominance (PAD) emotional state model (Mehrabian 1997) that employs 3 dimensions: pleasure (valence), arousal, and dominance. Other representations include control over the emotion (high control versus low control), and how constructive or obstructive the emotion is (Scherer 2005). As it is difficult to depict a 4-dimensional space, these representations are usually projected into a two-dimensional representation.

An advantage of dimensional models is that emotions can be represented in a space, and thus, it is possible to quantify the extent to which an emotion is similar to

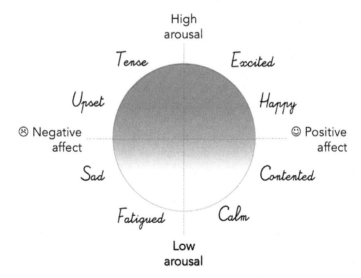

Fig. 14.1 A circumflex model of affect

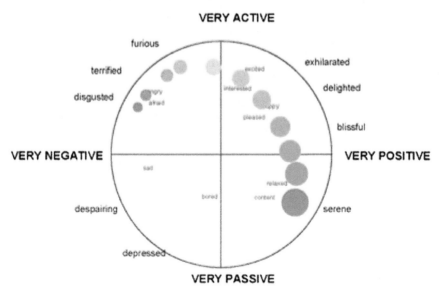

Fig. 14.2 Example of a FEELTRACE display during emotion tracking (reproduced with permission)

another emotion by measuring the distance between the two points that represent those emotions in the space. This can be useful when comparing emotional states, tracking changes in emotions, or measuring the quality of the annotation in corpora of emotions.

14.2.2   The Discrete Approach

Usually, there is no need to have such a fine-grained representation of emotion and it is sufficient to use a discrete approach in which a number of categories are considered and each unit (e.g., word or utterance) is classified into one of these categories. In this case, the first challenge is to determine which categories or labels to consider.

One approach is to employ repertoires of basic categories of emotion. These repertoires follow Darwin's theory that emotions are biologically determined and universal. Different authors have compiled an extensive body of research on how diverse cultures from all over the world consistently select the same labels to describe the same facial expressions, pictures, voices, or self-reports of emotion. However, some authors maintain that emotions are not universal and are to a great extent a cultural construct.

Well-known categories of emotion include the following: Ekman's six basic emotions—anger, disgust, fear, happiness, sadness, and surprise (Ekman 1999); or the eight basic emotions—trust, fear, surprise, sadness, disgust, anger, anticipation, and joy, described by Plutchik (2003). However, many authors create their own list of emotional terms depending on the application domain.

Some of these theories also have a dimensional component. For example, Plutchik's evolutionary theory of emotions can be represented as a cone, as shown in Fig. 14.3. In his model, there are 8 primary emotions that can appear with different degrees of intensity and that are in fact poles of the same axis (joy is opposite to sadness, fear to anger, anticipation to surprise, and disgust to trust). Moving from the center of the wheel to an extreme, the intensity of the emotion decreases (e.g., annoyance versus rage), and mixing two primary emotions results in a secondary emotion, e.g., fear and trust generate submission, and anticipation and joy generate optimism.

14.2.3   The Appraisal Approach

Other theories focus on appraisal as the main element. Different patterns of appraisal elicit different physiological and behavioral reactions. This approach is predominant in psychological and computational perspectives on emotion as it emphasizes the connection between emotion and cognition. In practical terms, appraisal models allow us not only to detect the emotion itself (e.g., being able to assign a label to what the user said), but also to determine what is the target of that emotion and the events, beliefs, desires, and/or intentions that provoked it. Computational appraisal models often provide elaborate mechanisms that encode the appraisal process in which significant events in the environment are evaluated to determine the nature of the emotional reaction.

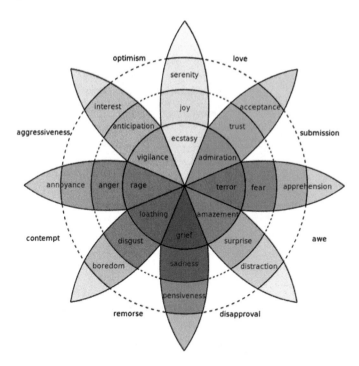

Fig. 14.3 Plutchik's wheel of emotions (*source* Wikimedia) (https://en.wikipedia.org/wiki/Contrasting_and_categorization_of_emotions#/media/File:Plutchik-wheel.svg Accessed 26 February 2016)

There are many computational appraisal models that follow an architecture where information flows like the circle proposed by Marsella et al. (2010), as shown in Fig. 14.4. In the figure, by agent we refer to either an artificial agent or the human user.

The main appraisal model used in affective computing is the Ortony, Clore, and Collins (OCC) model (Ortony et al. 1990). This model focuses on the cognitive structure of emotions, studying the eliciting conditions of emotions, and how users set their goals and behaviors accordingly. However, according to Gratch et al. (2009) the focus of OCC on cognitive components (appraisal dimension) and not on the overall emotion process results in very narrow computational models. Gratch et al. describe some alternatives that include other emotional elements such as somatic processes and behavioral responses, while others have proposed more comprehensive theories that not only encompass a wider range of emotional components (e.g., cognitions, somatic processes, behavioral tendencies, and responses) but also articulate basic process assumptions whereby emotions continuously influence and are influenced by cognition (Moors et al. 2013). Several authors have investigated these approaches within general agent architectures, for example, with the belief–desire–intention (BDI) and the biologically inspired cognitive architectures (BICA) models (Reisenzein et al. 2013; Hudlicka 2014).

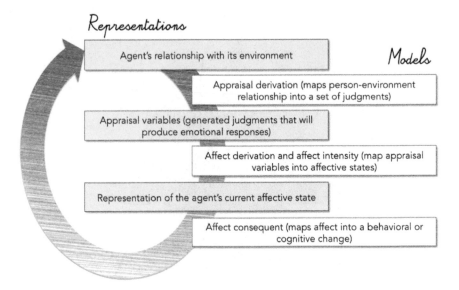

Fig. 14.4 Interpretation of Marsella's component model for computational appraisal models

14.3 Models of Personality

Personality can be defined as the characteristics of a person that uniquely influence their cognitions, motivations, and behaviors in different situations. Thus, personality is an important aspect to consider for understanding the user's behavior.

Digital personalities can be created with a consistent behavioral style that conveys the impression of having coherent characteristics of cognitions, motivations, behaviors, and emotions (Iurgel and Marcos 2007). Moreover, several studies have shown that users often assign personality to synthetic characters (Nass and Lee 2000; see also Trappl's TED talk[1]). Hence, there is a great interest in finding appropriate models to render consistent personalities that enable believable interactions with the agents.

Personality is usually modeled in terms of a space in which dimensions are based on psychological constructs called traits. Personality traits are consistent patterns of thoughts, feelings, or actions that remain stable during our lives and cause us to react consistently to the same situations over time. Trait psychologists have defined a taxonomy of five traits that, as in the case of dimensional models of emotion, have been shown to be replicable cross-culturally and to capture individual differences in personality. The "big five" traits, also known as the "OCEAN" model on account of their initials, are as follows: openness, conscientiousness, extroversion,

---

[1]https://www.youtube.com/watch?v=aixduAt3fL0. Accessed February 26, 2016.

agreeableness, and neuroticism (McCrae and John 1992). Each tuple or point in this space is usually addressed as a personality profile.

This model of personality traits has been widely adopted by the conversational agents community, although some authors provide their own features to represent personality-related phenomena in their systems, using either discrete approaches (choosing personality adjectives from a list), or using their own dimensional models that have been defined in a more ad hoc manner. For example, Leite et al. (2013) assess the friendliness of their robots using the dimensions: stimulation of companionship, help, intimacy, reliable alliance, self-validation, and emotional security.

Although the OCEAN model is frequently used as is to recognize the personality of users, usually only two of the five traits are used for rendering the system's behavior due to the complexity of modeling the traits. Thus, bi-dimensional models employ two main axes: introversion/extraversion and neuroticism/stability:

- Introversion/extraversion. Extraversion is being outgoing, talkative, and in need of external stimulation. Extraverts are under-aroused and bored and in need of stimuli, while introverts are over-aroused and need peace and calm.
- Neuroticism/stability. Neuroticism has high levels of negative affect such as depression and anxiety. Neurotic people, who have very low activation thresholds, are unable to inhibit or control their emotional reactions, and experience negative affect in the face of very minor stressors.

Although it would be possible to use a wide range of values within these axes, usually personalities are classified in the bi-dimensional model into four groups: stable extraverts (talkative, easy-going, and lively); unstable extraverts (excitable, impulsive, and irresponsible); stable introverts (calm, reliable, peaceful); and unstable introverts (quiet, reserved, pessimistic).

Another different dimension employed by some authors is psychoticism–socialization. Psychotic behavior is a key factor in tough-mindedness, non-conformity, and inconsideration. Examples of adjectives related to psychoticism are "aggressive," "egocentric," "manipulative," and "achievement-oriented."

14.3.1 The Detection of Personality

The personality of a user can be retrieved before interaction with the system as part of a process of creating a model of the user in order to fine-tune the agent. Thus in an adjustment phase, the user may be asked for personal data, such as age and gender, and also asked to answer a personality questionnaire. Then, the information collected is stored in the system, which adapts its behavior accordingly. Some of the most frequently used personality tests are as follows:

- The NEO-FFI test, Revised NEO Personality Inventory (NEO PI-R), in which there are several questions for each personality trait in the Big Five model.

- The Eysenck personality questionnaire (EPQ-R) that comprises three subscales: introversion/extraversion, neuroticism/stability, psychoticism/socialization, along with the Lie scale that was introduced in order to measure to what extent subjects attempted to control their scores.
- The Ten-Item Personality Inventory (TIPI). This test is especially interesting for the purpose of assessing the user's personality because, unlike the other tests, it is very short (10 items) and thus it is not as tedious for the user to respond to it.

Although this approach enables the creation of reliable models, it relies on the fact that the user who configured it at the beginning will be the only user of the system. Although this may be an appropriate assumption in some settings, e.g., a personal assistant on the user's mobile phone that will presumably always interact with the phone owner, this may not be the case for other application domains. The solution in these cases is to build an automatic personality recognizer.

Automatic personality detection has been widely studied recently in application domains related to text mining. For example, different studies have shown how to infer the user's personality from their behavior in social networks, including their writing style, number of friends, and even in terms of the number of likes and dislikes they give to different content (Ortigosa et al. 2014).

In the area of conversational interaction, in addition to text, acoustics can also be a good indicator of personality. Polzehl (2015) provides an extensive review of speech-based automatic personality estimation. This topic was also addressed in the INTERSPEECH 2012 Speaker Trait Challenge where participants were asked to determine the Big 5 personality profiles of speakers based on acoustic data. The annotated corpus used, the Speaker Personality Corpus (Mohammadi and Vinciarelli 2012), is freely available to the research community. The features and classifiers employed were similar to those that will be described for emotion recognition in Chap. 15 (Schuller et al. 2012).

14.3.2   Simulating Personality

If interactions with simulated agents are to be believable, users must be able to apply their models of human communication to the agents and the agents must have a convincing and intuitive behavior (Schonbrodt and Asendorpf 2011). Thus, it is important that these agents become recognizable individuals in order to have life-like interaction capabilities. There are two fundamental questions that will be addressed in the following subsections: which personality should be conveyed depending on the objectives of our system, and once we know the personality we would like to convey, what methods can be used to render it?

14.3.2.1 Which Personality to Render

Several studies have shown that people like personalities that are similar to their own, a phenomenon known as the "similarity-attraction principle." This principle also applies to personalities in conversational systems. For example, Nass and Lee (2000) showed that users are more attracted to voices that exhibit a similar personality to their own; users regard a TTS voice that exhibits a similar personality to their own as more credible; and users evaluate voice-based products more positively if they match their personality. Also similarity has been shown to increase the perceived agent's intelligence and competence. Thus, a popular approach for using personality in conversational interfaces is to adapt the agent's personality to match the users' personality in an attempt to foster the agent's likeability.

However, in some cases, the opposite principle applies. For example, in the case of human–robot interaction with a robotic pet, Lee et al. (2006) showed that participants preferred interacting with a robot with a personality that was complementary to their own and judged it as more intelligent and attractive.

Other authors, such as Mairesse and Walker (2010), propose tailoring the agent's personality according to the application domain. For example, in a tutoring system they suggest using extrovert and agreeable pedagogic agents, whereas for a psychotherapy agent it could be interesting to use more "disturbing" personalities. They also point out that the personality used by telesales agents should match the company's brand.

14.3.2.2 How to Render Personality

People consistently associate different verbal and non-verbal markers with certain speaker personality characteristics. For instance, the personality of the agent can be implemented by adjusting the linguistic and acoustic content of the system's interventions. Firstly, regarding language generation, Mairesse and Walker (2011) present the PERSONAGE system that has more than 25 parameters for tuning a natural language generator to generate personalities in the range extravert versus introvert. Some of the lessons learnt from this study are as follows:

- Sentences: Introverts communicate fewer content items in the same phrase, introduce fewer repetitions, and produce fewer restatements. However, they employ more complex syntactic structures. Introverts are more prone to negative content and, if an item with positive and negative content has to be conveyed, an introvert would focus more on the negative aspect and an extravert on the positive. Furthermore, introverts would use more negations (e.g., saying "it is not bad" instead of "it is good"). On the other hand, extraverts use more self-references and fewer unfilled pauses.
- Vocabulary: Extraverts use more social language and display better backchannel behaviors. Additionally, as opposed to introverts, they use more implicit language (e.g., "the food is good" instead of "the restaurant has good food") as well

as more informal language, and they also tend to exaggerate more. Introverts employ a richer vocabulary that is more formal and with a higher presence of hedging constructions.

Regarding speech synthesis, Nass and Lee (2000) present a study of TTS features with the following results:

- Speech rate: Extraverts speak more rapidly than introverts.
- Volume: Extraverts speak more loudly than introverts.
- Pitch: Extraverts speak with higher pitch and with more pitch variations.

More recent findings in the study of personality recognition can also be applied to speech synthesis. For example, Schuller and Batliner (2013) present an extensive description of the paralinguistics of personality that is relevant to TTS.

When rendering introvert versus extrovert personalities using the parameters described, it is necessary to take into account other possible effects on perception, for example:

- Speech rate: a slow speaking rate sounds cool; however, faster speakers appear more convincing, confident, intelligent, and objective.
- Pauses: fewer pauses and repeats, and a more dynamic voice give the impression of confidence.
- Volume: louder voices are judged to be friendlier.
- Pitch: high-pitched voices sound less benevolent and less competent, whereas low pitch raises the degree of apparent confidence. Besides, voices are evaluated as attractive when they are less monotonous and very clear with a low pitch.
- Pitch-range: speech with more variable intonation is evaluated as more benevolent.

For embodied conversational agents (ECAs) (see Chap. 15) that display gestures and facial expressions Neff et al. (2010) present a comprehensive list of features to differentiate the behaviors of introverts and extraverts, including:

- Body attitude: introverts lean backwards and extraverts lean forwards.
- Gesture amplitude, speed, direction, and rate: extraverts produce more gestures that are wider and face outwards, while introverts produce fewer gestures and they face inwards.
- Timing: the response latency of extraverts is shorter and their gestures are faster and smoother.
- Energy and persistence: extraverts are more energetic, but gestures from introverts are more persistent.

When using robots, another feature to be considered is their shape, which has been shown to have an effect on the robot's perceived personality (Hwang et al. 2013).

14.4 Making Use of Affective Behaviors in the Conversational Interface

There are a number of reasons conversational interfaces should incorporate information about affective behaviors. For example, as discussed in Chap. 10, dialog managers may have different inputs that enrich the knowledge that the system relies on in order to decide how to continue the interaction with the user. One of these information sources may be the recognition of the user's emotional state, which can be considered in isolation (as another input), or which can be included in some other higher-order component such as a user model or a model of the interaction context.

Bui et al. (2009, 2010) extend POMDPs with an affective component. Each state consists of the user's goal, emotional state, actions, and dialog state. However, their approach has been implemented only in a one-slot dialog. To make it more scalable, other authors such as Pittermann et al. (2009) provide semi-stochastic dialog models that have a stochastic component but also some rules, and the user's emotion is considered within the user model. In Griol et al. (2014), the user's perceived emotion is merged with the predicted user intention and input in a statistical dialog manager to enable it to decide the best system action and adapt the system's responses accordingly.

Apart from the selection of the system's next action, the adaptation of the system's behavior can be done in terms of tuning some of the dialog parameters, such as the style of prompting, the strategy, and frequency of system feedback, the confirmation strategy, the dialog initiative, and the dialog flow. Also multimodal output can be adapted to emotion (see Chap. 15).

In the following sections, we will briefly present the main interaction strategies and goals that can be generated as a result of recognizing the user's emotion. Although the interaction goals and strategies of conversational agents are very varied and depend to a great extent on the application domain, there are common goals such as making the interaction more fluid and satisfactory, fostering the acceptability, perceived social competence and believability of the agent, and keeping the user engaged during the interaction.

14.4.1 Acknowledging Awareness and Mirroring Emotion

A simple and effective strategy is to acknowledge that the agent knows that the user is displaying certain emotions. One way would be to explicitly show awareness of the user's emotions. This can be done in applications that are supposed to monitor the user's state (e.g., stress control applications that use biosignals to monitor the user), or as a strategy for social interaction (e.g., in a pedagogic system where the students are shown their recognized emotions during learning).

Another strategy is to align the emotional response of the agent with the recognized emotional state of the user, an approach known as mirroring. Mirroring is a sign of human empathy and interpersonal competence and can be an appropriate behavior to enable an agent to be perceived as socially competent.

Mirroring can be performed by any of the emotion synthesis strategies covered in Chap. 15. For example, by adjusting the pace and tone of the agent's synthetic voice to the user, or in terms of choice of vocabulary, head posture, gestures, and facial expressions. As discussed by Marinetti et al. (2011), interlocutors explicitly and implicitly convey and react to emotional information. This mutual adjustment can be deliberate or automatic, and that is why those agents that only react to explicit full-blown emotions seem artificial, as they miss the subtle mechanisms of this interpersonal process.

Different studies that address more complex mirroring strategies include more subtle mechanisms, as is the case with the social dynamics of laughter, in which the user's state is not copied but interpreted in different ways that affect the system's behavior at several levels. Also there are databases available to study the ways in which mimicry and synchronization take place between human interlocutors (Sun et al. 2011).

Schröder et al. (2012) demonstrated that it is possible to increase the acceptability of ECAs through the use of backchannel behaviors such as head nods, smiles, and expressions such as "wow" or "uh-huh." Also mimicry approaches have been shown to be successful in causing different social effects in users, e.g., contagious smiling. Agents that use effective mirroring strategies are perceived as being more empathic, and one of the consequences is an increase of rapport. For example, the Rapport Agent presented in Gratch and Marsella (2013) showed that its mimicry strategies had an impact on the users' feelings of rapport and embarrassment and also on the extent to which they engaged in intimate self-disclosure during the conversation.

14.4.2 Dealing with and Provoking the User's Emotions

In most application domains, it is important to not only acknowledge the user's emotional state but also to do something about it. In most approaches discussed in the literature, agents have been designed to avoid user frustration. Clear examples are tutoring systems, in which affective information can be used to choose the best pedagogic strategy to avoid frustration during learning. Several authors have explored this topic by attempting to frustrate users and then assess how they respond to agents with different degrees of empathic responses. Beale and Creed (2009) present a survey of these studies that highlight the potential for empathetic responses to enhance perceptions of agents and reduce feelings of frustration caused by a computer game.

However, there are also applications in which it is interesting to provoke negative emotional responses in users, as an uncomfortable experience can also make them reflect on their feelings and responses. In fact, as highlighted by Boehner et al. (2007), some authors have pointed out that the ultimate goal of affective systems should not be that the agent understands the user's emotions better, but that, thanks to the agent, the user becomes more aware of their own emotions. Sustained affective interactions can help users examine their emotional experiences over time. Attention to emotion has proven to be helpful for emotion regulation, and even labeling the affective states can help to reduce their ambiguity and facilitate coping, which can be important for diminishing dysfunctional behaviors such as drinking to handle negative affect. For example, FearNot!, developed during the EU projects Victec and eCircus, explored the use of negative ECAs for different purposes, including the treatment of phobias (Vannini et al. 2011).

14.4.3   Building Empathy

If agents are to be perceived as social actors by users, their behavior should be social, which implies being empathic and trustworthy, especially in application domains such as coaches, tutors, health care, and social companions.

Looije et al. (2010) build on the concepts of empathy and trust and divide empathy into three dimensions—complimentary, attentive, and compassionate—aspects that they take into account for a model of motivational interviewing for daily health self-management (harmonization of food, exercise, and medication). Similarly, Bickmore et al. (2010) present a relational agent for antipsychotic medication adherence for which the ability to conduct social and empathic chat as well as providing positive reinforcement and feedback is very important.

Moridis and Economides (2012) studied the implications of parallel and reactive empathy in tutoring agents. Parallel empathy describes a person displaying an emotional state similar to that of another individual, which expresses the ability to identify with the emotions of others. Reactive empathy provides insight for recovering from that emotional state. The authors showed that combining both types of empathy by displaying adequate emotional expressions moved students from a state of fear to a more neutral state. Similarly, McQuiggan and Lester (2007) present the CARE system, which supports empathetic assessment and interpretation. CARE induces a dual model of empathy. One component is used at runtime to support empathetic assessment and the other is used to support empathetic interpretation.

However, the ways in which empathy (especially parallel empathy) is portrayed have to be carefully considered. Beale and Creed (2009) showed that persuasive health messages appeared to be more effective (based on subjective measures) when they were presented with a neutral emotional expression as opposed to emotions that were consistent with the content of the presented message.

14.4.4 Fostering the User's Engagement

Engagement is a key aspect of human–computer interaction. In application domains where the agent is supposed to maintain long-term interactions with their users, a sustained interaction with an agent that always behaves in the same way would be likely to decrease user satisfaction over time.

In particular, humor may be of a great help. Dybala et al. (2009) have studied the influence of humorous stimuli and demonstrated that they enhance the positive involvement of users in conversation and their intentions to continue using the system. Also when users lose interest in the system this has a stronger effect than other negative indicators such as user frustration. For example, in tutoring systems boredom has been demonstrated to be more persistent and associated with poorer learning and less user satisfaction than frustration.

Engagement is influenced not only by the agent's behavior but is also determined by the model of the user. Pekrun et al. (2014) improved the engagement of students by considering their achievement goals and emotions. For example, they considered that in educational settings mastery goals are coupled with students' positive affect and enjoyment of learning and a negative link with anger and boredom. On the other hand, it has been shown that performance-approach goals are positively related to students' pride and hope, and performance-avoidance goals to their anxiety, shame, and sense of hopelessness. Similarly, Sellers (2013) provides insight into different layers of motivation and their coupled emotions for pedagogic systems.

Predictions of emotions that will happen in the future are also interesting as another source of information. Although the focus of conversational interfaces is to provide an immediate response, when the aim is to sustain long-term relations with the user the ability of the system to learn from a long history of previous interactions has become more important. The conversational nature of human–ECA interaction is beneficial for obtaining more information from the user, who is in some cases keener to self-disclose with an artificial interlocutor (Kang and Gratch 2014). Thus, there is more space for using sentiment analysis approaches to extract knowledge from large amounts of data. This is also relevant for producing a variety of behaviors and fostering engagement and trust. For these reasons, it could be beneficial to build long-term user models that include predictions of future user actions and affect.

14.4.5 Emotion as Feedback on the System's Performance

As discussed before, the emotional state of the user is also influenced by their interaction with the system, and some authors have employed dialog features such as the dialog duration and the degree of difficulty in obtaining pieces of information from the system as a predictor of negative user states (Callejas and López-Cózar 2008;

Callejas et al. 2011). Similarly, working in the other direction, the user's emotions can be used to predict the system's performance so that the system can adapt its behavior dynamically and avoid further malfunctions. This idea has been researched by several authors, for example, Schmitt and Ultes (2015).

14.5 Summary

Although affect and personality may seem unrelated to machines, scientists have shown that mechanisms to understand and exhibit emotion and personality in artificial agents are essential for human–computer interaction. This chapter has presented an overview of computational models of emotion and personality and the role they can play to improve different aspects of conversational interaction between humans and machines. The next chapter will build on these concepts and discuss how to endow artificial agents with the ability to understand and show emotional and rich expressive behaviors.

Further Reading
Sander and Scherer (2009) is a reference for non-specialists interested in emotions, moods, affect, and personality. It provides encyclopedic entries sorted in alphabetical order that cover the main affect-related concepts and the role that emotion plays in society, social behavior, and cognitive processes. For an introduction to emotion science from different perspectives including psychology and neuroscience, see Fox (2008). Scherer (2005) provides an excellent review of emotion theories from the perspective of computational models, including a review of emotion recognition. Marsella et al. (2010) and Marsella and Gratch (2014) examine computational models of emotion, while Trappl et al. (2002) introduces the components of human emotion and how they can be incorporated into machines as well as presenting interesting discussions between the contributing authors about the topics covered.

Exercise
On the shoulders of giants. These are some prominent researchers working on affective computing. Visit their Web pages to familiarize yourself with their latest advances, publications, and the projects they are involved in.

- **Rafael Calvo**[2] is a professor at the University of Sydney, ARC Future Fellow, director of the Positive Computing Lab, and co-director of the Software Engineering Group that focuses on the design of systems that support well-being in areas of mental health, medicine, and education.
- **Roddy Cowie**[3] is a professor at the School of Psychology at Queen's University Belfast. He has been Head of Cognitive and Biological Research Division, chair

---

[2]http://rafael-calvo.com/.

[3]http://www.qub.ac.uk/schools/psy/Staff/Academic/Cowie/. Accessed February 26, 2016.

of the School Research Committee, head of postgraduate research, director of research for the Emotion, Perception, and Individual Characteristics cluster.

- **Jonathan Gratch**[4] is a research professor of Computer Science and Psychology and Director or Virtual Human Research at the University of Southern California (USC) Institute for Creative Technologies and co-director of the USC Computational Emotion Group.
- **Eva Hudlicka**[5] is a principal scientist at Psychometrix Associates, which she founded in 1998 to conduct research in computational affective modeling and its applications in health care. Prior to founding Psychometrix in 1998, Dr. Hudlicka was a senior scientist at Bolt, Beranek and Newman in Cambridge, MA.
- **Stacy Marsella**[6] is a professor in the College of Computer and Information Science with a joint appointment in Psychology. Prior to joining Northeastern, he was a research professor in the Department of Computer Science at the University of Southern California, and a research director at the Institute for Creative Technologies.
- **Ana Paiva**[7] is an associate professor in the Department of Computer Science and Engineering of Instituto Superior Técnico from the Technical University of Lisbon. She is also the group leader of GAIPS, a research group on agents and synthetic characters at INESC-ID.
- **Paolo Petta**,[8] head of the Intelligent Agents and New Media group at the Austrian Research Institute for Artificial Intelligence.
- **Rosalind W. Picard**[9] is founder and director of the Affective Computing Research Group at the Massachusetts Institute of Technology (MIT) Media Lab, co-director of the Media Lab's Advancing Wellbeing Initiative, and faculty chair of MIT's Mind + Hand + Heart Initiative. She has co-founded Empatica, Inc., creating wearable sensors and analytics to improve health, and Affectiva, Inc. delivering technology to help measure and communicate emotion.
- **Klaus Scherer** is a professor emeritus at the University of Geneva. He founded and directed the Centre Interfacultaire en Science Affectives and the Swiss Center for Affective Sciences.[10]
- **Mark Schröder**[11] is a senior researcher at DFKI GmbH in Saarbrücken, Germany, and has acted as coordinator and Project leader of very important emotion-related EU research projects. He is the leader of the DFKI Speech Group, and chair of the W3C Emotion Markup Language Incubator Group.

---

[4]http://people.ict.usc.edu/∼gratch/. Accessed February 2016.

[5]https://www.cics.umass.edu/faculty/directory/hudlicka_eva. Accessed February 26, 2016.

[6]http://www.ccs.neu.edu/people/faculty/member/marsella/. Accessed February 26, 2016.

[7]http://gaips.inesc-id.pt/∼apaiva/Ana_Paiva_Site_2/Home.html. Accessed February 26, 2016.

[8]http://www.ofai.at/∼paolo.petta/. Accessed February 26, 2016.

[9]http://web.media.mit.edu/∼picard/. Accessed February 26, 2016.

[10]http://www.affective-sciences.org/user/scherer. Accessed February 26, 2016.

[11]http://www.dfki.de/∼schroed/index.html. Accessed February 26, 2016.

- **Robert Trappl**[12] is head of the Austrian Research Institute for Artificial Intelligence in Vienna, which was founded in 1984. He is a professor emeritus of Medical Cybernetics and Artificial Intelligence at the Center for Brain Research, Medical University of Vienna. He was a full professor and head of the department of Medical Cybernetics and Artificial Intelligence, University of Vienna, for 30 years.

References

Beale R, Creed C (2009) Affective interaction: how emotional agents affect users. Int J Hum-Comput Stud 67:755–776. doi:10.1016/j.ijhcs.2009.05.001

Bickmore TW, Puskar K, Schlenk EA, Pfeifer LM, Sereika SM (2010) Maintaining reality: relational agents for antipsychotic medication adherence. Interact Comput 22(4):276–288. doi:10.1016/j.intcom.2010.02.001

Boehner K, DePaula R, Dourish P, Sengers P (2007) How emotion is made and measured. Int J Hum-Comput 65(4):275–291. doi:10.1016/j.ijhcs.2006.11.016

Bui TH, Poel M, Nijholt A, Zwiers J (2009) A tractable hybrid DDN–POMDP approach to affective dialogue modeling for probabilistic frame-based dialogue systems. Nat Lang Eng 15 (2):273–307. doi:10.1017/S1351324908005032

Bui TH, Zwiers J, Poel M, Nijholt A (2010) Affective dialogue management using factored POMDPs. In: Babuška R, Groen FCA (eds) Interactive collaborative information systems: studies in computational intelligence. Springer, Berlin, pp 207–236. doi:10.1007/978-3-642-11688-9_8

Callejas Z, López-Cózar R (2008) Influence of contextual information in emotion annotation for spoken dialogue systems. Speech Commun 50(5):416–433. doi:10.1016/j.specom.2008.01.001

Callejas Z, Griol D, López-Cózar R (2011) Predicting user mental states in spoken dialogue systems. EURASIP J Adv Signal Process 1:6. doi:10.1186/1687-6180-2011-6

Cowie R, Douglas-Cowie E, Savvidou S, McMahon E, Sawey M, Schröder M (2000) FEELTRACE: an instrument for recording perceived emotion in real time. In: Proceedings of the International Speech Communication Association (ISCA) workshop on speech and emotion. Newcastle, Northern Ireland, pp 19–24

Dybala P, Ptaszynski M, Rzepka R, Araki K (2009) Activating humans with humor—a dialogue system that users want to interact with. IEICE T Inf Syst E92-D(12):2394–2401. doi:10.1587/transinf.E92.D.2394

Ekman P (1999) Basic emotions. In: Dalgleish T, Power MJ (eds) Handbook of cognition and emotion. Wiley, Chichester, pp 45–60. doi:10.1002/0470013494.ch3

Fox E (2008) Emotion science: cognitive and neuroscientific approaches to understanding human emotions. Palgrave Macmillan, Basingstoke

Gratch J, Marsella S (eds) (2013) Social emotions in nature and artifact. Oxford University Press, Oxford. doi:10.1093/acprof:oso/9780195387643.001.0001

Gratch J, Marsella S, Petta P (2009) Modeling the cognitive antecedents and consequences of emotion. Cogn Syst Res 10(1):1–5. doi:10.1016/j.cogsys.2008.06.001

Griol D, Molina JM, Callejas Z (2014) Modeling the user state for context-aware spoken interaction in ambient assisted living. Appl Intell 40(4):749–771. doi:10.1007/s10489-013-0503-z

---

[12]http://www.ofai.at/~robert.trappl/. Accessed February 26, 2016.

Hudlicka E (2014) Affective BICA: challenges and open questions. Biol Inspired Cogn Archit 7:98–125. doi:10.1016/j.bica.2013.11.002

Hwang J, Park T, Hwang W (2013) The effects of overall robot shape on the emotions invoked in users and the perceived personalities of robot. Appl Ergon 44(3):459–471. doi:10.1016/j. apergo.2012.10.010

Iurgel IA, Marcos AF (2007) Employing personality-rich virtual persons. New tools required. Comput Graph 31(6):827–836. doi:10.1016/j.cag.2007.08.001

Johnson G (2009) Emotion, theories of. Internet Encycl Philos http://www.iep.utm.edu/emotion/

Kang S-H, Gratch J (2014) Exploring users' social responses to computer counseling interviewers' behavior. Comput Hum Behav 34:120–130. doi:10.1016/j.chb.2014.01.006

Lee KM, Peng W, Jin S-A, Yan C (2006) Can robots manifest personality? An empirical test of personality recognition, social responses, and social presence in human-robot interaction. J Commun 56(4):754–772. doi:10.1111/j.1460-2466.2006.00318.x

Leite I, Pereira A, Mascarenhas S, Martinho C, Prada R, Paiva A (2013) The influence of empathy in human–robot relations. Int J Hum-Comput St 71(3):250–260. doi:10.1016/j.ijhcs.2012.09.005

Looije R, Neerincx MA, Cnossen F (2010) Persuasive robotic assistant for health self-management of older adults: design and evaluation of social behaviors. Int J Hum-Comput St 68(6):386–397. doi:10.1016/j.ijhcs.2009.08.007

Mairesse F, Walker MA (2010) Towards personality-based user adaptation: psychologically informed stylistic language generation. User Model User-Adap 20(3):227–278. doi:10.1007/s11257-010-9076-2

Mairesse F, Walker MA (2011) Controlling user perceptions of linguistic style: trainable generation of personality traits. Comput Linguist 37(3):455–488. doi:10.1162/COLI_a_00063

Marinetti C, Moore P, Lucas P, Parkinson B (2011) Emotions in social interactions: unfolding emotional experience. In: Cowie R, Pelachaud C, Petta P (eds) Emotion-oriented systems, cognitive technologies. Springer, Berlin, pp 31–46. doi:10.1007/978-3-642-15184-2_3

Marsella S, Gratch J (2014) Computationally modeling human emotion. Commun ACM 57 (12):56–67, doi:10.1145/2631912

Marsella S, Gratch J, Petta P (2010) Computational models of emotion. In: Scherer KR, Bänziger T, Roesch EB (eds) Blueprint for affective computing: a source book. Oxford University Press, Oxford, pp 21–41

McCrae RR, John OP (1992) An introduction to the five-factor model and its applications. J Pers 60(2):175–215. doi:10.1111/j.1467-6494.1992.tb00970.x

McQuiggan SW, Lester JC (2007) Modeling and evaluating empathy in embodied companion agents. Int J Hum-Comput St 65(4):348–360. doi:10.1016/j.ijhcs.2006.11.015

Mehrabian A (1997) Comparison of the PAD and PANAS as models for describing emotions and for differentiating anxiety from depression. J Psychopathol Behav 19(4):331–357. doi:10.1007/BF02229025

Mohammadi G, Vinciarelli A (2012) Automatic personality perception: prediction of trait attribution based on prosodic features. IEEE T Affect Comput 3(3):273–284. doi:10.1109/T-AFFC.2012.5

Moors A, Ellsworth P, Scherer KR, Frijda NH (2013) Appraisal theories of emotion: state of the art and future development. Emot Rev 5(2):119–124. doi:10.1177/1754073912468165

Moridis CN, Economides AA (2012) Affective learning: empathetic agents with emotional facial and tone of voice expressions. IEEE T Affect Comput 3(3):260–272. doi:10.1109/T-AFFC.2012.6

Nass C, Lee KM (2000) Does computer-generated speech manifest personality? An experimental test of similarity-attraction. In: Proceedings of the SIGCHI conference on human factors in computing systems (CHI'00). The Hague, Netherlands, 1–6 April 2000:329–336. doi:10.1145/332040.332452

Neff M, Wang Y, Abbott R, Walker M, (2010) Evaluating the effect of gesture and language on personality perception in conversational agents. In: Allbeck J, Badler N, Bickmore T, Pelachaud C, Safonova A (eds) Intelligent virtual agents 6356. Springer, Berlin, pp 222–235. doi:10.1007/978-3-642-15892-6_24

Ortigosa A, Carro RM, Quiroga JI (2014) Predicting user personality by mining social interactions in facebook. J Comput Syst Sci 80(1):57–71. doi:10.1016/j.jcss.2013.03.008

Ortony A, Clore GL, Collins A (1990) The cognitive structure of emotions. Cambridge University Press, Cambridge

Pekrun R, Cusack A, Murayama K, Elliot AJ, Thomas K (2014) The power of anticipated feedback: effects on students' achievement goals and achievement emotions. Learn Instr 29:115–124. doi:10.1016/j.learninstruc.2013.09.002

Picard RW (1997) Affective computing. The MIT Press, Cambridge

Pittermann J, Pittermann A, Minker W (2009) Handling emotions in human-computer dialogues. Springer Science & Business Media, Netherlands. doi:10.1007/978-90-481-3129-7

Plutchik R (2003) Emotions and life: perspectives from psychology, biology, and evolution. American Psychological Association, Washington, DC

Polzehl T (2015) Personality in speech: assessment and automatic classification. Springer, New York. doi:10.1007/978-3-319-09516-5

Reisenzein R, Hudlicka E, Dastani M, Gratch J, Hindriks KV, Lorini E, Meyer J-JC (2013) Computational modeling of emotion: toward improving the inter- and intradisciplinary exchange. IEEE T Affect Comput 4(3):246–266. doi:10.1109/t-affc.2013.14

Sander D, Scherer KR (eds) (2009) The Oxford companion to emotion and the affective sciences. Oxford University Press, Oxford

Scherer K (2005) What are emotions? And how can they be measured? Soc Sci Inform 44(4):695–729. doi:10.1177/0539018405058216

Schmitt A, Ultes S (2015) Interaction quality: assessing the quality of ongoing spoken dialog interaction by experts—and how it relates to user satisfaction. Speech Commun 74:12–36. doi:10.1016/j.specom.2015.06.003

Schonbrodt FD, Asendorpf JB (2011) The challenge of constructing psychologically believable agents. J Media Psychol-GER 23(2):100–107. doi:10.1027/1864-1105/a000040

Schröder M, Bevacqua E, Cowie R, Eyben F, Gunes H, Heylen D, ter Maat M, McKeown G, Pammi S, Pantic M, Pelachaud C, Schuller B, de Sevin E, Valstar M, Wöllmer M (2012) Building autonomous sensitive artificial listeners. IEEE T Affect Comput 3(2):165–183. doi:10.1109/T-AFFC.2011.34

Schuller B, Batliner A (2013) Computational paralinguistics: emotion, affect and personality in speech and language processing. Wiley, Chichester. doi:10.1002/9781118706664

Schuller B, Steidl S, Batliner A, Nöth E, Vinciarelli A, Burkhardt F, van Son R, Weninger F, Eyben F, Bocklet T, Mohammadi G, Weiss B (2012) The INTERSPEECH 2012 speaker trait challenge. In: Proceedings of the 13th annual conference of the international speech communication association (Interspeech 2012), Portland, 9–13 Sept 2012. http://www.isca-speech.org/archive/interspeech_2012/i12_0254.html

Sellers M (2013) Toward a comprehensive theory of emotion for biological and artificial agents. Biol Inspired Cogn Archit 4:3–26. doi:10.1016/j.bica.2013.02.002

Sun X, Lichtenauer J, Valstar M, Nijholt A, Pantic M (2011) A multimodal database for mimicry analysis. In: D'Mello S, Graesser A, Schuller B, Martin J-C (eds) Affective computing and intelligent interaction, Lecture Notes in Computer Science. Springer, Berlin, pp 367–376. doi:10.1007/978-3-642-24600-5_40

Trappl R, Petta P, Payr S (eds) (2002) Emotions in humans and artifacts. MIT Press, Cambridge

Vannini N, Enz S, Sapouna M, Wolke D, Watson S, Woods S, Dautenhahn K, Hall L, Paiva A, André E, Aylett R, Schneider W (2011) "FearNot!": a computer-based anti-bullying-programme designed to foster peer interventions. Eur J Psychol Educ 26(1):21–44. doi:10.1007/s10212-010-0035-4

Chapter 15
Affective Conversational Interfaces

Abstract In order to build artificial conversational interfaces that display behaviors that are credible and expressive, we should endow them with the capability to recognize, adapt to, and render emotion. In this chapter, we explain how the recognition of emotional aspects is managed within conversational interfaces, including modeling and representation, emotion recognition from physiological signals, acoustics, text, facial expressions, and gestures and how emotion synthesis is managed through expressive speech and multimodal embodied agents. We also cover the main open tools and databases available for developers wishing to incorporate emotion into their conversational interfaces.

15.1 Introduction

Building on the overview of approaches to affect, emotion, and personality presented in Chap. 14, this chapter discusses how these features can be incorporated into conversational interfaces to make them more believable and more expressive. The first section looks at the Emotion Markup Language (EmotionML), a recommendation of the W3C for annotating features of emotion. Next, we provide a detailed discussion of the processes of emotion recognition, looking at the phases of data collection and annotation, learning, and optimization, and at the behavioral signals that are used to recognize emotion, including physiological signals, paralinguistic features in speech and text, facial expressions, and gestures. This is followed by an overview of the synthesis of emotion. For each of the different aspects discussed, we provide a list of various tools that are available for developers.

15.2   Representing Emotion with EmotionML

The Emotion Markup Language (EmotionML)[1] is a recommendation of the W3C published by the Multimodal Interaction Working Group. At the time of writing, the latest version is 1.0, published in 2014. Emotion Markup Language 1.0 is designed to be practically applicable and based on concepts from research in affect and emotion. EmotionML can be used to represent emotions and related concepts following any of the approaches described in Chap. 14.

The root element of an EmotionML file is `<emotionml>`, which may contain different `<emotion>` elements that represent the annotated emotions. Depending on the model used to represent emotions, the `<emotion>` element may include a `dimension set`, `category set`, or `appraisal set` attribute indicating the dimensional space, list of categories, or appraisal model, respectively. For each of these representations, different tags may be nested inside the `<emotion>` element.

The `<dimension>` element can be used for dimensional models. The attributes required are a name for the dimension and a value, and optionally, a confidence can be assigned to the annotation. Each `<emotion>` that is annotated may have as children as many `<dimension>` elements as are required by that space. For example, in the tridimensional space of pleasure (valence), arousal, and dominance (PAD), there would be three dimension elements. As anger can be characterized by low pleasure, high arousal, and high dominance, it would be represented as shown in Code 15.1.

It is possible to define the dimension set manually or to use previously defined models that can be found in the document Vocabularies for EmotionML.[2] In this document, there are lists of emotion vocabularies that can be used with EmotionML that follow scientifically valid inventories corresponding to categories (e.g., Ekman's Big Six), dimensions (e.g., Mehrabian's PAD), appraisals (e.g., the Ortony, Clore, and Collins (OCC) model of emotion), and action tendencies (e.g., Frijda's action tendencies).

```
<emotionml version="1.0"
xmlns="http://www.w3.org/2009/10/emotionml">
<emotion dimension-set="http://www.w3.org/TR/emotion-
voc/xml#pad-dimensions">
  <dimension name="pleasure" value="0.2/>
  <dimension name="arousal" value="0.8"/>
  <dimension name="dominance" value="0.8/>
</emotion>
</emotionml>
```

Code 15.1 Representing anger in a dimensional space with EmotionXML

---

[1]https://www.w3.org/TR/emotionml/. Accessed February 27, 2016.
[2]http://www.w3.org/TR/emotion-voc/. Accessed March 1, 2016.

The <category> element can be used for discrete models in which emotion is assigned a name attribute. As shown in Code 15.2, an item has been tagged as "anger," but in this case, it is not represented as a point in a space but as a category chosen from a catalog called "big6."

Here, we can see how to define the vocabulary manually, though we could have used a predefined catalog as in the previous example using <emotion category-set = "http://www.w3.org/TR/emotion-voc/xml#big6">. Optionally, it is possible to include a confidence attribute indicating the annotator's confidence that the annotation for the category is correct.

The <appraisal> element allows the use of appraisal models. The only required element is the name, but again we can also specify a value and a confidence attribute. For example, according to Scherer, anger entails appraising an event as incongruent with one's goals and values and intentionally caused. Note that, as discussed earlier in Chap. 14, with the appraisal model we are not interested in representing anger, but rather a situation that could be appraised as anger (Code 15.3).

```
<emotionml version="1.0"
xmlns="http://www.w3.org/2009/10/emotionml">

<!-- Vocabulary definition -->
<vocabulary type="category" id="big6">
  <item name="anger"/>
  <item name="disgust"/>
  <item name="fear"/>
  <item name="happiness"/>
  <item name="sadness"/>
  <item name="surprise"/>
</vocabulary>

<!-- Sample annotation of an item -->
<emotion category-set="#big6">
    <category name="anger"/>
</emotion>

</emotionml>
```

Code 15.2 Representing anger as an emotion category with EmotionXML

```
<emotionml version="1.0"
xmlns="http://www.w3.org/2009/10/emotionml">
<emotion appraisal-set="http://www.w3.org/TR/emotion-
voc/xml#scherer-appraisals">
  <appraisal name="self-compatibility" value="0.1"/>
  <appraisal name="cause-intentional" value="0.9"/>
</emotion>
</emotionml>
```

Code 15.3 Representing appraisal of anger with EmotionXML

Usually, it is necessary to tag more than one emotional item, and EmotionML facilitates different mechanisms to include time stamps and durations. For example, it is possible to indicate the absolute starting and finishing time or a starting time and duration. In the following example, we see that surprise starts at moment 1268647334000 and ends at 1268647336000, while anger starts at 1268647400000 and lasts 130 ms (Code 15.4).

To indicate relative durations, it is also possible to include an identifier for the <emotion> element. For example, we can say that anger starts 66,000 ms after surprise (Code 15.5).

Also, the <trace> element can be used to represent a periodic sampling of the value of an emotion (Code 15.6).

```
<emotion category-set="http://www.w3.org/TR/emotion-
voc/xml#big6" start="1268647334000" end="1268647336000">
  <category name="surprise"/>
</emotion>

<emotion category-set="http://www.w3.org/TR/emotion-
voc/xml#big6" start="1268647400000" duration="130">
  <category name="anger"/>
</emotion>
```

Code 15.4 Including time stamps and durations in EmotionXML

```
<emotion id="referenceSurprise" category-
set="http://www.w3.org/TR/emotion-voc/xml#big6"
start="1268647334000" end="1268647336000">
  <category name="surprise"/>
</emotion>

<emotion category-set="http://www.w3.org/TR/emotion-
voc/xml#big6" time-ref-uri="#referenceSurprise" offset-
to-start="66000">
  <category name="surprise"/>
</emotion>
```

Code 15.5 Including relative durations in EmotionXML

```
<emotion category-set="http://www.w3.org/TR/emotion-
voc/xml#big6">
  <category name="anger">
  <trace freq="10Hz" samples="0.1 0.1 0.15 0.2 0.2 0.25
0.25 0.25 0.3 0.3 0.35 0.5 0.7 0.8 0.85 0.85"/>
  </category>
</emotion>
```

Code 15.6 Representing a sampled emotion in EmotionXML

According to its specification, EmotionML is conceived as a multipurpose language that can be used for manual annotation of corpora, as a standard language for the output of emotion recognizers, or to specify the emotional behavior generated by automated systems. In its specification, there are also different examples of how it can be used in combination with other compatible languages such as Extensible Multimodal Annotation Markup Language (EMMA), Synchronized Multimedia Integration Language (SMIL), or Speech Synthesis Markup Language (SSML).

15.3   Emotion Recognition

The processes involved in building an emotion recognizer are shown in Fig. 15.1. In this section, we focus on the following phases: data collection and annotation, learning, and optimization.

During the *data collection* phase different signals are recorded from the user and preprocessed to eliminate noise and other phenomena that may degrade them. The question here is which information can be obtained from the user that is a reliable source of emotional information and how to acquire it. Emotion recognition can be performed using any of the input modalities of the conversational interface (e.g., detecting emotion in the user's voice or facial expression) or using a combination of them. It can also take into account the appraisal mechanisms in users and the effect that the interaction itself may have on their emotional responses.

However, raw signals are not appropriate inputs for a classifier. They must be sampled and different features and statistics are usually computed in order to make

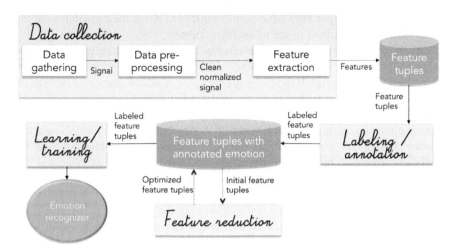

Fig. 15.1 Processes involved in building an emotion recognizer

them suitable for processing. Thus, the features are not only the raw values computed from the input signal (e.g., heart rate or voice volume), but statistical measures (e.g., heart rate variance and average volume). Sometimes there is a classification process in order to obtain meaningful features to be entered into the emotion recognizer. For example, from the recorded video, we may just focus on the mouth and from the mouth apply a classifier to determine whether the user is smiling or not. The unit being used for classification can also have an important impact on the performance of the classifier. Schuller and Batliner (2013) present a discussion of the advantages and disadvantages of different units.

Once we have obtained a database with all the recognition units (e.g., system utterances) represented as feature vectors, the database must be annotated to assign an emotion to each of the units. The *annotation* procedure depends on how the data was collected. The data can be obtained from acted emotions, elicited emotions, or spontaneous naturally occurring emotions. Acted data by professionals can be appropriate for some modalities although they miss some of the subtleties of emotional response production that cannot be consciously produced. For other signals, such as physiological data, databases of acted emotions are not suitable. With respect to elicited emotions, it is important to avoid inducing emotions different from the target emotion and eliminating the chances of inducing several emotions. There are some widespread emotion elicitation methods using pictures, films, music, and personal images (Calvo et al. 2014). Some authors have tuned video games or produced faulty versions of systems to induce negative emotions in users. Spontaneous emotions are the most natural, but they demand a complex emotion annotation process in order to obtain a reliable database. Also, they may have the drawback that not all emotions are frequent in all application domains and usually databases of spontaneous emotions are unbalanced.

Once annotated, the emotional database is used to *train a pattern recognition algorithm* that from a feature vector generates a classification hypothesis for an emotion. Different algorithms have been compared to check their suitability for this task, as will be described in the following sections.

As the overall idea is to recognize emotional states automatically from patterns of input features, there must be a process of feature design and optimization to keep only the relevant features. If too many features are considered, they may mislead the recognition process and slow it down to the point that it cannot be computable online (while the user is interacting with the system). On the other hand, the feature set must contain all the relevant features so that recognition is reliable. Sometimes *feature selection* is done offline, and on other occasions, algorithms are used that automatically select the best features while the system is operating.

The process used to recognize the user's emotion while interacting with a system is shown in Fig. 15.2.

Emotion recognition is a vast research area. In the following sections, we will describe the different sources of information that can be used in the recognition of emotion along with some discussion of the challenges involved.

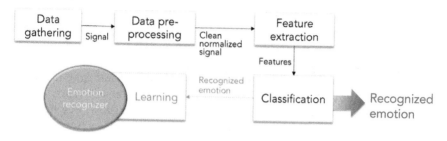

Fig. 15.2 The process of emotion recognition

15.3.1 Emotion Recognition from Physiological Signals

Different emotional expressions result in changes in autonomic activity, producing physiological signals that can be measured and used for emotion recognition. This activation affects different areas. Jerritta et al. (2011) classify them into cardiovascular system, electrodermal activity (EDA), respiratory system, muscular system, and brain activity and present a detailed table of measures. In this section, we will describe some of the most relevant of these.

The advantage of physiological signals is that they can be collected continuously to check for changes or patterns. The signals are robust against social artifacts that can hide emotions (e.g., a polite smile when the person is tense). As explained in Jerritta et al. (2011), even if a person does not overtly express his/her emotion through speech, gestures, or facial expression, a change in physiological patterns is inevitable and detectable because the sympathetic nerves of the autonomous nervous system become activated when a person is positively or negatively excited.

On the other hand, recording physiological signals requires sensors that, although in most cases they are not invasive, can be intrusive and uncomfortable for users. As discussed in Chap. 13, there are an increasing number of devices that can be connected to computers and smartphones and that provide sensing capabilities. However, developers will have to find a balance between the wearability and price of the equipment and its reliability as a source of information for recognizing emotion, since the cheap sensing devices that most users would find in stores are not sufficiently reliable in most cases and sensors that are reliable are expensive and mostly only used in laboratory conditions.

15.3.1.1 The Cardiovascular System

A heartbeat is a series of electrical impulses, involving depolarization and repolarization of the muscle whose electrical waveforms can be recorded. For example, electrocardiography (ECG) detects the electrical activity of the heart through electrodes attached to the outer surface of the skin and reflects emotional states such as tension or stress.

With every beat, the volume of blood is pushed and the generated pulse wave travels from the heart to all regions of the body. Blood volume pulse (BVP), also called heart photoplethysmography (PPG), sensors detect blood flow by using infrared light through the tip of a finger and measuring how much light is reflected.

The typical features for emotion classification obtained with the techniques described are heart rate variability (HRV), respiratory sinus arrhythmia (RSA), cardiac output, interbeat interval (IBI), and blood pressure (BP). Using these features, it is possible to differentiate mainly valence (distinguishing between positive and negative emotions), but it is also possible to recognize mental stress and effort. In this way, decreasing heart rate is a cue of relaxation and happiness, and increasing HRV is a sign of stress, anger, fear, and frustration.

Anger increases diastolic BP to the greatest degree, followed by fear, sadness, and happiness, and can be distinguished from fear by larger increases in blood pulse volume, while an increased interbeat interval can be used to detect amusement and sadness.

15.3.1.2 Electrodermal Activity

EDA measures the skin's ability to conduct electricity, which reflects changes in sympathetic nervous systems due to emotional responses and is specifically correlated with arousal (e.g., it usually increases for amusement and sadness as found in emotion elicitation using films). It is usually measured in terms of skin conductance (SC) and galvanic skin response (GSR).

Sensors for SC and GSR, which are usually placed on fingers, are based on applying a small voltage to the skin to measure its conductance or resistance. As SC depends on the activity of the sweat glands, it is important to consider the perspiration component when measuring these features. Sometimes skin temperature is used to measure the temperature at the surface of the skin, usually also on the fingers. It is important to take into account that the temperature varies depending on where the sensor is placed on the user's body and even on the time of day or the activity.

15.3.1.3 The Respiratory System

Typical features related to the respiratory system are breaths per minute, respiration volume, and relative breath amplitude. These features measure how deep and fast the user breathes and can be gathered with a sensor incorporated in a band fixed to the user's chest that accounts for chest expansions. An increasing respiration rate is related to anger and joy and a decreased respiration rate with relaxation and bliss. Respiratory cues can then be used to recognize arousal. However, it is important to consider that shocking events may cause the user's respiration to cease for a moment. Also, this can be indicative of negative valence as negative emotions may cause irregular respiratory patterns.

Breathing is linked to cardiac features and is also related to talking. These correlations must be considered in multimodal conversational interfaces in order to avoid interdependencies.

15.3.1.4 The Muscular System

A frequent method for gathering information from the muscular system is electromyography (EMG), which measures the electric impulses generated by muscles during their activity: the higher the amplitude of the electric signal, the higher the power of muscular contraction, although the signal can be very subtle, and thus, it is mainly used for strong emotions.

Although it can be useful to distinguish facial expressions, facial EMG requires fixing electrodes on the user's face, which can be intrusive, and thus, other approaches based on video are usually employed to recognize emotions from facial expressions and gestures (see further Sect. 15.3.3).

15.3.1.5 Brain Activity

Brain activity is usually measured by means of electroencephalography (EEG) and brain imaging methods such as positron emission tomography. EEG measures the electrical voltages generated by neurons when they fire. There are different frequency subsets: high beta (20–40 Hz), beta (15–20 Hz), sensorimotor rhythm (13–15 Hz), alpha (8–13 Hz), theta (4–8 Hz), and delta (2–4 Hz). The meaning of the signals gathered and their relations to emotional states are described in Bos (2006).

EEG is still not very well suited for practical implementations because of the high sensitivity to physiological artifacts such as eye blinks and electrostatic artifacts.

15.3.1.6 Classification Based on Physiological Cues

Usually, classification based on physiological cues makes use of a combination of the features described in order to obtain patterns for a certain emotion. For example, Jang et al. (2014) report that the responses of the autonomous system for fear comprise broad sympathetic activation including cardiac acceleration, increased myocardial contractility, vasoconstriction, and electrodermal activity (EDA).

It is interesting to note that the accuracy of arousal discrimination is usually higher than that for valence. The reason might be that the change in the arousal level corresponds directly to the intensity of activities such as sweat glands and BP, which is straightforward to measure with single features, while the valence differentiation of emotion requires a multifactor analysis (Kim and André 2008).

With respect to classification accuracy, Kim and André (2008) report results from early studies that already attained over 80 % accuracy. Picard et al. (2001) obtained a recognition accuracy of over 80 % on average with a linear approach; Nasoz et al. (2003) achieved an emotion classification accuracy of 83 %; and Haag et al. (2004) classified arousal and valence separately using a neural network classifier and obtained recognition accuracy rates of 96.6 and 89.9 %, respectively. Kim et al. (2004) obtained a classification ratio of over 78 % for three emotions (sadness, stress, and anger) and over 61 % for four emotions (sadness, stress, anger, and surprise) by adopting support vector machines as a pattern classifier. In all these approaches, the emotion database used was acted or elicited, which may make the results more difficult to replicate in spontaneous settings.

A more recent study by Jerritta et al. (2011) presents a detailed survey of more than 40 studies on physiological emotion recognition with accuracies ranging from 66 to 95 % using support vector machines, linear discriminant analysis, Bayesian networks, Hidden Markov models, k-nearest neighbors, and neural networks, with databases containing from 1 to 154 subjects.

15.3.1.7 Open Tools for the Analysis of Physiological Signals

The Augsburg Biosignal Toolbox (AuBT)[3] is a toolbox written in MATLAB and developed at the University of Augsburg. The toolbox can be used to analyze physiological signals by extracting their features, automatically selecting the relevant features, and using these features to train and evaluate a classifier. AuBT includes two corpora: a corpus containing physiological data of a single user in four different emotional states (Augsburg database of biosignals (AuDB)) and a corpus containing physiological data recorded from a single user under varying stress DRIving under VArying WORKload (DRIVAWORK).

AuDB was collected by recording ECG, EMG, SC, and respiratory features of one participant while listening to music in order to induce one of the emotions of joy, anger, sadness, and pleasure. There were 25 separate sessions on different days with a total of 200 min of data.

DRIVAWORK contains recordings of ECG, EMG, SC, temperature, BVP, and respiratory features along with audio and video recordings of participants in a simulated car drive. It contains a total of 15 h from 24 participants where relaxed and stressed states were elicited by giving the participants different tasks on top of a driving task.

---

[3]https://www.informatik.uni-augsburg.de/de/lehrstuehle/hcm/projects/tools/aubt/. Accessed February 27, 2016.

15.3.2   Emotion Recognition from Speech

For conversational interfaces, the user's spoken input is probably the most relevant source of emotional information in that it encodes the message being conveyed (the textual content) as well as how it is conveyed (paralinguistic features such as tone of voice).

15.3.2.1   Paralinguistic Features

Many acoustic features can be obtained from the speech signal, although there is no single approach for classifying them. Batliner et al. (2011) distinguish segmental and suprasegmental features.

Segmental features are short-term spectral and derived features, including mel-frequency cepstral coefficients (MFCCs), linear predictive coding (LPC), and wavelets. Suprasegmental features model prosodic types such as pitch, intensity duration, and voice quality. Features can be represented as raw data or they can be normalized, standardized, and presented as statistics (means, averages, etc.).

The main groups of acoustic features used for emotion recognition are listed below. Usually, for each of these groups, different features are computed, including statistics such as minimum, maximum, variance, mean, and median.

- **Intensity (energy)**. Intensity is the physical energy of the speech signal and models the loudness of a sound as perceived by the human ear.
- **Duration**. Duration models temporal aspects of voiced and unvoiced segments. It can be computed over the whole signal or on higher-order phonological units, e.g., words, to be correlated with their linguistic content.
- **Zero Crossing Rate (ZCR)**. ZCR counts the number of times the speech signal changes its sign and thus at some point equals zero. It is useful to tell whether a speech signal is voiced (low ZCR) or not.
- **Pitch/Fundamental frequency**. The fundamental frequency F0 is very representative of emotion, as human perception is very sensitive to changes in pitch.
- **Linear Prediction Cepstral Coefficients (LPCCs)**. Spectral features represent phonetic information. Their extraction can be based on LPC. The main idea of linear prediction is that the current speech sample can be predicted from its predecessors, i.e., it can be approximated by a linear combination of previous samples.
- **Mel-Frequency Cepstral Coefficients (MFCCs)**. MFCCs are among the most widely used speech features for automatic speech processing including speech and speaker recognition. They are computed by transforming the signal into a cepstral space. Coefficient 0 describes the signal energy. Coefficients 1–12 (approximately) describe mainly the phonetic content, and higher-order coefficients describe more the vocal tract and thus speaker characteristics.

Table 15.1 Common effects of emotion in speech features of Western languages (based on Väyrynen 2014)

	Anger	Fear	Joy	Sadness	Disgust	Surprise
Speech rate	>	≫	> or <	<	⋘	>
Pitch average	⋙	⋙	≫	<	⋘	>
Pitch range	≫	≫	≫	<	>	>
Pitch changes	Abrupt	Normal	Smooth up	Down	Down terminal	High
Intensity	>	=	>	<	<	>
Voice quality	Breathy	Irregular	Breathy	Resonant	Grumbled	Breathy
Articulation	Tense	Precise	Normal	Slurring	Normal	

The symbols ">," "≫," and "⋙" represent increase and symbols "<," "≪," and "⋘" decrease, while "=" indicates no perceived change

- **Formants**. Cepstral coefficients are very widespread features for speech processing, but they have a poor performance with noisy speech. Thus, to handle real-life speech, they can be supplemented with formant parameters. Formants are used to model changes in the vocal tract shape and they vary according to the spoken content, in particular formants F1 and F2 and their bandwidths.
- **Wavelets** represent a multilevel analysis of time, energy, and frequencies of a speech signal and account for its sharp transitions and drifts.
- **Voice Quality**. Voice quality features are based on acoustical models of the vocal folds. They model jitter, shimmer, and further microprosodic events.

For a review of emotionally relevant features and extraction techniques, see (Batliner et al. 2011; Cowie and Cornelius 2003; Ververidis and Kotropoulos 2006). A summary of commonly associated emotion effects in relation to normal speech is shown in Table 15.1.

15.3.2.2 Classification of Paralinguistic Features

Since 2009, the INTERSPEECH Conference, organized by the International Speech Communication Association (ISCA), has held Computational Paralinguistics Challenges. Not all editions have focused on emotion, but they have provided an opportunity to share databases, to replicate and compare results, and to compile the best features and algorithms to be used for classification. All subchallenges allow contributors to use their own features and machine learning algorithms, although participants adhere to the definition of training, development, and test sets. Thus, results are directly comparable, and it is easy for the interested reader to check which approaches have obtained the best results in each challenge.[4]

---

[4]http://compare.openaudio.eu/. Accessed February 27, 2016.

15.3.2.3 Open Tools for the Analysis of Paralinguistic Features

There are several tools—most of them open source—that provide the algorithms to perform acoustic analysis and visualization, as well as tools for scripting and classification.

Praat phonetics software, developed at the University of Amsterdam, is an open-source software package for the analysis of speech (Boersma and Weenink 2016). Praat implements algorithms to perform the main phonetic measurement and analysis procedures, including working with waveforms and spectrograms, measuring pitch, pulses, harmonics, formants, intensity, and sound quality parameters. Praat also features graphic representations and statistics and allows users to create their own scripts and communicate with other programs. On its Web page[5], it is possible to find information about how to use it to compute the features described previously. There is also a Praat Users Group in yahoo[6] and conversations about Praat in other communities of programmers such as Stack Overflow.[7]

EmoVoice/Open SSI. The Open Social Signal Interpretation framework (Open SSI)[8] offers tools to record, analyze, and recognize human behavior in real time, including gestures, mimics, head nods, and emotional speech (Wagner et al. 2013). EmoVoice was developed in the Human-Centered Multimedia Lab in the University of Augsburg and has been used by several EU-funded projects related to affective interaction. EmoVoice is integrated into the Social Signal Interpretation (SSI) framework and provides modules for real-time recognition of emotions from acoustics. The modules include speech corpus creation, segmentation, feature extraction, and online classification. The phonetic analysis used by EmoVoice relies on the algorithms provided by Praat.

openSMILE. The Speech and Music Interpretation by Large-space Extraction (SMILE) tool also provides general audio signal processing, feature extraction, and statistics as in the previously described tools. Its input/output formats are compliant with other widespread tools for machine learning such as the Hidden Markov Toolkit, Waikato Environment for Knowledge Analysis (WEKA), and the Library for Support Vector Machines (LibSVM). openSMILE was started at the Technical University of Munich by Florian Eyben, Martin Wöllmer, and Björn Schuller (Eyben et al. 2013) and is now maintained by audEERING and distributed free of charge for research and personal use.[9]

Databases. In order to train emotion recognizers, there is a need for emotionally labeled corpora. There are different corpora that have been released under varying licenses. The Association for the Advancement of Affective Computing compiles the main databases and tools and is constantly updated. For example, they have the

---

[5]http://www.fon.hum.uva.nl/praat/. Accessed February 27, 2016.

[6]https://uk.groups.yahoo.com/neo/groups/praat-users. Accessed February 27, 2016.

[7]http://stackoverflow.com/questions/tagged/praat. Accessed February 27, 2016.

[8]http://hcm-lab.de/projects/ssi/. Accessed February 27, 2016.

[9]http://www.audeering.com/research/opensmile. Accessed February 27, 2016.

HUMAINE, Belfast Naturalistic, and Geneva Vocal Emotion Expression Stimulus databases.[10] We recommend readers to check this Web page as it contains information about projects, journals and conferences, researchers, tools, and databases related to affective computing. Also, the European Language Resources Association (ELRA) has a catalog of spoken, written, and multimodal resources, some of them related to emotion.[11]

15.3.2.4 Extracting Affective Information from Text

In conversational interfaces, the user's spoken input is translated into text by means of an automatic speech recognizer. The text is used to extract the semantics of the message conveyed and to compute the most adequate system response. However, the text also carries information about the user's emotional state. This is encoded in the words and grammatical structure. For example, saying "as you wish" is not the same as saying "do what the hell you want."

There are many techniques for extracting affective information from text. Most of these involve applying techniques that are widely used for ASR and SLU to this new classification task. For example, emotion recognition from text uses preprocessing stages that are common to techniques used in SLU, such as stemming (separating lexemes from morphemes in order to avoid a dimensionality problem, especially in highly inflective languages) and stopping (removing non-relevant words). For the processing of affect, non-linguistic vocalizations such as sighs, yawns, laughs, and cries are important and are usually included as vocabulary.

The main approaches used are bag of words, n-grams, rule models, and semantic analysis. Of these, bag of words and n-grams are widely used because of their simplicity.

Bag of words. The main idea behind this approach is that words have an affective charge and some words are more frequent in expressions produced under certain emotional states than others. In this approach, the emotional salience of each word in the vocabulary to be considered is computed in a training corpus that is emotionally annotated. Then, when a new user utterance is ready to be processed, an overall emotional score is computed taking into account the most representative emotional category for each word in the utterance.

Sometimes we can also use already prepared vocabularies so that we just have to count the number of appearances of each word to compute the probability of each emotion being considered. For example, "hell" is very likely to appear under an anger setting, and thus, "do what the hell you want" can we considered as anger, while "as you wish" could be considered neutral as the vocabulary employed would have a low probability of being related to any affect category.

---

[10]http://emotion-research.net/toolbox/toolbox_query_view?category=Database. Accessed February 27, 2016.

[11]http://catalog.elra.info/. Accessed February 27, 2016.

N-grams. Sometimes considering words in isolation does not provide accurate results, as it may be necessary to account for the relation between the different words in the phrase. For example, if the user says "this is fun as hell!", "hell" should not be considered an indicative of anger. Using n-grams (see Chap. 8), we could account for how the preceding structure "fun as" changes the polarity of "hell" from negative to positive. In current work, mostly unigrams and bigrams (and very rarely trigrams) have been employed for emotion recognition, e.g., Polzin and Waibel (2000).

Rule-based approaches. These approaches are based on expert knowledge of the topic and have been extensively used in opinion mining, usually to determine the polarity of opinion (whether it is positive or negative). For example, saying that the battery of a wearable is rechargeable is usually categorized as a positive opinion. This is, however, a tricky method in application domains that are very variable, for example, saying that a smartphone is big might have been negative some years ago but is positive now (and may be negative again in the future).

The rules can also be applied to quantify the effects of connectors and qualifiers in combination with a bag-of-words approach. Thus, if the word "funny" indicates positive polarity, "extremely funny" should have even a more positive value, and a rule-based approach can be used to indicate explicitly how much higher. Similarly, if we say someone is "friendly and kind," this is more positive than just "friendly" or just "kind."

The rules may also change depending on user models. For example, in a tutoring system, if a student says that he or she finds a problem difficult, it may be bad for a student who is having problems with the subject and who may become frustrated, but good for a student who is doing well and who likes challenges.

Linguistic analysis. Emotion can also be obtained from text by means of a semantic analysis using the techniques described in Chap. 8.

Usually, techniques such as these are used for sentiment analysis (see the discussion on the difference between the semantic analysis and affective interaction communities in Clavel and Callejas (2016)). Sentiment analysis is becoming very popular for opinion mining (e.g., for companies to control conversations about their brands in social networks). Although the bag-of-words and n-gram approaches can also be employed in this area, it is very relevant to detect which is the object of the opinion/affect, something that is also very important for emotion recognition when using appraisal approaches.

The conversational context. Irrespective of the approach used, the information derived from the current user utterance must be interpreted in the context of the ongoing dialog. Techniques used to evaluate the contribution of the semantics of the input to the conversation and how the system response is computed are described in Chap. 10. With respect to the affect-related interpretation, it is possible to use the conversational context to compute the probability of each emotional state of the user more reliably. For example, if the system has been faulty, this may cause a negative user state (Callejas et al. 2011).

15.3.2.5 Open Tools for Extracting Emotional Information from Text

Tools for processing natural language text and that can also be used for extracting emotional information from text were described in Chap. 9, for example, the Natural Language Toolkit (NLTK)[12] and Apache OpenNLP.[13]

Specialized databases. The following are lexical resources for sentiment analysis and/or opinion mining. Sentiment lexicons are the most crucial resource for most sentiment analysis algorithms.

- **SentiWordNet**.[14] SentiWordNet assigns to each synset of WordNet three sentiment scores: positivity, negativity, and objectivity (Baccianella et al. 2010).
- **Affective Norms for English Words** (ANEW).[15] This dataset provides normative emotional ratings for a large number of words in the English language in terms of pleasure, arousal, and dominance.
- Opinion and sentiment lexicons and other resources by **Bing Liu**, author of books and scientific papers on sentiment analysis (Liu 2015).[16]
- **General Inquirer Home Page**.[17] The Harvard General Inquirer is a lexicon attaching syntactic, semantic, and pragmatic information to part-of-speech tagged words.
- **Sentiment in finance and accounting**. Words appearing in documents from 1994 to 2014. The dictionary reports count statistics, proportion statistics, and nine sentiment category identifiers (e.g., negative, positive, uncertainty, litigious, modal, constraining) for each word.[18]
- **Creating your own sentiment lexicon**. Sometimes, especially when working in a very specific application domain, it is necessary to build a specific sentiment lexicon. This is a very demanding task, but it is possible to expand existing resources to facilitate this process. In Feldman (2013), there is a description of how to do this from WordNet by using a vocabulary of seed adjectives and introducing synonyms with "sentiment consistency."

15.3.3 Emotion Recognition from Facial Expressions and Gestures

Some authors have identified facial expressions as the most important clue for emotion detection, and in fact, emotion recognition from facial features is one of the research

---

[12]http://www.nltk.org/. Accessed February 27, 2016.

[13]https://opennlp.apache.org/. Accessed February 27, 2016.

[14]http://sentiwordnet.isti.cnr.it/. Accessed February 27, 2016.

[15]http://csea.phhp.ufl.edu/media.html#bottommedia. Accessed February 27, 2016.

[16]https://www.cs.uic.edu/~liub/FBS/sentiment-analysis.html. Accessed February 27, 2016.

[17]http://www.wjh.harvard.edu/~inquirer/. Accessed February 27, 2016.

[18]http://www3.nd.edu/~mcdonald/Word_Lists.html. Accessed February 27, 2016.

topics with a longer trajectory in the area (Ekman 1999). There are two approaches for facial expression analysis: message based and sign based (Calvo et al. 2014).

Message-based analysis is based on the assumption that the face "is the mirror of the soul" and that it displays a representation of a person's emotional state. Some authors have provided evidence of facial expressions that signal a reduced number of basic emotions that are recognizable across cultures. Darwin described facial expressions for more than 30 emotions, and the work by Ekman is quite paradigmatic on "universal" basic emotions (Ekman 2003; Ekman and Rosenberg 2005). There are even studies of homologous emotion recognition from facial expressions in primates.

However, many other authors are not comfortable with the assumption of message-based measurement and believe that interpreting the meaning of an expression depends on the context. For example, the same expression may indicate different emotional states depending on the context in which it was produced and the person who produced it. Also, facial expressions could be posed, and thus, there would be no correspondence between the real emotional state of the user and their facial expression.

Sign-based analysis is more similar to the speech signal analysis described earlier. The idea is to obtain relevant features from corpora of annotated facial expressions and by means of a machine learning approach to learn patterns that show the relation between the feature tuples and the annotated emotion.

There are different methods employed to discretize facial expressions into relevant inputs for classification. The most relevant is the Facial Action Coding System (FACS) (Ekman and Rosenberg 2005), which deconstructs facial expressions into action units (AUs) chosen from a repertoire of more than 40 indicating their presence or absence or their intensity (see some examples in Fig. 15.3).

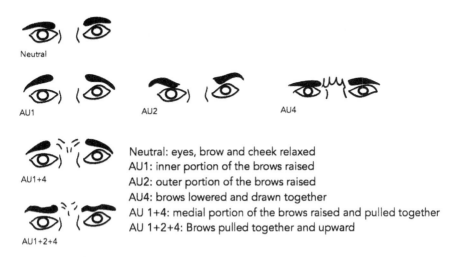

Neutral: eyes, brow and cheek relaxed
AU1: inner portion of the brows raised
AU2: outer portion of the brows raised
AU4: brows lowered and drawn together
AU 1+4: medial portion of the brows raised and pulled together
AU 1+2+4: Brows pulled together and upward

Fig. 15.3 Sample action units

Once the AUs have been detected, a classification process can be used to determine the emotion, although some authors have also detected mappings that allow the use of rule-based approaches.

15.3.3.1 The Facial Expression Recognition Process

The process of facial expression recognition is shown in Fig. 15.4:

Face detection. Face detectors are reviewed in Zhang and Zhang (2010). The main challenge here is to overcome events that make image analysis difficult, as the user's face is not captured in its totality or in the correct position for an optimal feature extraction. For instance, head motion, partial coverage of the face (e.g., if the user puts their hand in front of their face or there is another obstacle between the camera or the face), and non-frontal poses.

Face normalization. There are individual differences in head shape, skin color, facial proportion as well as effects of the spatial face position that can be reduced by converting the face detected to a canonical size and orientation.

Facial feature detection and tracking. The features typically used for emotion recognition from facial expressions are based on the local spatial position or displacement of specific points and regions of the face:

- Position and shape features (also called geometric features) that account for shapes (e.g., eyebrows) and positions (e.g., edges of the mouth).
- Motion features that account for the movement of facial muscles (e.g., optical flow or dynamic models of specific regions).
- Appearance features that represent changes in skin texture (e.g., wrinkles).

Geometric features refer to facial landmarks such as the eyes or brows. They can be represented as points, a connected face mesh, active shape model, or face component shape parameterization.

As described in Calvo et al. (2014), we can further divide geometric features into sparse (e.g., eyes or eye corners) or dense (e.g., the contours of the eyes and other permanent facial features). An advantage of the latter is that they provide information from which to infer a 3D pose. To track a dense set of facial features, active

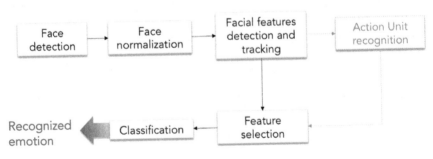

Fig. 15.4 The process of facial expression recognition

Fig. 15.5 Schema of an
AAM mesh

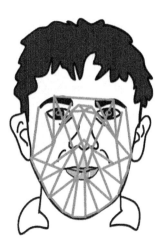

appearance models (AAMs) are often used that describe shape by a 2D triangulated mesh. In particular, the coordinates of the mesh vertices define the shape and the vertex locations correspond to a source appearance image from which the shape is aligned (see Fig. 15.5).

Motion features include optical flow and dynamic textures or motion history images (MHI). These methods all encode motion in a video sequence.

Appearance features represent changes in skin texture such as wrinkling and deepening of facial furrows and pouching of the skin. Many techniques for describing local image texture have been proposed. A major challenge is that lightning conditions affect texture. Biologically inspired appearance features, such as Gabor wavelets or magnitudes, are more robust.

AU recognition. The features described can be directly used to recognize an emotional state or can be used to recognize action units that are then used as a basis for emotion recognition (i.e., *Features* $\rightarrow$ *Emotion* or *Features* $\rightarrow$ *AUs* $\rightarrow$ *Emotion*).

Feature selection. As happened with speech and physiological signals, there can be many features derived from facial expressions that could be used for classification, but it is important to reduce dimensionality. The same methods as in the previous cases apply (e.g., component analysis, bootstrapping), but there are also some that are specific to visual features, including eigenmaps and locality preserving projections.

15.3.3.2 Classification of Emotions from Facial Expressions

Most approaches use supervised learning with previously annotated data. For a review, see Ryan et al. (2009). Calvo et al. (2014) discuss two approaches to supervised learning:

1. Static modeling—typically posed as a discriminative classification problem in which each video frame is evaluated independently.
2. Temporal modeling—frames are segmented into sequences and typically modeled with a variant of dynamic Bayesian networks (e.g., Hidden Markov models, conditional random fields).

Temporal dynamics also help the study of transitions between emotions. In conversational interfaces, temporal models are also interesting for coping with the effect of the movement of the facial muscles while talking and how they may interfere with one another.

15.3.3.3   Emotion Recognition from Gestures

Although gestures convey important affective information, they have not been exploited much yet. The process followed is the same as in Fig. 15.5, but with other features focusing on the body instead of on the face.

Usually, to avoid interpersonal differences of body shape and other aspects, work focuses on different abstract representations:

- Skeleton. A representation of the skeleton is used. In order to do this, either professional equipment or generally available commercial applications such as Microsoft Kinect have been used.
- Silhouette and blobs. For example, hand blobs from which motion features are extracted (acceleration, fluidity, symmetry, duration). See, for example, the work by Castellano et al. (2010).

15.3.3.4   Tools for Recognizing Facial Expressions and Gestures

ANVIL.[19] ANVIL is a free video annotation tool. It offers multilayered annotation based on a user-defined coding scheme. During coding, the user can see color-coded elements on multiple tracks in time alignment. Some special features are cross-level links, non-temporal objects, coding agreement analysis, 3D viewing of motion capture data, and a project tool for managing whole corpora of annotation files (Kipp 2012).

MUMIN annotation model. Implemented in various annotation tools, this model deals with communicative nonverbal behaviors such as facial expressions, head movements, hand gestures, body postures, and gaze. The MUMIN coding scheme, developed in the Nordic Network on Multimodal Interfaces, is intended as a general instrument for the study of gestures (in particular, hand gestures and facial displays) in interpersonal communication, focusing on the role played by multimodal expressions for feedback, turn management, and sequencing (Allwood et al. 2008).

---

[19]http://www.anvil-software.org/. Accessed February 27, 2016.

Databases

- **Danish first encounter NOMCO corpus**[20] (Paggio and Navarretta 2011).
- **Cohn–Kanade database.**[21] The Cohn–Kanade AU-coded facial expression database is for research in automatic facial image analysis and synthesis and for perceptual studies. Cohn–Kanade is available in two versions. The first version comprises 486 sequences from 97 posers, and the second includes both posed and non-posed (spontaneous) sequences.
- The **MMI facial expression database**[22] (Pantic et al. 2005). This database consists of over 2900 videos and high-resolution still images of 75 subjects annotated for the presence of AUs in videos (event coding) and partially coded on the frame level, indicating for each frame whether an AU is in either the neutral, onset, apex, or offset phase.
- A complete overview of publicly available data sets that can be used in research on automatic facial expression analysis is provided in Pantic and Bartlett (2007).

15.4 Emotion Synthesis

Emotion synthesis is based to a large extent on the same features described for emotion recognition, so we will focus mainly on the tools and resources available.

There is an extensive body of work that shows that humans assign human characteristics to artificial interlocutors. Especially relevant are the experimental results achieved by Nass who shows that we often assign emotional content to synthetic voices and treat conversational systems as social counterparts (Nass and Lee 2000; Nass and Yen 2012).

15.4.1 Expressive Speech Synthesis

In the case of speech synthesis, the same parameters described in Sect. 15.3.2 can be applied to color a message conveyed with emotion. This area of study is known as expressive speech synthesis (ESS).

There are several approaches to ESS. On the one hand, it is possible to modify naturally synthesized speech based on the prosodic rules that generated the desired expression. That is, once we have a series of rules that determine which parameters to change to synthesize emotional speech, we tweak those parameters to convey the desired emotion (e.g., make it faster and louder in order to sound angry).

---

[20]http://metashare.cst.dk/repository/browse/danish-first-encounters-nomco-corpus/6f4ee05644421 1e2b2e00050569b00003505d6478d484ae2b75b737aab697e99/. Accessed February 27, 2016.

[21]http://www.pitt.edu/~emotion/ck-spread.htm. Accessed February 27, 2016.

[22]http://mmifacedb.eu/. Accessed February 27, 2016.

On the other hand, we could use recordings corresponding to the target emotion that are already available in a database. To gain flexibility, the recordings can consist of small units that can correspond to the target emotion or to other emotions that are blended to generate new styles.

The procedure followed in both cases is the same as described for TTS (see Chap. 5), but accepting the target emotion as another input. Additional details on work addressing each of these techniques can be found in the comprehensive reviews by Govind and Prasanna (2012), Schröder (2009), and van Santen et al. (2008).

15.4.1.1   Tools

The same tools described in Sect. 15.3.2 apply here. For the specific case of speech synthesis, **EmoFilt**[23] is a very interesting tool for readers who want to experiment with ESS, as it shows very clearly how to generate emotions from a neutral voice by configuring the synthesis parameters. EmoFilt is an open-source program based on the free-for-non-commercial-use MBROLA synthesis engine (Burkhardt 2005).

15.4.2   Generating Facial Expressions, Body Posture, and Gestures

Embodied conversational agents (ECAs) are able to display facial expressions and gestures (see Chap. 14). Based on the earlier discussion of emotion perception in facial expression and gestures (Sect. 15.3.3), it is possible to generate static snapshots of emotional expression (e.g., by generating the relevant action units). However, expressive behaviors for conversational interfaces involve not only choosing the appropriate features but also deciding how they are realized and especially what are the dynamic qualities of the signals generated.

With respect to facial expressions, different approaches have been used in the literature, from interpolating discrete emotions to fuzzy logic, or by superposing different facial areas corresponding to different emotions. For gestures and body movements, different dynamic models are used based on features that correspond to temporal, spatial, power, and fluidity aspects. A comprehensive description can be found in the following references (Pelachaud 2009; Niewiadomski et al. 2013).

In addition to the synthesis of these general qualities, given that the generation of multimodal expressive behavior encompasses many details, it is possible to find very complex research on narrower aspects, such as emotional eye movement and gaze or smiling.

---

[23]http://emofilt.syntheticspeech.de/. Accessed February 27, 2016.

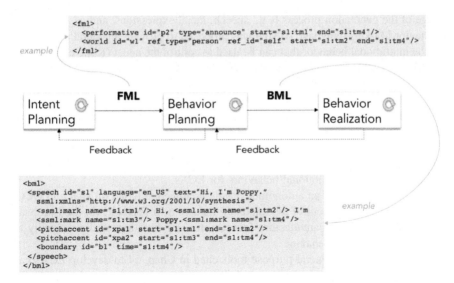

```
                    <fml>
                       <performative id="p2" type="announce" start="s1:tm1" end="s1:tm4"/>
                       <world id="w1" ref_type="person" ref_id="self" start="s1:tm2" end="s1:tm4"/>
  example           </fml>
```

```
<bml>
  <speech id="s1" language="en_US" text="Hi, I'm Poppy."
    ssml:xmlns="http://www.w3.org/2001/10/synthesis">
    <ssml:mark name="s1:tm1"/> Hi, <ssml:mark name="s1:tm2"/> I'm               example
    <ssml:mark name="s1:tm3"/> Poppy.<ssml:mark name="s1:tm4"/>
    <pitchaccent id="xpa1" start="s1:tm1" end="s1:tm2"/>
    <pitchaccent id="xpa2" start="s1:tm3" end="s1:tm4"/>
    <boundary id="b1" time="s1:tm4"/>
  </speech>
</bml>
```

Fig. 15.6 The SAIBA model

15.4.2.1 Tools for Generating Facial Expressions, Body Posture, and Gestures

SAIBA, FML, and BML. SAIBA is a model for unifying multimodal behavior generation for ECAs. It consists of three stages: intent planning, behavior planning, and behavior realization. Figure 15.6 shows the general architecture and an example of the FML and BML files generated.[24]

The intent planner decides the communicative intention, the behavior planner schedules the communicative signals, and finally the behavior realizer realizes the behaviors scheduled to generate the corresponding animation. As shown in Fig. 15.6, the interface between stages (1) and (2)—intent planning and behavior planning—describes communicative and expressive intent without any reference to physical behavior. This information (e.g., the agent's current goals, emotional state, and beliefs) can be specified with the Functional Markup Language (FML),[25] which provides a semantic description that accounts for the aspects that are relevant and influential in the planning of verbal and nonverbal behaviors (Cafaro et al. 2014).

The interface between stages (2) and (3)—behavior planning and behavior realization—describes multimodal behaviors as they are to be realized by the final

---

[24]The example code shown is from the SEMAINE project: http://semaine.opendfki.de/wiki/FML. Accessed February 27, 2016.

[25]http://secom.ru.is/fml/. Accessed February 27, 2016.

stage of the generation process (e.g., speech, facial expressions, and gestures). The Behavior Markup Language (BML) was proposed to provide a general description of the multimodal behavior that can be used to control the agent (Kopp et al. 2006).

Alma.[26] Alma is a computational model of real-time affect for virtual characters. It contains appraisal rules for emotion, mood, and personality to control the physical behavior of ECAs. It also provides a CharacterBuilder tool based on the AffectML language. It has been programmed in Java, and its code is available in GitHub.

EMA (Marsella and Gratch 2009). EMA is another computational model of emotion, based on the time dynamics of emotional reactions that can be used to generate naturalistic emotional behaviors for ECAs.

Fatima.[27] Fatima is an autonomous agent architecture based on BDI and OCC, initially developed to control the minds of the agents in FearNot! (in the European project VICTEC). The architecture focuses on using emotions and personality to influence the agent's behavior.

Other tools. The general-purpose tools cited in Chap. 14 to develop ECAs can also be employed to render emotional behaviors.

15.4.3   The Uncanny Valley

Many experiments have demonstrated that a higher degree of human likeness increases the appeal of agents and robots. However, when building very realistic agents, there is the danger of falling into the so-called *uncanny valley* (Kätsyri et al. 2015). As shown in Fig. 15.7, as the agent becomes more sophisticated, it is more familiar and better accepted by users until it reaches a point when it becomes disturbing because it is very real but still not as natural as would be expected (that is the valley), and then as they become more human-like, the acceptability increases again.

The generation of rich expressive behaviors may lead to expectations in users that, when not addressed, may negatively affect their perception of the system. Ben Mimoun et al. (2012) present an interesting discussion of the reasons for the failure of ECAs, including an exaggeration of expectations. They present some solutions, such as explaining clearly to users the limitations of the agent and its functionality.

As discussed in Mathur and Reichling (2016), humans seem to appear to infer trustworthiness from affective cues (e.g., subtle facial expressions) that are known to contribute to human–human social judgments, and thus, affective interaction is a key to developing believable and likeable conversational interfaces.

---

[26]http://alma.dfki.de. Accessed February 27, 2016.

[27]http://sourceforge.net/projects/fearnot/files/FAtiMA/FAtiMA/. Accessed February 27, 2016.

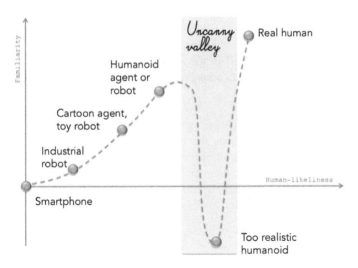

Fig. 15.7 Interpretation of the Uncanny Valley

15.5 Summary

Endowing conversational interfaces with the ability to display believable and expressive behaviors involves modeling and representing information from physiological signals, acoustic and paralinguistic features of speech and text, facial expressions, and gestures. Features of emotion can be marked up using the W3C Emotion Markup Language (EmotionML). The processes of implementing emotion recognition and synthesis include stages of data collection and annotation, learning, and optimization. There is a wide range of tools and databases available to developers who wish to incorporate emotional behaviors into conversational interfaces.

Further Reading

Schuller and Batliner (2013) give a complete survey of paralinguistics, including interesting aspects that we have not covered, such as the difference between acted versus spontaneous, felt versus perceived, intentional versus instinctual, universal versus culture-specific, and many details on emotion modeling such as type and segmentation units, features, balancing, partitioning, and laboratory versus life approaches. There is also a full chapter on corpus engineering, corpora and benchmarks, and a "hands-on" practical tutorial with openSMILE.

Petta et al.'s study (2011) is a collection of 41 chapters describing the HUMAINE project, funded by the European Commission. Calvo et al. (2014) present a comprehensive survey of affective computing, while Gratch and Marsella (2013) cover social aspects of emotion processing.

Exercises

1. **Not that easy**! Put yourself in the agent's shoes and take a test[28] of emotion recognition (if you take it in a language that you cannot speak, it will give you an even better perspective).
2. **A world of opportunities**. Go over the tools and databases that have been presented throughout the chapter and get familiarized with them, as they are open and provide instructions that will enable you to easily develop simple emotion recognizers and synthesizers.
3. **What a feeling**! Follow the demos created by Christopher Potts for the Sentiment Symposium Tutorial[29] to see how sentiment analysis works using natural language processing.

References

Allwood J, Cerrato L, Jokinen K, Naravetta C, Paggio P (2008) The MUMIN coding scheme for the annotation of feedback, turn management and sequencing phenomena. Lang Resour Eval 41(3/4):273–287. doi:10.1007/s10579-007-9061-5

Baccianella S, Esuli A, Sebastiani F (2010) SentiWordNet 3.0: an enhanced lexical resource for sentiment analysis and opinion mining. In: International conference on language resources and evaluation (LREC2010). European Language Resources Association (ELRA), Malta, 17–23 May 2010, pp 2200–2204

Batliner A, Schuller B, Seppi D, Steidl S, Devilliers L, Vidrascu L, Vogt T, Aharonson V, Amir N (2011) The automatic recognition of emotions in speech. In: Cowie R, Pelachaud C, Petta P (eds) Emotion-oriented systems. Springer Berlin Heidelberg, Berlin, Heidelberg, pp 71–99. doi:10.1007/978-3-642-15184-2_6

Ben Mimoun MS, Poncin I, Garnier M (2012) Case study—embodied virtual agents: an analysis on reasons for failure. J Retail Consum Serv 19(6):605–612. doi:10.1016/j.jretconser.2012.07.006

Boersma P, Weenink D (2016) Praat: doing phonetics by computer. http://www.fon.hum.uva.nl/praat/

Bos DO (2006) EEG-based emotion recognition; the influence of visual and auditory stimuli. http://hmi.ewi.utwente.nl/verslagen/capita-selecta/CS-Oude_Bos-Danny.pdf

Burkhardt F (2005) Emofilt: the simulation of emotional speech by prosody-transformation. In: Proceedings of the 9th European conference on speech communication and technology (Interspeech2005), Lisbon, Portugal, 4–8 Sept 2005, pp 509–512. http://www.isca-speech.org/archive/interspeech_2005/i05_0509.html

Cafaro A, Vilhjálmsson HH, Bickmore T, Heylen D, Pelachaud C(2014) Representing communicative functions in SAIBA with a unified function markup language. In: Bickmore T, Marsella S, Sidner C (eds) Intelligent virtual agents. Springer International Publishing, Switzerland, pp 81–94. doi:10.1007/978-3-319-09767-1_11

Callejas Z, Griol D, López-Cózar R (2011) Predicting user mental states in spoken dialogue systems. EURASIP J Adv Signal Process 1:6. doi:10.1186/1687-6180-2011-6

---

[28]http://www.affective-sciences.org/content/exploring-your-ec. Accessed February 27, 2016.

[29]http://sentiment.christopherpotts.net/. Accessed February 27, 2016.

Calvo RA, D'Mello S, Gratch J, Kappas A (eds) (2014) The Oxford handbook of affective computing, 1st edn. Oxford University Press, Oxford. doi:10.1093/oxfordhb/9780199942237. 001.0001

Castellano G, Leite I, Pereira A, Martinho C, Paiva A, McOwan PW (2010) Affect recognition for interactive companions: challenges and design in real world scenarios. J Multimodal User Interfaces 3(1–2):89–98. doi:10.1007/s12193-009-0033-5

Clavel C, Callejas Z (2016) Sentiment analysis: from opinion mining to human-agent interaction. IEEE Trans Affect Comput 7(1):74–93. doi:10.1109/TAFFC.2015.2444846

Cowie R, Cornelius R (2003) Describing the emotional states that are expressed in speech. Speech Commun 40(1–2):5–32. doi:10.1016/S0167-6393(02)00071-7

Ekman P (1999) Basic emotions. In: Dalgleish T, Power MJ (eds) Handbook of cognition and emotion. Wiley, Chichester, pp 45–60. doi:10.1002/0470013494.ch3

Ekman P (2003) Emotions revealed: recognizing faces and feelings to improve communication and emotional life, 1st edn. Times Books, New York

Ekman P, Rosenberg EL (eds) (2005) What the face reveals: basic and applied studies of spontaneous expression using the facial action coding system (FACS), 2nd edn. Oxford University Press, Oxford. doi:10.1093/acprof:oso/9780195179644.001.0001

Eyben F, Weninger F, Gross F, Schuller B (2013) Recent developments in openSMILE, the munich open-source multimedia feature extractor. In: Proceedings of the 21st ACM international conference on multimedia (MM'13), Barcelona, Spain, 21–25 Oct 2013, pp 835–838. doi:10.1145/2502081.2502224

Feldman R (2013) Techniques and applications for sentiment analysis. Commun ACM 56(4):82. doi:10.1145/2436256.2436274

Govind D, Prasanna SRM (2012) Expressive speech synthesis: a review. IJST 16(2):237–260. doi:10.1007/s10772-012-9180-2

Gratch J, Marsella S (eds) (2013) Social emotions in nature and artifact. Oxford University Press, Oxford 10.1093/acprof:oso/9780195387643.001.0001

Haag A, Goronzy S, Schaich P, Williams J (2004) Emotion recognition using bio-sensors: first steps towards an automatic system. In: André E, Dybkjær L, Minker W, Heisterkamp P (eds) Affective dialogue systems. Springer Berlin Heidelberg, New York, pp 36–48. doi:10.1007/978-3-540-24842-2_4

Jang E-H, Park B-J, Kim S-H, Chung M-A, Park M-S, Sohn J-H (2014) Emotion classification based on bio-signals emotion recognition using machine learning algorithms. In: Proceedings of 2014 international conference on information science, Electronics and Electrical Engineering (ISEEE), Sapporo, Japan, 26–28 April 2014, pp 104–109. doi:10.1109/InfoSEEE.2014.6946144

Jerritta S, Murugappan M, Nagarajan R, Wan K (2011) Physiological signals based human emotion recognition: a review. In: 2011 IEEE 7th international colloquium on signal processing and its applications (CSPA), Penang, Malaysia, 4–6 March 2011, pp 410–415. doi:10.1109/CSPA.2011.5759912

Kätsyri J, Förger K, Mäkäräinen M, Takala T (2015) A review of empirical evidence on different uncanny valley hypotheses: support for perceptual mismatch as one road to the valley of eeriness. Front Psychol 6:390. doi:10.3389/fpsyg.2015.00390

Kim J, André E (2008) Emotion recognition based on physiological changes in music listening. IEEE Trans Pattern Anal 30(12):2067–2083. doi:10.1109/TPAMI.2008.26

Kim KH, Bang SW, Kim SR (2004) Emotion recognition system using short-term monitoring of physiological signals. Med Biol Eng Comput 42(3):419–427. doi:10.1007/BF02344719

Kipp M (2012) ANVIL: a universal video research tool. In: Durand J, Gut U, Kristofferson G (eds) Handbook of corpus phonology. Oxford University Press, Oxford. doi:10.1093/oxfordhb/9780199571932.013.024

Kopp S, Krenn B, Marsella S, Marshall AN, Pelachaud C, Pirker H, Thórisson KR, Vilhjálmsson H (2006) Towards a common framework for multimodal generation: the behavior markup language. In: Gratch J, Young M, Aylett R, Ballin D, Olivier P (eds) Intelligent virtual agents. Springer International Publishing, Switzerland, pp 205–217. doi:10.1007/11821830_17

Liu B (2015) Sentiment analysis: mining opinions, sentiments, and emotions. Cambridge University Press, New York. doi:10.1017/CBO9781139084789

Marsella SC, Gratch J (2009) EMA: a process model of appraisal dynamics. Cogn Syst Res 10 (1):70–90. doi:10.1016/j.cogsys.2008.03.005

Mathur MB, Reichling DB (2016) Navigating a social world with robot partners: a quantitative cartography of the Uncanny Valley. Cognition 146:22–32. doi:10.1016/j.cognition.2015.09. 008

Nasoz F, Alvarez K, Lisetti CL, Finkelstein N (2003) Emotion recognition from physiological signals using wireless sensors for presence technologies. Cogn Technol Work 6(1):4–14. doi:10.1007/s10111-003-0143-x

Nass C, Lee KM (2000) Does computer-generated speech manifest personality? An experimental test of similarity-attraction. In: Proceedings of the SIGCHI conference on human factors in computing systems (CHI'00), The Hague, Netherlands, 1–6 April 2000, pp 329–336. doi:10. 1145/332040.332452

Nass C, Yen C (2012) The man who lied to his laptop: what we can learn about ourselves from our machines. Penguin Group, New York

Niewiadomski R, Hyniewska SJ, Pelachaud C (2013) Computational models of expressive behaviors for a virtual agent. In: Gratch J, Marsella S (eds) Social emotions in nature and artifact. Oxford University Press, Oxford, pp 143–161. doi:10.1093/acprof:oso/ 9780195387643.003.0010

Paggio P, Navarretta C (2011) Head movements, facial expressions and feedback in danish first encounters interactions: a culture-specific analysis. In: Stephanidis C (ed) Universal access in human-computer interaction users diversity. Springer Berlin Heidelberg, New York, pp 583– 590. doi:10.1007/978-3-642-21663-3_63

Pantic M, Bartlett MS (2007) Machine analysis of facial expressions. In: Delac K, Grgic M (eds) Face recognition. I-Tech Education and Publishing, Vienna, Austria, pp 377–416. doi:10. 5772/4847

Pantic M, Valstar MF, Rademaker R, Maat L (2005) Web-based database for facial expression analysis. In: IEEE International conference on multimedia and expo (ICME), Amsterdam, The Netherlands, 6–8 July 2005, pp 317–321. doi:10.1109/ICME.2005.1521424

Pelachaud C (2009) Modelling multimodal expression of emotion in a virtual agent. Philos Trans R Soc B Biol Sci 364(1535):3539–3548. doi:10.1098/rstb.2009.0186

Petta P, Pelachaud C, Cowie R (eds) (2011) Emotion-oriented systems: the Humaine handbook. Springer, Berlin Heidelberg. doi:10.1007/978-3-642-15184-2

Picard RW, Vyzas E, Healey J (2001) Toward machine emotional intelligence: analysis of affective physiological state. IEEE Trans Pattern Anal 23(10):1175–1191. doi:10.1109/34. 954607

Polzin TS, Waibel A (2000) Emotion-sensitive human-computer interfaces. In: International speech communication association (ISCA) tutorial and research workshop on speech and emotion. Newcastle, Northern Ireland, UK, pp 201–206

Ryan A, Cohn JF, Lucey S, Saragih J, Lucey P, De La Torre F, Rossi A (2009) Automated facial expression recognition system. In: 43rd annual international Carnahan conference on security technology, Zurich, Switzerland, 5–8 Oct 2009, pp 172–177. doi:10.1109/CCST.2009. 5335546

Schröder M (2009) Expressive speech synthesis: past, present, and possible futures. In: Tao J, Tan T (eds) Affective information processing. Springer, London, pp 111–126. doi:10.1007/ 978-1-84800-306-4_7

Schuller B, Batliner A (2013) Computational paralinguistics: emotion, affect and personality in speech and language processing. Wiley, Chichester, UK. doi:10.1002/9781118706664

Van Santen J, Mishra T, Klabbers E (2008) Prosodic processing. In: Benesty J, Sondhi MM, Huang Y (eds) Springer handbook of speech processing. Springer, Berlin Heidelberg, pp 471– 488. doi:10.1007/978-3-540-49127-9_23

Väyrynen E (2014) Emotion recognition from speech using prosodic features. Doctoral Dissertation, University of Oulu, Finland. http://urn.fi/urn:isbn:9789526204048

Ververidis D, Kotropoulos C (2006) Emotional speech recognition: resources, features and methods. Speech Commun 48(9):1162–1181. doi:10.1016/j.specom.2006.04.003

Wagner J, Lingenfelser F, Baur T, Damian I, Kistler F, André E (2013) The social signal interpretation (SSI) framework: multimodal signal processing and recognition in real-time. In: Proceedings of the 21st ACM international conference on Multimedia (MM'13), Barcelona, Spain, 21–25 Oct 2013, pp 831–834 doi:10.1145/2502081.2502223

Zhang C, Zhang Z (2010) A survey of recent advances in face detection. Microsoft TechReport MSR-TR-2010-66. http://research.microsoft.com/apps/pubs/default.aspx?id=132077

Chapter 16
Implementing Multimodal Conversational Interfaces Using Android Wear

Abstract When they first appeared, conversational systems were developed as speech-only interfaces accessible usually via landline phones. Currently, they are employed in a wide variety of devices such as smartphones and wearables, with different input and output capabilities. Traditional speech-based multimodal interfaces were designed for Web and desktop applications, but current devices pose particular restrictions and challenges for multimodal interaction that must be tackled differently. In this chapter, we discuss these issues and show how they can be solved practically by building several apps for smartwatches using Android Wear that demonstrate the different alternatives available.

16.1 Introduction

Android Wear extends the technology of Android to wearables, enabling synchronization with the smartphone to send and receive notifications and data between applications, provide access to sensors and hardware on the wearable, and support the creation of appropriate layouts and voice actions. Google provides regularly updated information about how to build apps for wearables with Android Wear.[1]

Android Wear builds its user interface (UI) model around two functions: suggest and demand. With these functions, either the device suggests useful and timely information to the user, or the user explicitly demands certain information or actions from the device.

The suggest function is based on the context stream: users do not have to launch applications to check for updates, and they can simply glance at a vertical list of cards showing useful information (see Chap. 13 for more detail on cards). Users swipe vertically to navigate from card to card and horizontally from right to left to obtain further information on a card, or they press buttons to perform actions on the card. They swipe from left to right to dismiss a card so that it is removed from the screen until it has useful information to display (see Fig. 16.1).

---

[1]http://developer.android.com/intl/es/training/building-wearables.html. Accessed February 22, 2016.

M. McTear et al., *The Conversational Interface*,
DOI 10.1007/978-3-319-32967-3_16

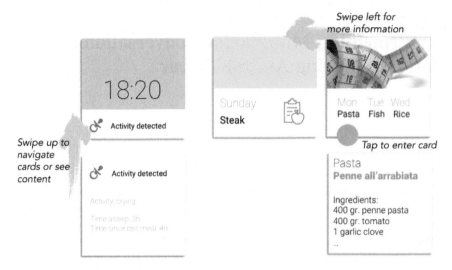

Fig. 16.1 Interaction with Android cards using swiping gestures

Fig. 16.2 Sample cue card
Google and the Google logo
are registered trademarks of
Google Inc., used with
permission

The demand function is based on the cue card. When the user cannot find the desired information in the context stream, they can tap on the background of the home screen or say "Ok Google" to show the cue card. This card is a kind of menu with a list of options that the user can select on the screen by swiping up and tapping, or by voice (the options shown in Fig. 16.2 are in fact voice commands).

Apart from the suggest and demand options, Android Wear devices can run full-screen apps that are shown on top of the main stream for which the interaction pattern is defined by the developer (though there are some guidelines that can be followed to make these apps more usable[2]).

---

[2]http://developer.android.com/intl/es/design/wear/structure.html. Accessed February 22, 2016.

We will focus on issues related to the multimodal interface. Firstly, we will describe the novelties of these graphical UIs with respect to traditional UIs, including layouts, cards, and watch faces. Then, we will describe how to add voice actions and spoken notifications. These are the functionalities that have been explicitly addressed in Android Wear. However, we are also interested in conversational capabilities, so we will go a step further to explain how to create conversational multimodal interfaces using the available technology.

You can download the code corresponding to the examples in this chapter from the folder *chapter16* of the ConversationalInterface[3] repository.

16.2   Visual Interfaces for Android Wear

Wearables use similar layouts to handheld Android devices, but their specific properties must be considered when designing wearable apps as they may not be portable from a smartphone to a smartwatch or even from a square smartwatch to a round smartwatch.[4]

Also, the information shown by the wearable must be processable at a glance and actionable, that is, it should be easy to generate commands and work with the information displayed. They must also be attractive and fit the screen of the wearable. Table 16.1 shows the primary components of the Android Wear UI.

The actions that can be used to navigate between these components are tapping, pressing with a finger for some seconds, swiping, pressing with the palm of the hand, or speaking. The design of application patterns is based on combining the elements in Table 16.1 with these navigation actions and involves creating a card map with a story of the valid actions that will enable navigation between cards, as shown in Fig. 16.3.

Android Wear provides developers with tutorials and materials[5] including

- Specification (size, position, color, etc) of the primary Android Wear UI components (e.g., peek cards, text notifications, actions, etc.).
- Sample notification and application patterns.
- Specific cards for sample apps.
- Specification of watch faces.
- Layout templates.
- Sticker sheets, icons, typographies, and color palettes.

---

[3]http://zoraidacallejas.github.io/ConversationalInterface/. Accessed March 2, 2016.

[4]http://developer.android.com/intl/es/training/wearables/ui/layouts.html. Accessed February 22, 2016.

[5]http://developer.android.com/intl/es/design/downloads/index.html#Wear. Accessed February 22, 2016.

Table 16.1 Primary Android Wear UI components

Component	Elements	Example
Notification card	Background photograph Text string App icon Action button	
Action button	Text string App white icon	
Confirmation	Animation Caption	
Picker (1D or 2D)	Options a list of cards	

(continued)

Table 16.1 (continued)

Component	Elements	Example
Speech entry	Hint text	Speak a writer or book name

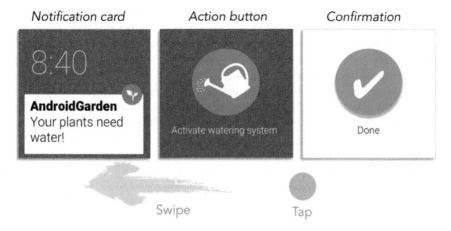

Fig. 16.3 Sample interface design for a gardening app

16.3   Voice Interfaces for Android Wear

Using voice input in Android Wear is relatively straightforward, as the interaction has been devised to be mainly oral. There are two options:

- Using voice actions. Google distinguishes two types of voice action: system provided and app provided.[6] System-provided actions are predefined voice actions that come with the Wear platform, while app-provided actions are declared by the programmer.
- Using the general-purpose speech recognition capabilities described in Chap. 6.

---

[6]http://developer.android.com/intl/es/training/wearables/apps/voice.html. Accessed February 22, 2016.

16.3.1 System-Provided Voice Actions

These are voice actions that are supported by the Android Wear platform. To use them, we must follow three simple steps:

1. Define an intent filter.[7]
2. Handle the intent in our app.
3. Update our app completion status.

As an example, we present the `MorningCoffee` app that communicates with an intelligent coffee machine to prepare coffee for the time requested by the user. This app shows how to program a custom treatment for a system-defined voice action (set an alarm). Instead of setting the alarm as usual, when this system action is launched, the `MorningCoffee` app is executed. To build the app on your device, follow the instructions in the tutorials of the "Further Reading" section.

`MorningCoffee` is a native app, that is, it is directly installed in the wearable and thus can be directly executed in the smartwatch by selecting it in the menu. Alternatively, it can be invoked through a system-defined command to set the alarm. As shown in Fig. 16.4, when the user chooses `MorningCoffee` in the menu of the smartwatch, they can select the time to set the coffee alarm and then they obtain a message that the coffee will be ready by that time. Alternatively, the user can go to the microphone in the smartwatch and speak a command such as "set an alarm for 7 am," and the same processing is done and the same final message is obtained.

The app was built following the three steps mentioned earlier. To build it from scratch in Android Studio, be sure that you select that the target Android device is "Wear" rather than "Phone or Tablet," as these are selected by default.

Firstly, in the `Manifest` file, we define an intent filter for the `SET_ALARM` action and specify that the activity `CoffeeActivity` will process it (Code 16.1).

As we have associated the `SET_ALARM` intent with our app, the first time that you try to set an alarm using a voice action the smartphone will prompt you to choose how you are going to manage that action in the future, with the built-in `Alarm` app, or with `MorningCoffee`.

There is a list of the voice intents supported by the Android Wear platform with their respective actions, categories, extras, and mime types.[8] These include call a taxi, take a note, set an alarm, set timer, start and stop watch, start and stop bike ride/run/workout, show heart rate, and show step count.

---

[7]See the most common intents here: http://developer.android.com/intl/es/guide/components/Intents-common.html. Accessed February 22, 2016.

[8]http://developer.android.com/intl/es/training/wearables/apps/voice.html. Accessed February 22, 2016.

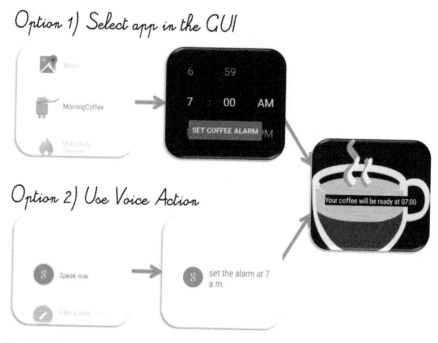

Fig. 16.4 Screenshots of the MorningCoffee app

```
<activity android:name=".CoffeeActivity"

  android:label="@string/app_name">
  <intent-filter>
   <action android:name="android.intent.action.SET_ALARM" />
   <category android:name="android.intent.category.DEFAULT" />
  </intent-filter>
</activity>
```

Code 16.1 Fragment of the AndroidManifest.xml file of the MorningCoffee app

We handle the intent in CoffeeActivity where we just obtain the hour and minute of the alarm (which are the extras of the intent) and show a message in a TextView. The default value if no extras are indicated is 7:00 am (Code 16.2).

The CoffeeActivity class would suffice to manage the voice action, but if the MorningCoffee app is selected from the smartphone menu, it will do nothing, as it will not be the result of an intent to set an alarm. That is why in the MainActivity class, we have included the code necessary to start the intent to set the alarm from a GUI in which the user selects the time and presses a button (Code 16.3).

```
TextView textView = (TextView) findViewById(R.id.text);
Intent intent = getIntent();


if (AlarmClock.ACTION_SET_ALARM.equals(intent.getAction())) {
 if (intent.hasExtra(AlarmClock.EXTRA_HOUR)) {
  int hour = intent.getIntExtra(AlarmClock.EXTRA_HOUR, 7);
  int mins = intent.getIntExtra(AlarmClock.EXTRA_MINUTES, 0);
  //**** The communication with the smart coffee machine will
  go here with the parameters hour and mins ;)
  textView.setText("Your coffee will be ready at " +

  String.format("%02d", hour) +":"+

  String.format("%02d", mins));
 } else
  textView.setText("Your coffee will be ready");
}
```

Code 16.2 Fragment of the CoffeeActivity.java file of the MorningCoffee app

```
//Reads the hour selected in the interface

TimePicker t = (TimePicker) findViewById(R.id.timePicker);
int hour, mins;
if (Build.VERSION.SDK_INT >= 23) {
  hour = t.getHour();
  mins = t.getMinute();
} else {
  hour = t.getCurrentHour();
  mins = t.getCurrentMinute();
}


//Creates and starts the intent to set an alarm.

It is managed by the CoffeeActivity class
Intent intent = new Intent(AlarmClock.ACTION_SET_ALARM);
if (intent.resolveActivity(getPackageManager()) != null) {
  intent.putExtra(AlarmClock.EXTRA_HOUR, hour);
  intent.putExtra(AlarmClock.EXTRA_MINUTES, mins);
  startActivity(intent);
}
```

Code 16.3 Fragment of the MainActivity.java file of the MorningCoffee app

16.3.2 Developer-Defined Voice Actions

There are several different ways in which developers can use their own voice actions. We will discuss three of these:

- Voice actions to start apps.

```
<application>
<activity android:name="MainActivity"

    android:label="MyPrettyApp">
 <intent-filter>
  <action android:name="android.intent.action.MAIN" />
  <category android:name="android.intent.category.LAUNCHER"/>
 </intent-filter>
</activity>
</application>
```

Code 16.4 Fragment of the Manifest of an app that can be started with a voice command

- Notification-related voice actions.
- General-purpose voice actions.

Voice actions to start apps

Users can start apps directly with the command "Start X" where X is the name of the activity. In order to provide this functionality, the developer must register for a "start" action indicating a label for the activity that will be started, as shown in the following code, where the specified intent filter recognizes "Start MyPrettyApp" and launches MainActivity (Code 16.4).

Notification-related voice actions

Notifications usually offer the possibility of providing a response. For example, when you are notified about an e-mail, you should be able to respond to it. On a wearable where there is no keyboard, the user can take the phone and type a response, but it is perhaps more appropriate to respond directly on the device (e.g., smartwatch), either by tapping on a list of predefined responses or by speaking directly to the device. In some cases, there will not be a list of responses, for example, for responding to an e-mail.

In order to receive notifications correctly on your smartwatch, you must make sure that you have given the Android Wear app on your smartphone the appropriate permissions and that the wearable accepts notifications and is not in theater mode.[9]

The CookingNotifications app (Fig. 16.5) shows how to build different types of notifications. This time the app is not installed on the smartwatch but instead on the smartphone, although the notifications will appear on both devices and the actions to respond to them are different.

When a button is clicked, a notification is issued. The first button issues a simple notification that only shows an icon, a title, and a descriptive text (Fig. 16.6). This is done in the ShowSimpleNotification method of MainActivity.java.

---

[9]You can check how to do this here: https://support.google.com/androidwear/answer/6090188?hl=en. Accessed February 22, 2016.

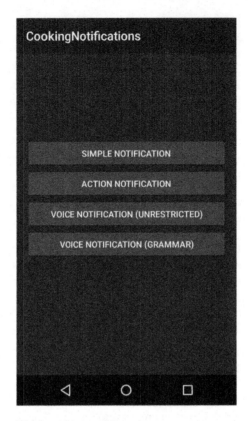

Fig. 16.5 Screenshot of the CookingNotifications app

Fig. 16.6 Simple notification—behavior triggered by the "SIMPLE NOTIFICATION" button (CookingNotifications app)

```
private void showSimpleNotification(){
  int notificationId = 1;


  //Building notification layout
  NotificationCompat.Builder notificationBuilder =
    new NotificationCompat.Builder(this)
                 .setSmallIcon(R.mipmap.cook)
                 .setContentTitle("Time for lunch!")
                 .setDefaults(Notification.DEFAULT_ALL)

                 .setContentText("Your lunch is ready");
  //Creating an instance of the NotificationManager service
  NotificationManagerCompat notificationManager =
          NotificationManagerCompat.from(this);

  //Building the notification and issuing it with the

  notification manager
  notificationManager.notify(notificationId,

    notificationBuilder.build());
}
```

Code 16.5 Creating and issuing a simple notification (fragment of `MainActivity.java` in the `CookingNotifications` app)

As can be observed in Code 16.5, we first create the layout for the notification indicating the icon, title, and description, then we instantiate the notification manager service and issue the notification.

The same code can be used to issue notifications with attached actions just by including the `Intent` associated with the particular action (Code 16.6).

The result is a notification with a small button to perform an action on both the smartphone and the smartwatch as shown in Fig. 16.7. On the phone, the notification presents a button that when clicked directs to a recipes Web page. On the smartwatch, the notification appears and can be swiped to show the action button, which when pressed opens the recipes Web page on the smartphone.

We can also create actions that receive voice input in response to a notification using the `RemoteInput` class as explained here.[10] To do so, we proceed as in the previous example, replacing the URL action by the voice action created in Code 16.7. As can be observed, we create a `PendingIntent` that when finished is processed in `SecondActivity`.

---

[10]http://developer.android.com/intl/es/training/wearables/notifications/voice-input.html. Accessed February 22, 2016.

```
// Build an intent for an action to open a url
Intent urlIntent = new Intent(Intent.ACTION_VIEW);
Uri uri = Uri.parse("http://www.reciperoulette.tv/");
urlIntent.setData(uri);
PendingIntent pendingIntent =
     PendingIntent.getActivity(this, 0, urlIntent, 0);

//Building notification layout
NotificationCompat.Builder notificationBuilder =
  new NotificationCompat.Builder(this)

    .setSmallIcon(R.mipmap.cook)
    .setContentTitle("Cooking tips")

    .setContentText("New recipe available")

    .setDefaults(Notification.DEFAULT_ALL)

    .setAutoCancel(true)

    .setContentIntent(pendingIntent)

    .addAction(R.mipmap.cook, "Check recipe", pendingIntent);
```

Code 16.6 Fragment of the showActionNotification method that attaches an action Intent to a notification (in MainActivity.java, CookingNotifications app)

Fig. 16.7 Notification with an action—behavior triggered by the "ACTION NOTIFICATION" button (CookingNotifications app)

Unlike in the previous example with the URL action, voice actions can only be performed on the wearable, so they are attached to the notification specifying that it is an action specific to the wearable with WearableExtender (Code 16.8).

As can be observed in Fig. 16.8, the notification shown on the smartphone is simple and does not have actions, whereas the notification shown on the wearable allows a voice response.

```
// Creates an intent for the reply action
Intent replyIntent = new Intent(this, SecondActivity.class);
PendingIntent replyPendingIntent =
        PendingIntent.getActivity(this, 0, replyIntent,
                PendingIntent.FLAG_UPDATE_CURRENT);


RemoteInput remoteInput = new
RemoteInput.Builder(EXTRA_VOICE_REPLY)
        .setLabel(replyLabel).build();


//Creates the reply action and adds the remote input
NotificationCompat.Action action =
   new NotificationCompat.Action.Builder(

       R.mipmap.conversandroid,
      "Tell us what you think", replyPendingIntent)
                .addRemoteInput(remoteInput)
                .build();
```

Code 16.7 Fragment of the getVoiceAction method that creates the voice action (in MainActivity.java, CookingNotifications app)

```
NotificationCompat.Builder notificationBuilder =
   new NotificationCompat.Builder(this)

   .setSmallIcon(R.mipmap.cook)

   .setContentTitle("New recipe")
   .setContentText("There is a tasty new recipe")

   .setDefaults(Notification.DEFAULT_ALL

   .setAutoCancel(true)
   .extend(new NotificationCompat.WearableExtender()

   .addAction(action));
```

Code 16.8 Fragment of the showVoiceNotification method that attaches the voice action to the notification (in MainActivity.java, CookingNotifications app)

Unlike the methods described in Chap. 6, there is the possibility to specify a recognition grammar defined in XML, which is usually added to res/values/strings.xml (Code 16.9).

To add it to the speech input, we would have code as in Code 16.7, but adding the recognition choices to the RemoteInput object (Code 16.10).

As is observed in Fig. 16.9, now the user is presented with the options in a cue card that are defined in the grammar and they can select them either by speaking the command or by tapping on it in the GUI.

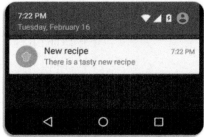

SMARTPHONE

WEARABLE

Tell us what you
think about it

Speak now

Draw emoji

Swipe left   Tap   [ ] Speak

Fig. 16.8 Notification with unrestricted voice response—behavior triggered by the "voice notification (unrestricted)" button (`CookingNotifications` app)

```
<string-array name="grammar">
    <item>No</item>
    <item>Yes, for lunch</item>
    <item>Yes, for dinner</item>
</string-array>
```

Code 16.9 Recognition grammar (fragment of `strings.xml` in the app `CookingNotifications`)

To obtain and process the user's response, we will use the activity declared in the reply action's intent (`SecondActivity`) and obtain the recognized text using `getResultsFromIntent()`. For the code to work, the activity must be declared in the application's `Manifest` (Code 16.11). The results of processing the voice input in the `CookingNotifications` app are shown in Fig. 16.10.

```
String replyLabel = "Do you want to try it today?";
String[] replyChoices =
    getResources().getStringArray(R.array.grammar);

(...)

RemoteInput remoteInput =

    new RemoteInput.Builder(EXTRA_VOICE_REPLY)
        .setLabel(replyLabel)
        .setChoices(replyChoices)
        .build();
```

Code 16.10 Adding a recognition grammar to a RemoteInput (fragment of the getGrammarAction method in MainActivity.java, CookingNotifications app)

Fig. 16.9 Notification with grammar-based voice response—behavior triggered by the "VOICE NOTIFICATION (GRAMMAR)" button (CookingNotifications app)

```
Intent intent = getIntent();
TextView textView = (TextView) findViewById(R.id.textView);

Bundle remoteInput =

  RemoteInput.getResultsFromIntent(intent);
if (remoteInput != null)
  textView.setText(remoteInput.getCharSequence

   (MainActivity.EXTRA_VOICE_REPLY));
```

Code 16.11 Processing the recognized spoken input (fragment of `SecondActivity.java`, `CookingNotifications` app)

Fig. 16.10 Result of processing the voice input (`CookingNotifications` app)

General-purpose speech recognition

The speech input and output capabilities described in Chap. 6 are also available for Android Wear. This makes speech interaction in Android very powerful and easy to port between devices. To show it, we present the `WriteBack` app, a native app (installed in the wearable) that uses the `VoiceActivity` class presented in Chap. 6 to process spoken interaction with the smartwatch.

This app is very similar to `TalkBack` (Chap. 6), as it presents a button that the user can press to initiate speech recognition and the result is presented back to the user. However, instead of being synthesized back, it is just shown as text in a `TextView` (Fig. 16.11). This is because most smartwatches still do not have TTS capabilities, but the TTS methods provided by `VoiceActivity` could be implemented as in `TalkBack` for wearables with TTS capabilities.

Fig. 16.11 Interacting with the `WriteBack` app

We have kept the app simple to emphasize how we have been able to port the speech processing mechanisms that we implemented previously from a smartphone to a smartwatch without introducing changes in the code of the `VoiceActivity` class and with just some minor edits in the `MainActivity` class.

We leave it as an exercise for the reader to compare the `MainActivity` class in `WriteBack` and `TalkBack` and to verify that minimum editing has been required. The edits include: 1) a different `onCreate` method that in `WriteBack` initializes the components in the watch GUI with `WatchViewStub`, although even that piece of code is autogenerated when creating the new Android Wear project and 2) replacing synthesized messages with texts in the GUI. In order for the code to work, remember to include the required permissions in the `Manifest` file.

16.4   Summary

Wearable devices present new opportunities for spoken multimodal interfaces, as due to their size restrictions their operation requires interfaces that are attractive, simple, and easily actionable. In this chapter, we have explained how to build these types of interface with Android Wear and have presented three apps for smartwatches that illustrate the main alternatives offered by this technology: using predefined system voice actions, using voice actions provided by our apps, and using general-purpose Android speech recognition mechanisms.

Further Reading
To build an app in your device, couple your wearable with your smartphone and activate the debugging modes. You can find detailed instructions in the following tutorials:

- http://developer.android.com/intl/es/training/wearables/apps/creating.html. Accessed February 22, 2016

- https://software.intel.com/en-us/android/articles/android-wear-through-adb. Accessed February 22, 2016
- http://www.howtogeek.com/125769/how-to-install-and-use-abd-the-android-debug-bridge-utility/. Accessed February 22, 2016

Bear in mind that in some systems (e.g., some Mac operating systems), it is necessary to write `adb connect 127.0.0.1:4444` instead of `localhost:4444`.

Part IV
Evaluation and Future Directions

Chapter 17
Evaluating the Conversational Interface

Abstract The evaluation of conversational interfaces is a continuously evolving research area that encompasses a rich variety of methodologies, techniques, and tools. As conversational interfaces become more complex, their evaluation has become multifaceted. Furthermore, evaluation involves paying attention not only to the different components in isolation, but also to interrelations between the components and the operation of the system as a whole. This chapter discusses the main measures that are employed for evaluating conversational interfaces from a variety of perspectives.

17.1 Introduction

The evaluation of a conversational interface usually takes place either during the development stage and/or just before it is released to the public. Evaluation may fulfill different purposes—for example, to compare a system with previous versions in order to assess the adequacy of changes; to compare different systems; or to predict system behavior.

When the evaluation is done cyclically over a certain system, the benchmark is the system being developed and comparisons are made of the system's performance at several points during the development process. At the time of the initial assessment, there is no reference system against which a comparison can be made. In this case, an a priori estimation of the operation of the system is usually carried out during the specification phase and subsequent evaluations are made to assess deviations from the expected behavior.

In comparative assessment, a system may be evaluated with respect to another system with the same features but using a different technology, or with respect to a completely different system or technology. This evaluation has been used in important projects such as the DARPA Communicator (Walker et al. 2002).

Performance metrics are useful as they allow the weak points of the system to be detected and suggest ways to overcome these, especially at the early stages of development. However, they do not necessarily produce relevant information about

© Springer International Publishing Switzerland 2016
M. McTear et al., *The Conversational Interface*,
DOI 10.1007/978-3-319-32967-3_17

the quality of the system. As suggested by Engelbrecht (2012), quality involves users of a system making comparisons of the perceived qualities of the system against its desired qualities. For this reason, quality can only be measured by taking into account the opinions of users.

Bernsen et al. (1998) make a three-way distinction between performance evaluation, which measures the performance of the system and its components in terms of quantitative parameters; diagnostic evaluation, which detects design and implementation errors; and adequacy evaluation, which describes how well the system and its components fit their purpose and meet the needs and expectations of users.

Traditionally, the criteria that have been used for evaluation have been divided into objective and subjective metrics:

- Objective metrics are computed from logs of the interactions of users with the system, such as the duration of the dialog or the word error rate (WER), which considers the number of substituted, deleted, and inserted words in the output of the speech recognizer.
- Subjective metrics elicit the opinions of users about some aspect of quality, such as the intelligibility of the synthesized speech.

This classification is very widespread, although it is not always accurate to say that performance metrics are "objective" as they may involve judgments by human subjects (Möller 2005). For example, expert evaluators are required to calculate WER, as they must listen to the real input of the user and compare it with the recognizer's hypothesis in order to calculate the number of errors. Thus, instead of the subjective versus objective distinction, some authors make a distinction between quality judgments (subjective metrics) and interaction parameters, where the latter can be measured instrumentally (e.g., the duration of the dialog), or calculated by experts (e.g., WER).

Other important concerns are the objects of the evaluation (e.g., the whole system or a certain component), the environment in which it will take place (e.g., controlled laboratory conditions or field study), and the life cycle phase in which it takes place (e.g., a prototype or the final fully operative version). These issues will be described in detail in the following sections.

17.2   Objective Evaluation

First, we will describe how systems are evaluated in terms of their overall performance and the performance of their components; that is, how systems are evaluated "objectively" using interaction parameters.

17.2.1 Overall System Evaluation

There have been several attempts to create a full list of metrics to be used for the evaluation of conversational interfaces. Some authors have proposed catalogs of important aspects to be considered for evaluation. For instance, Dybkjaer and Bernsen (2000) proposed the following list:

> modality appropriateness, input recognition adequacy, naturalness of user speech, output voice quality, output phrasing adequacy, feedback adequacy, adequacy of dialog initiative, naturalness of the dialog structure, sufficiency of task and domain coverage, sufficiency of the system's reason capabilities, sufficiency of interaction guidance, error handling adequacy, sufficiency of adaptation to user differences, number of interaction problems, user satisfaction.

As can be observed, these are quite broad categories rather than particular metrics, and measuring each of the criteria proposed would involve considering a mixture of interaction parameters and user judgments.

In the late 1990s, various comprehensive evaluation frameworks were developed for use within the scientific community. The Expert Advisory Group on Language Engineering Standards (EAGLES) proposed a list of metrics that were applied and interpreted following an innovative framework.[1] This framework provided guidelines on how to carry out the evaluation and how to make the results available in such a way that they could be easily interpretable and comparable. In the DISC project, best practice guidelines were proposed that complemented the EAGLES proposal by using life cycle development methodologies (Dybkjaer et al. 1998). Dybkjaer et al. (2004) and López-Cózar and Araki (2005) provide detailed reviews of this work and Möller et al. (2007) present a review of the de facto criteria extracted from all these studies and an example of their usage to evaluate a particular dialog system.

The most popular methodology for performing overall system evaluation is PARAdigm for DIalogue Evaluation System (PARADISE) (Walker et al. 1997, 1998). This method models performance as a weighted function of the following:

> task success (exact scenario completion), dialog efficiency (task duration, system turns, user turns, total turns), dialog quality (word accuracy, response latency), and user satisfaction (sum of TTS performance, ease of task, user expertise, expected behavior, and future use).

The application of PARADISE to the evaluation of a dialog system requires dialog corpora extracted from controlled experiments in which users have to evaluate satisfaction on a scale after they have interacted with the system.

Other authors have focused on how to obtain and study speech corpora to compute evaluation measures. These are frequently large corpora extracted from system usage or from human–human dialogs. In the case of human–human corpora, human behavior can be used as a baseline to compare against the system's behavior.

---

[1] http://www.ilc.cnr.it/EAGLES/browse.html. Accessed March 2, 2016.

Despite its age, Gibbon et al. (1997) is an interesting reference on the statistical analysis of data corpora for system evaluation.

Some organizations also focus on the study and definition of assessment corpora and techniques, for example, the International Committee for Co-ordination and Standardisation of Speech Databases (COCOSDA[2]), which supports the development of spoken language resources and procedures for the purpose of building and/or evaluating spoken language technology, and the European Language Resources Association (ELRA[3]), which focuses on the collection and distribution of linguistic resources.

17.2.2 Component Evaluation

Each of the components of the conversational interface can also be evaluated independently using specific evaluation metrics. Next, we will describe how the speech recognition, spoken language understanding, dialog management, natural language generation, and text-to-speech synthesis components may be evaluated.

17.2.2.1 Automatic Speech Recognition

The National Institute of Standards and Technology (NIST) has conducted evaluations of speech technologies since the mid-1980s. Training and development datasets have been provided for different speech domains, including conversational telephone speech, broadcast speech, air travel planning kiosk speech, and meeting speech, and speech has been recorded under various noise conditions, with varied microphones, and in English, Arabic, and Mandarin.[4] Three evaluation tasks are included:

- Speech-to-text transcription, in which the spoken words are transcribed automatically using ASR technology and the transcripts are evaluated against transcripts of the same speech data by human transcribers.
- "Who spoke when" diarization—annotation of the transcript of a meeting to indicate when each participant speaks but without outputting speaker names or identifying speakers.
- Speaker attributed speech-to-text, in which the spoken words are transcribed and also associated with a speaker.

The most recent evaluation was held in 2007 (Fiscus et al. 2008).

---

[2]http://www.cocosda.org/. Accessed February 29, 2016.

[3]http://www.elra.info/en/. Accessed February 29, 2016.

[4]http://itl.nist.gov/iad/mig/publications/ASRhistory/index.html. Accessed February 29, 2016.

The primary metric for speech-to-text evaluation is the WER. WER is calculated by comparing the recognized text against a reference, such as a transcription by a human expert, using the following formula in which error types are substitutions (S), deletions (D), and insertions (I), and N represents the total number of words:

$$WER = 100\frac{(S+D+I)}{N}\%$$

(17.1)

Generally, WER in commercial ASR systems has improved over time but it is important to distinguish between different speech domains, some of which are more challenging than others, as well as other factors such as microphone and noise factors that can have a significant effect on accuracy.

Other related measures are Word Accuracy (WA), defined as the contrary to WER (1-WER), Word Insertion Rate, Word Substitution Rate, and Word Deletion Rate.

17.2.2.2 Spoken Language Understanding

Evaluation of the SLU component involves comparing the output of SLU with a reference representation from a test set. The most commonly used metrics are as follows:

- Sentence accuracy: the percentage of correct syntactic or semantic representations.

$$\%fc = 100 \times \frac{\text{num. of sentences correctly represented}}{\text{total number of sentences}}$$

(17.2)

- Concept error rate (also known as slot error rate): the percentage of incorrectly identified concepts or slots. This can also be expressed in terms of concept accuracy rate:

$$\%P_f = 100 \times \frac{\text{num. of correct concepts in the hypothesis}}{\text{num. of semantic concepts in the hypothesis}}$$

(17.3)

- Slot precision/recall/F1 score: the precision/recall score for slots, often combined as the F1 score. Precision corresponds to the ratio of correctly identified slots over all slots detected, whereas recall is the ratio of identified slots over all slots that should have been identified. Thus, precision reflects the system's ability to reject incorrect answers while recall measures its ability to find as many correct answers as possible. F1 is a method for measuring the balance between precision and recall since if you try to improve recall you frequently obtain a lower score for precision, and vice versa.

$$\text{Precision} = \frac{\text{num. of reference slots correctly detected by SLU}}{\text{num. of slots detected by SLU}} \quad (17.4)$$

$$\text{Recall} = \frac{\text{num. of reference slots correctly detected by SLU}}{\text{num. of total reference slots}} \quad (17.5)$$

$$F_1 = \frac{2 \times (\text{Precision} \times \text{Recall})}{\text{Precision} + \text{Recall}} \quad (17.6)$$

As in the case of the ASR component, it is necessary to label each semantic concept generated by the SLU component as correct or wrong (incorrect concepts, inserted, deleted, and substituted) in order to compute these metrics.

In a number of evaluations, it has been found that the performance of statistical models is at least competitive with that of handcrafted models. For example, Henderson and Jurčíček (2012) found that the statistical parsers developed in the CLASSiC project (Lemon and Pietquin 2012) outperformed the handcrafted Phoenix semantic parser when parsing ASR output from the TownInfo dataset. Similar results have been reported in many other studies—see chapters in (Tur and de Mori 2011).

17.2.2.3 Dialog Management

A number of different statistical metrics have been proposed for the evaluation of the DM strategy (Scheffler and Young 2001; Schatzmann et al. 2005). These may be divided into three groups:

- High-level features of the dialog: average length of the dialog (number of dialog turns), average number of actions per dialog turn, proportion of user versus system talk, and ratio of user versus system actions.
- Style of the dialog: frequency of different speech acts, ratio of goal-directed actions versus grounding actions versus dialog formalities versus misunderstandings, and user cooperativeness (proportion of slot values provided when requested).
- Success rate and efficiency of the dialog goal achievement rates and goal completion times.

Sometimes, the evaluation is carried out by comparing the dialogs generated by the dialog manager under study with respect to a desired behavior, e.g., human–human conversations in the same application domain, or with respect to other systems or versions of the same system.

In addition to these metrics, some other measures can be used specifically for the purposes of comparison, such as perplexity and distance measures. Another method is to use the Common Answer Specification (CAS) protocol that compares the system's chosen response with a canonical response in a database. This method allows an automatic evaluation once the reference responses have been specified and a labeled dialog corpus is available. It also makes a direct comparison between systems easier. However, the CAS assessment procedure is rather limited as it is done on the sentence level and it does not account for partially correct answers.

Perplexity is a commonly used evaluation metric in the field of statistical language modeling for testing how well a given model predicts the sequence of words in a given test dataset. It is a useful metric for determining whether the dialogs contain similar action sequences (or dialog state sequences). The definition of perplexity (PP) is based on the per-action (per state) entropy H representing the amount of non-redundant information provided by each new action (state) on average.

$$PP = 2^{\widehat{H}} \tag{17.7}$$

The latter can be approximated as follows:

$$H = -\frac{1}{m} \log_2 P(a_1, a_2, \ldots, a_m) \tag{17.8}$$

where $P(a_1, a_2, \ldots, a_m)$ is the probability estimate assigned to the action sequence $a_1, a_2, \ldots, a_m$ by a dialog model (Young 2002).

Another interesting method for measuring similarity between dialogs is presented in Cuayahuitl et al. (2005). The authors propose training HMMs to compute the similarity between two dialog corpora in terms of the distance between the two HMMs. The assumption is that the smaller the distance between the two HMMs, the greater the similarity between the two corpora. To compute the distance between two HMMs, the authors propose the symmetrized Kullback-Leibler divergence, which is defined as:

$$D(P, Q) = \frac{D_{KL}(P||Q) + D_{KL}(Q||P)}{2} \tag{17.9}$$

where D_{KL} is the distance between the probability distributions P and Q.

Williams (2008) proposes the divergence between the distributions of dialog scores between different corpora as a measure of the quality of the dialogs. The normalized Cramer-von Mises divergence is proposed for evaluating and rank-ordering dialogs. The dialog manager maintains a representation of the state of the dialog in a process called dialog state tracking (DST). Numerous techniques have been proposed for dialog state tracking; however, direct comparisons between these methods have not been possible because past studies use different domains and system components for ASR, SLU, and DM.

The Dialog State Tracking Challenge (DSTC) has addressed this problem by providing a corpus of 15 K human–computer dialogs in a standard format, along with a suite of 11 evaluation metrics (Williams et al. 2013). The challenge received a total of 27 entries from 9 research groups. This was a similar challenge to the Spoken Dialog Challenge 2010, which investigated how different spoken dialog systems perform on the same task (Black et al. 2010). The results showed considerable variation both between systems and between the control and live tests. However, even though the systems were quite different in their designs, similar correlations were observed between WER and task completion for all the systems.

17.2.2.4 Natural Language Generation

Evaluation of texts produced by natural language generation (NLG) systems has used the following methodologies (Reiter and Belz 2009):

- Task-based evaluation
- Evaluation by humans
- Automatic evaluation

Task-based evaluation measures the impact of the generated texts on task performance by end users—for example, whether persuasive texts actually have an effect on human behavior by persuading users to stop smoking (Reiter et al. 2003) or by convincing them by means of evaluative arguments that something is desirable or right (Carenini and Moore 2006). While task-based evaluations provide useful feedback, they are expensive to conduct in terms of time and money, and they depend on the goodwill of the subjects performing the tasks.

Evaluations by humans involve human judges rating texts using ordinal scales or comparing different versions of a text. Commonly used metrics include the number of times the user requires the system to repeat the response provided by the system, user response time, the number of times the user does not provide a response, and the number of out of vocabulary words. Judges can also be asked to edit the generated texts and then the generated and edited texts can be compared to see what changes have been suggested. Evaluations by humans are quicker and less costly to conduct than task-based evaluations. We will see how user judgments can be considered for evaluating conversational interfaces in Sect. 17.3.

In automatic evaluation, generated texts are compared to reference texts authored by humans. This is a method used in other areas of natural language processing, such as machine translation. The metrics used include string-edit distance, tree similarity, and the Bilingual Evaluation Understudy (BLEU) metrics used widely in machine translation evaluation. As Reiter and Belz (2009) found in an extensive investigation of the validity of various metrics for the automatic evaluation of NLG systems, automatic evaluation was most reliable and useful for comparing the linguistic quality of generated texts, i.e., the realization stage, but not for higher levels such as content or discourse structure, i.e., the document planning and microplanning stages.

Evaluation of statistical approaches to NLG usually involves comparisons of a baseline system with a system based on reinforcement learning (see Chap. 10), looking at objective measures such as task completion and subjective measures such as user satisfaction. For example, in the study described in Lemon et al. (2010)

the trained information presentation strategy outperformed a baseline system that used conventional hand-coded prompts in terms of task completion rate.

17.2.2.5 Text-to-Speech Synthesis

Evaluation of TTS is usually done by human listeners who make judgments about the quality of the synthesized speech using metrics for intelligibility and naturalness. In recent evaluations, a common speech dataset has been used—for example, in the Blizzard challenge, which has been running annually since 2005. The basic idea behind the Blizzard challenge is to take an agreed speech dataset and build a synthetic voice using the data that can synthesize a prescribed set of test sentences. The performance of each synthesizer is evaluated using listening tests[5]—see also Bennett (2005) and Black and Tokuda (2005).

Jekosch (2005) provides a systematic review of voice and speech quality perception, including a model of speech quality measurements and the issues that must be controlled, such as the sound source, acoustic conditions, and type of speech context. Some studies of the evaluation of synthesized speech put special emphasis on cognitive and perceptual factors in the listener. For example, the study by Delogu et al. (1998) highlights speech perception, memory, and attention.

17.2.3 Metrics Used in Industry

The metrics described in the preceding sections are also used in industry to measure the quality of their voice user interfaces (VUI), to compare them to those of their competitors, and to evaluate the appropriateness of any changes that are made in the voice interface of their own companies. James Larson presents a comprehensive list of criteria for measuring effective VUIs,[6] while Jason Brome presents some informal best practices for VUI design.[7] Suhm (2008) presents an interesting industrial perspective on the design of VUIs based on the limitations of the current technologies (e.g., ASR and SLU errors) and of the users (e.g., limitations of working memory). Also included is a survey of the industry know-how acquired through the evaluation of commercial systems, and a description of how to mine call recordings to obtain a variety of information and to detect problems.

Commercial systems have some particular requirements for evaluation given that a common objective is to reduce costs and ensure user satisfaction. There are a number of objective metrics that can be collected from the data of user–system

---

[5]http://festvox.org/blizzard/. Accessed February 29, 2016.

[6]http://www.speechtechmag.com/Articles/Editorial/Feature/Ten-Criteria-for-Measuring-Effective-Voice-User-Interfaces-29443.aspx. Accessed February 28, 2016.

[7]http://help.voxeo.com/go/help/xml.vxml.bestprac.vui. Accessed February 28, 2016.

calls. *Time-to-task* measures the amount of time that it takes to start engaging in a task after any instructions and other messages provided by the system. The *correct transfer rate* measures whether the customers are correctly redirected to the appropriate human agent, while the *containment rate* measures the percentage of calls not transferred to human agents and that are handled by the system. This metric is useful in determining how successfully the system has been able to reduce the costs of customer care through automation. The converse of the containment rate is the *abandonment rate* that measures the percentage of callers who hang up before completing a task with an automated system.

New applications in industry such as conversational interfaces have brought new challenges for evaluation. For example, personal assistants such as Cortana, Siri, or Google Now are used mainly to perform Web searches, so specific evaluation metrics and methods have been developed to estimate search satisfaction (White 2016). Jiang et al. (2015) present a study with Cortana where they analyze search behavior and its associated satisfaction levels and their relation to search outcomes and search effort.

17.3 Subjective Evaluation

As discussed before, the quality of a conversational interface is a perceptual event that compares whether what the users perceive is what they expected or desired. Thus, even if a system performs well from an "objective" perspective, it may not meet the users' expectations and thus be judged to be of low quality. In order to avoid this, it is important to assess the perception of users about the system after interacting with it. In this way, a holistic evaluation of the system should also consider the impressions of users of different dimensions of quality. Usually, this is done using a single process in which the opinions of users about the system are gathered through questionnaires after the interaction in order to check whether their expectations were matched.

The Subjective Assessment of Speech System Interfaces (SASSI) questionnaire is a widely used test for all types of speech interfaces. SASSI uses Likert scales to obtain quantitative measures of the users' agreement with 34 statements related to six factors: system response accuracy, likeability, cognitive demand, annoyance, habitability, and speed (Hone and Graham 2000). SASSI was adopted by the speech community as a basis for subjective evaluation and has been extended and now recommended for evaluating telephone-based speech interfaces by the International Telecommunication Union (ITU-T) in its Recommendation P.851.[8] This recommendation describes methods for conducting subjective evaluations of telephone-based spoken dialog systems and establishes quality dimensions that can also be employed to evaluate other types of conversational interface. The

---

[8]https://www.itu.int/rec/T-REC-P.851-200311-I/en. Accessed February 28, 2016.

recommendation is comprehensively documented in terms of the different questionnaires that may be employed, covering the following:

- The user's background: personal information (e.g., age and gender), task-related information (e.g., how often a user uses the system to perform a task and their motivation), and system-related information (e.g., previous experience with speech interfaces).
- Individual interaction: information provided from the system (e.g., availability and consistency), speech input/output (e.g., perceived system understanding and perceived intelligibility), the system's interaction behavior (e.g., flexibility and congruence with expectation), the perceived system personality (e.g., friendliness and politeness), the impression on the user (e.g., pleasantness and cognitive demand), and perceived task fulfillment (e.g., task success and reliability).
- The user's overall impression of the system (e.g., overall impression, perceived usability, and expected future use).

Another frequently used questionnaire is AttrakDiff (Hassenzahl 2001). Although it is not designed specifically for conversational interfaces and has been used in many areas (e.g., Web site evaluation), it can be employed to assess the perceived quality of speech interfaces according to its subscales for system attractiveness, pragmatic quality, hedonic quality-stimulation, and hedonic quality-identity.

Various subjective metrics are used in industry to measure the quality of a user's experience with a VUI, including caller satisfaction (see Sect. 17.3.1), ease of use, quality of audio output, and perceived first-call resolution rate. Perceived first-call resolution rate measures the extent to which a caller is able to achieve their goals on the first call as opposed to having to call back on more than one occasion. More generally, since an interaction with an automated system is often the first contact that customers have with a company, the VUI should be consistent in its behaviors and maintain the value of the company's brand.

17.3.1 Predicting User Satisfaction

The questionnaire approach is very useful for obtaining information from users. However, it is a time consuming and costly process and it may be difficult to get real system users to answer a questionnaire after interacting with the system. Also asking real users to leave a judgment after interaction can bias the evaluation toward the experience of users with a particular profile (those who are willing to respond to a satisfaction questionnaire). This is why subjective evaluation is usually performed with recruited subjects, but the results obtained with them may not always be translatable to real settings with the final users (in Sect. 17.4, we will discuss in more detail the differences between laboratory and field studies and between recruited users and end users).

One solution is to complement real field studies with automatic methods that create predictions of user satisfaction based on factors related to the system's performance, characteristics of the user, and the environmental and contextual conditions in which the interaction takes place. Figure 17.1 shows a summary of some factors that can be considered in relation to user satisfaction. Using this scheme, it is possible to automatically assess whether the user is satisfied with the interaction. In order to learn such models, different approaches can be employed. In the following subsections, we will describe some of the approaches that are used widely in the conversational interfaces community.

17.3.1.1 Prediction of User Satisfaction Using PARADISE

The PARADISE framework mentioned earlier has also been used to develop models of user satisfaction prediction from dialog data based on the weighted linear combination of different measures (Walker et al. 2000). The goal of this evaluation method was to maximize user satisfaction by maximizing task success and minimizing interaction costs (see, Fig. 17.2).

Costs are quantified using different measures of efficiency and quality. The weights of each measure are computed via a multivariable linear regression in which user satisfaction is considered as the dependent variable, and task success, efficiency, and quality measures are considered as independent variables. User satisfaction is predicted by means of the cost of task success and several other costs associated with the interaction:

$$\text{User satisfaction} = (aN(\text{task success})) - \sum_{i=1}^{N} w_i N(\text{costs of the dialog}) \quad (17.10)$$

where the distributions of the measures related to the success of the task and the costs of the dialog are normalized to a normal distribution with mean = 0 and variance = 1.

The measures of the success of the task most commonly used are the Kappa Factor (K) and the task completion rate. The Kappa Factor was proposed in the initial formulation of the PARADISE model. This factor is calculated from a confusion matrix showing the values of the attributes exchanged between the user and the system, so that the main diagonal of the matrix indicates the cases in which the system correctly recognized and understood the information provided by the user. The following expression is used:

$$K = \frac{P(A) - P(E)}{1 - P(E)} \quad (17.11)$$

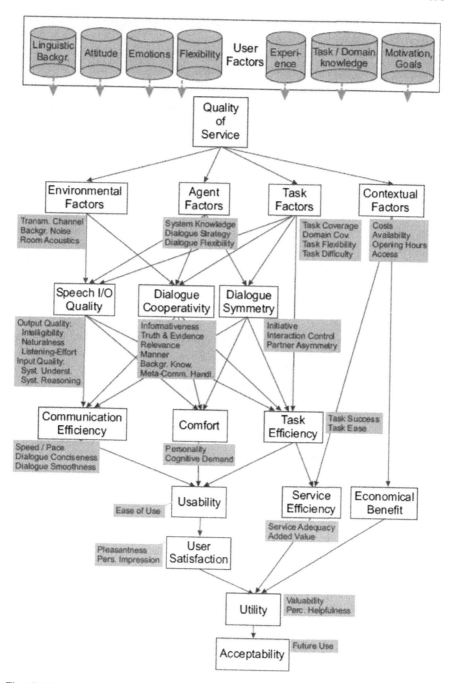

Fig. 17.1 Factors that intervene in models of user satisfaction (from Möller 2005, Fig. 2.9, reproduced with the permission from Springer)

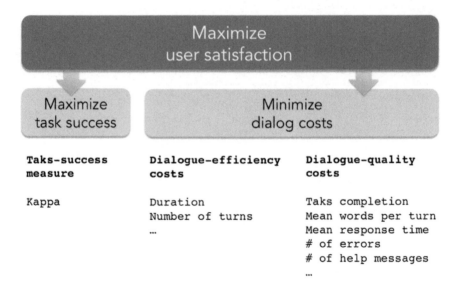

Fig. 17.2 The PARADISE framework

where $P(A)$ is the probability of the system correctly recognizing and understanding the information provided by the user, and $P(E)$ estimates the number of times the system could provide a correct response by chance, calculated by:

$$P(E) = \sum_{i=1}^{n} \left(\frac{t_i}{T}\right)^2 \qquad (17.12)$$

where t_i is the sum of attempts in the column i and T is the total sum of attempts.

The task completion rate is the percentage of times that the system successfully fulfills the users' requests. The dialogs have to be manually transcribed and labeled in order to obtain this measure.

The most important efficiency measures are as follows:

- The average time required to complete a task.
- The average time per turn.
- The average number of turns per task.
- The minimum number of turns or time required to complete a task.
- The types of confirmations strategies that are used.
- The number of words correctly recognized per turn.

These measures can be calculated considering all the dialogs or only the successful ones.

The most important measures for evaluating the quality of the system are as follows:

- The recognition rate (percentage of correctly recognized words).
- The rate of correct semantic concepts (percentage of semantic concepts correctly generated by the SLU module).
- The percentage of errors successfully corrected (efficiency of the techniques used for error detection and correction).
- The response time of the system (the time used by the system to recognize and understand the words spoken by the user).
- The response time of the user (the time spent by the user to provide a response).
- The number of times the user does not provide an answer.
- The number of times the user requests repetition.
- The number of times the user requests help.
- The number of times the user interrupts the system.

The generality of PARADISE as an assessment framework for conversational interfaces relies on the fact that it decouples the requirements of the task and the behavior of the interface, comparing dialog strategies, measuring the quality of complete dialogs and subdialogs, specifying the specific contributions to the overall performance, and comparing interfaces by normalizing the complexity of the task. The main drawbacks of the model include the excessive coupling between user satisfaction and usability, the complexity of predicting user satisfaction from the information recorded in the log files of the system, the difficulty in interpreting the questionnaires, and usage limited to controlled experiments (and not usually with real users). The PARADISE framework has also been enhanced to enable the evaluation of multimodal dialog systems. For example, it was used in the SmartKom Project to create the PROMISE framework (Beringer et al. 2002).

17.3.1.2 Other Models for Predicting User Satisfaction

Just as the availability of corpora and new statistical learning models have fostered the use of statistical approaches for the development of the different components of the conversational interface, so also statistical models have become common for the purposes of evaluation.

The main idea behind these models is to build a corpus of user–system interactions that incorporates the values of the interaction parameters as well as user responses to opinion questionnaires. Then, a machine-learning approach is used to learn the relation between the different parameters with the aim of obtaining a model that can predict system quality from a set of metrics that can be calculated automatically during the operation of the system. For example, Möller et al. (2008) presented prediction models based on two databases corresponding to the BoRIS and INSPIRE systems and used interaction metrics to estimate judgments related to perceived quality and usability. In order to do so, they used different models including linear regression, decision trees, and neural networks. Similarly, Yang et al. (2012) used collaborative filtering to predict user evaluations of unrated dialogs assuming that they will be similar to the ratings received by similar dialogs.

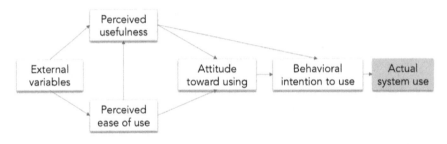

Fig. 17.3 The technology acceptance model

Models of prediction have also provided a basis for newly derived metrics such as interaction quality (IQ) that delivers complementary information with respect to user satisfaction (Schmitt and Ultes 2015).

In addition to the prediction of user satisfaction, other general purpose models have been used to measure acceptance and to predict system usage—for example, the technology acceptance model (TAM) that predicts, among other factors, system usage in terms of perceived usefulness and perceived ease of use (Davis et al. 1989) (see, Fig. 17.3).

17.4 Evaluation Procedures

Apart from the metrics employed, there are other factors that are relevant to the evaluation process, such as differences between laboratory and field or real-world settings, the different subjects that can participate in the evaluation (recruited vs. simulated users), and whether the evaluated system is a complete or an incomplete prototype.

17.4.1 Evaluation Settings: Laboratory Versus Field

In laboratory interactions, subjects are asked to interact with the system in accordance with predefined scenarios. Scenarios describe a particular task that the user has to achieve by interacting with the system—for example, book a certain flight. Usually, these tasks are categorized in terms of the different topics or functionality that the system can deal with, and within each category in terms of the final objective of the dialog. Thus, they allow maximum control over the task carried out by the test subjects while at the same time covering a wide range of possible situations (and possible problems) in the interaction.

Compared to field evaluations, laboratory interactions are cheaper, easier to achieve, and permit control of all the variables, thus ensuring that the effect of a certain change in a system is due to that particular improvement and no other uncontrolled variable. Scenarios make it easier to compare results obtained in different dialogs and to control the influence of the task on the experiments.

In some cases, more relaxed scenarios can be used in which the objective is not so accurately defined but the domain of the conversation is restricted. This is the case with companion agents, where sometimes the conversation does not match a particular objective but it still relates to a certain topic.

A problem with laboratory settings is that the scenarios may differ from the tasks that a user would have selected in a non-predefined interaction. In contrast, field evaluation is based on real users interacting with the final system spontaneously. Field tests involve evaluations of real system–user interactions in which the user employs the system freely without following predefined scenarios created by the evaluators. This is the best way to obtain a valid judgment on some system characteristics such as its helpfulness or difficulties in achieving a certain goal, since with predefined scenarios there is no real motivation and the user's opinion may not reflect what would happen in a real-life situation.

Field evaluations are not replicable as the interaction context is highly variable. This can also be their main advantage as they gather results from different users (differences in gender, voice, knowledge, experience of using the system), talking on different devices, and in different environments. As the results obtained from field tests are robust to this heterogeneity, they are more relevant for predicting the real behavior of systems.

To study the implications of using field tests, some authors have focused on non-restricted evaluation studies. This is the case with the Let's Go system that was evaluated using interactions of real users who phoned the system to get information about bus schedules (Raux et al. 2005). The evaluation involved reporting the results of interaction parameters. The corpus acquired was open for research purposes and has been used and augmented by many research groups. Schmitt et al. (2012) present a parameterization that includes new metrics, including emotion.

17.4.2 Wizard of Oz

In prototyping development life cycles, the different stages of system development (especially design, implementation, and testing) are performed iteratively, generating prototypes that are increasingly more complex. This implies that the iteration of some components may not be complete or even implemented at all but it is still interesting to perform an evaluation of the prototype. This is solved using the Wizard of Oz (WOZ) technique.

In WOZ, users believe that they are interacting with a real system when in reality a human (the wizard) simulates all or a part of it. In order for this technique to be effective, it must be very carefully planned as the wizard cannot improvise but has to behave just like the system would behave. In order to do that, the prospective system behavior that is specified in the current design must be clear and the wizard must be adequately trained to execute it. To make users believe that they are interacting with a system, the wizard needs to be able to select the system's response in real time so that it is synthesized appropriately for the user.

Despite its complexity, WOZ is a good method for performing evaluations of the system-in-the-loop, that is, early tests prior to implementation, thus saving the cost of having to revise an already-implemented system if it turns out to be unacceptable due to issues of performance or usability (Bernsen et al. 1998).

However, it is important to pay attention to possible misleading effects resulting from the superior capabilities of a human wizard compared with a conversational agent. For example, the design of the dialog manager may lead the system to behave in a way that the wizard knows is erroneous. It is important that the wizard adheres to the design rather than making use of their own knowledge and capabilities. This is not easy to do, which is why WOZ is usually automated as much as possible so as to constrain the behavior of the wizard.

In some settings, for example, when the human wizard replaces the ASR or SLU components, different error-generation approaches have been applied to introduce errors at a rate that simulates the performance of the components as it would be when simulated by the wizard.

17.4.3   Test Subjects

When discussing the difference between laboratory and field conditions, we were also implicitly distinguishing between recruited and real users. Recruited users are subjects who are asked to participate in the evaluation (usually for some incentive). The users recruited should be representative of the target population that represents the final users of the system and they must be sufficiently numerous to provide evaluation results that are statistically significant.

Users can also be recruited to gather evaluation corpora from which to compute models, e.g., satisfaction prediction models. In both cases, achieving a high number of users poses a challenge for developers, especially in academia where it is difficult to recruit some population groups. We will focus on two ways to mitigate this problem: crowdsourcing and user simulation.

17.4.3.1   Crowdsourcing

Crowdsourcing is a rather novel approach that involves recruiting users through the Internet by means of a service created for that effect. Traditionally, academics have used students as test subjects, but student populations may not be representative of the final user groups in all contexts. Collecting user judgments with crowdsourcing makes it possible to have a large number of dialogs evaluated by a large population in a short period of time. This allows the results to be more representative and the process to be more cost effective.

Crowdsourcing services act as intermediaries between a body of workers and the creators of tasks (developers). Developers create well-defined tasks and define a profile of the workers that may be assigned to it (e.g., age, gender, education, language, and experience). Workers are paid for their responses to the assignments and developers can approve or reject their results. The most widespread crowd-sourcing platform within the speech community is Amazon Mechanical Turk,[9] but some other platforms are also being employed, such as CrowdFlower,[10] microWorkers,[11] and Crowdee,[12] a spin-off of the Department of Quality and Usability at the Technical University of Berlin.

Eskenazi et al. (2013) present a detailed study of crowdsourcing for speech processing, describing the benefits of this recruitment procedure and also the issues that developers should consider. For example, there may be bots, malicious indi-viduals, and well-intentioned individuals, such as non-native speakers, who are not well enough qualified or suitable for the task. As a partial solution, in some plat-forms it is possible to check the ratio of responses that were accepted for a certain worker and establish a threshold for accepting workers for your evaluations.

Different studies show that evaluation results for conversational interfaces using crowdsourcing are reliable. For example, Jurčíček et al. (2011) compared the results of Mechanical Turk with results obtained from evaluators recruited locally in their university (Cambridge, UK) and found that the results were consistent and indistinguishable.

As crowdsourcing has become popular, new authors are providing interesting insights on how to avoid cheaters in evaluations, how to perform evaluations that require spoken interaction with a system within the crowdsourcing platform, and how to control the quality of the results. For example, at the INTERSPECH 2015 conference there was a special session on "Advanced Crowdsourcing for Speech and Beyond" where experts discussed these topics.[13]

17.4.3.2 User Simulation

Systems can also be evaluated automatically, that is, by creating software that interacts with the conversational interface as if it were a real user. The idea is to learn possible user responses to the system's interventions from a corpus of real user data (e.g., from previous versions of the system or from human–human con-versations in the same domain), or to generate rules that produce user responses for each system intervention. These rules can be as detailed as desired, from learning

---

[9]https://www.mturk.com/mturk/welcome. Accessed on February 28, 2016.

[10]http://www.crowdflower.com/. Accessed on February 28, 2016.

[11]https://microworkers.com/. Accessed on February 28, 2016.

[12]https://www.crowdee.de/en/. Accessed on February 28, 2016.

[13]The papers presented at the special session are accessible here: http://www.isca-speech.org/archive/interspeech_2015/#Sess074. Accessed February 29, 2016.

general user models to simulating users that behave differently according to user profiles. In this way, user simulators make it possible to generate a large number of dialogs in a very simple way, reducing the time and effort needed for the evaluation of a conversational interface each time it is modified.

Eckert et al. (1997) identified the following advantages of user simulation:

- It allows the automatic evaluation of a large number of dialogs without expensive manual investigation.
- As there is less manual work, the results are less error-prone.
- The characteristics of different user populations can be easily modeled.
- The same user model can be used to perform a comparative evaluation of other competing systems.
- The simulations can also be used for other purposes such as optimizing dialog strategies.

It is possible to classify different approaches with regard to the level of abstraction at which they model the conversation between the system and the simulated user, which is related to whether the integrated system or only some of its components are to be evaluated. For example, to evaluate the dialog manager, user simulators are frequently built at the level of intentions by generating dialog acts rather than actual spoken text (see Chap. 10). This approach is very popular and has been used with different learning techniques, including reinforcement learning (Schatzmann et al. 2006), n-gram models, and other statistical approaches (Griol et al. 2014).

Modeling interaction at the intention level avoids the need to reproduce the enormous variety of speech signals and word sequences that can be produced during interactions with the system. However, for a holistic system evaluation it is necessary to test the performance of components such as ASR and SLU. For example, López-Cózar et al. (2007) employ a corpus of possible user utterances for each semantic representation that is input to the speech recognizer, thus evaluating the whole interaction cycle. These approaches can be complemented by channel simulation techniques to generate noise and realistic ASR errors (Jung et al. 2009).

17.5 Summary

Evaluating conversational interfaces encompasses a variety of methodologies and tools. Evaluation examines not only whether systems operate correctly and are usable but also whether they can engage in a believable manner in social communication. Evaluation can take place during the development of a system as well as when the system is due for public release.

Criteria that are used for evaluation can be either objective metrics that are computed from logs of the interactions of users with the system, or subjective metrics based on the opinions of users about the system. Different metrics are used for the different components of a system.

In conducting evaluation, it is important to consider differences between laboratory and real-world settings as well as whether the test subjects have been recruited specially for the evaluation or are real users of the system. Laboratory-based evaluations with recruited users allow the evaluation to be more tightly controlled but have the drawback that they might not reflect usage of the system by real users. However, evaluations in the field with real users can produce highly variable results, as it is not easy to control the interactions. For these reasons, other methods such as simulated users and crowdsourcing have become more prevalent in recent years.

Further Reading

There are many other aspects that are involved in system evaluation that it is not possible to cover here, including:

- *Multimodal conversation.* Kühnel (2012) presents a systematic review of how the quality aspects of multimodal interactive systems can be quantified, including multimodal behavior as well as output and input aspects.
- *System personality and social interaction.* Callejas et al. (2014) present an overview of models and approaches for the evaluation of personality based mainly on assessing whether the personality perceived by users matches the personality that the development team intended to reproduce, also covering the effect of the users' own personality on their perceptions.
- *Human–robot interaction evaluation.* Sim and Loo (2015) provide a discussion of the strengths and weaknesses of the main assessment and evaluation methodologies that includes a very wide range of models covering aspects such as friendship, personality matching, empathy, acceptance, and task performance.
- *Embodied conversational agents (ECAs).* The evaluation of ECAs includes, among other things, metrics related to appearance and behavioral believability, social behaviors, domain knowledge, agency, responsiveness, reliability, and final visuals and interaction mechanisms (Ruttkay and Pelachaud 2004).

References

Bennett C (2005) Large scale evaluation of corpus-based synthesizers: results and lessons from the Blizzard challenge 2005. In: Proceedings of the 9th European conference on speech communication and technology (Interspeech'2005—Eurospeech), Lisbon, Portugal, 4–8 Sept 2005, pp 105–108. http://www.isca-speech.org/archive/interspeech_2005/i05_0105.html

Beringer N, Kartal U, Louka K, Schiel F, Türk U (2002) PROMISE: a procedure for multimodal interactive system evaluation. In: Proceedings of the LREC workshop on multimodal resources and multimodal systems evaluation, Las Palmas, Spain, 1 June 2002, pp 77–80. http://www.lrec-conf.org/proceedings/lrec2002/pdf/50.pdf

Bernsen NO, Dybkjær H, Dybkjær L (1998) Designing interactive speech systems: from first ideas to user testing. Springer, London. doi:10.1007/978-1-4471-0897-9

Black A, Tokuda K (2005) The Blizzard challenge—2005: evaluating corpus-based speech synthesis on common datasets. In: Proceedings of Interspeech'2005—Eurospeech, 9th european conference on speech communication and technology, Lisbon, Portugal, 4–8 Sept 2005, pp 77–80. http://www.isca-speech.org/archive/interspeech_2005/i05_0077.html

Black A, Burger S, Langner B, Parent G, Eskenazi M (2010) Spoken dialog challenge 2010. In: Proceedings of IEEE spoken language technology workshop (SLT), Berkeley, California USA, 12–15 Dec 2010, pp 448-453. doi:10.1109/SLT.2010.5700894

Callejas Z, Griol D, López-Cózar R (2014) A framework for the assessment of synthetic personalities according to user perception. Int J Hum-Comput Stud 72:567–583. doi:10.1016/j.ijhcs.2014.02.002

Carenini G, Moore JD (2006) Generating and evaluating evaluative arguments. Artif Intell 170:925–952. doi:10.1016/j.artint.2006.05.003

Cuayáhuitl H, Renals S, Lemon O, Shimodaira H (2005) Human-computer dialogue simulation using Hidden Markov models. In: Proceedings of the IEEE automatic speech recognition and understanding workshop (ASRU'05), San Juan, Puerto Rico, 27 Nov–1 Dec 2005, pp 290–295. doi:10.1109/ASRU.2005.1566485

Davis FD, Bagozzi RP, Warshaw PR (1989) User acceptance of computer technology: a comparison of two theoretical models. Manage Sci 35:982–1003. doi:10.1287/mnsc.35.8.982

Delogu C, Conte S, Sementina C (1998) Cognitive factors in the evaluation of synthetic speech. Speech Commun 24:153–168. doi:10.1016/S0167-6393(98)00009-0

Dybkjaer L, Bernsen NO (2000) Usability issues in spoken language dialogue systems. Nat Lang Eng 6(3–4):243–271. doi:10.1017/s1351324900002461

Dybkjaer L, Bernsen NO, Carlson R, Chase L, Dahlbäck N, Failenschmid K, Heid U, Heisterkamp P, Jönsson A, Kamp H, Karlsson I, Kuppevelt J, Lamel L, Paroubek P, Williams D (1998) The DISC approach to spoken language systems development and evaluation. In: Proceedings of the first international conference on language resources and evaluation, Granada, Spain, 28–30 May 1998, pp 185–189

Dybkjaer L, Bernsen NO, Minker W (2004) Evaluation and usability of multimodal spoken language dialogue systems. Speech Commun 43(1–2):33–54. doi:10.1016/j.specom.2004.02.001

Eckert W, Levin E, Pieraccini R (1997) User modeling for spoken dialogue system evaluation. In: IEEE workshop on automatic speech recognition and understanding, Santa Barbara, CA, 14–17 Dec 1997, pp 80–87

Engelbrecht K-P (2012) Estimating spoken dialog system quality with user models. Springer Science & Business Media, Berlin. doi:10.1007/978-3-642-31591-6

Eskenazi M, Levow G-A, Meng H, Parent G, Suendermann D (eds) (2013) Crowdsourcing for speech processing: applications to data collection, transcription, and assessment. Wiley, Chichester. doi:10.1002/9781118541241

Fiscus JG, Ajot J, Garofolo JS (2008) The rich transcription 2007 meeting recognition evaluation. In: Stiefelhagen R, Bowers R, Fiscus J (eds) Multimodal technologies for perception of humans. Springer, Berlin, pp 373–389. doi:10.1007/978-3-540-68585-2_36

Gibbon D, Moore R, Winski R (1997) Handbook of standards and resources for spoken language systems. Walter de Gruyter, Berlin

Griol D, Callejas Z, López-Cózar R, Riccardi G (2014) A domain-independent statistical methodology for dialog management in spoken dialog systems. Comput Speech Lang 28:743–768. doi:10.1016/j.csl.2013.09.002

Hassenzahl M (2001) The effect of perceived hedonic quality on product appealingness. Int J Hum-Comput Interact 13:481–499. doi:10.1207/S15327590IJHC1304_07

Henderson J, Jurčíček F (2012) Data-driven methods for spoken language understanding. In: Lemon O, Pietquin O (eds) Data-driven methods for adaptive spoken dialogue systems: computational learning for conversational interfaces. Springer, New York, pp 19–38. doi:10.1007/978-1-4614-4803-7_3

Hone KS, Graham R (2000) Towards a tool for the subjective assessment of speech system interfaces (SASSI). Nat Lang Eng 6:287–303. doi:10.1017/S1351324900002497

Jekosch U (2005) Voice and speech quality perception: assessment and evaluation. Springer, Berlin. doi:10.1007/3-540-28860-0

Jiang J, Awadallah AH, Jones R, Ozertem U, Zitouni I, Kulkarni RG, Khan OZ (2015) Automatic online evaluation of intelligent assistants. In: Proceedings of the 23rd international conference on World Wide Web (WWW '15), Florence, Italy, 18–22 May 2015, pp 506–516. http://www.www2015.it/documents/proceedings/proceedings/p506.pdf

Jung S, Lee C, Kim K, Jeong M, Lee GG (2009) Data-driven user simulation for automated evaluation of spoken dialog systems. Comput Speech Lang 23(4):479–509. doi:10.1016/j.csl.2009.03.002

Jurčíček F, Keizer S, Gašić M, Mairesse F, Thomson B, Yu K, Young S (2011) Real user evaluation of spoken dialogue systems using Amazon Mechanical Turk. In: Proceedings of the 12th annual conference of the international speech communication association (Interspeech 2011), Florence, Italy, 27–31 Aug 2011, pp 3061–3064. http://www.isca-speech.org/archive/interspeech_2011/i11_3061.html

Kühnel C (2012) Quantifying quality aspects of multimodal interactive systems. Springer, Berlin. doi:10.1007/978-3-642-29602-4

Lemon O, Pietquin O (eds) (2012) Data-driven methods for adaptive spoken dialog systems: computational learning for conversational interfaces. Springer, New York. doi:10.1007/978-1-4614-4803-7

Lemon O, Janarthanam S, Rieser V (2010) Statistical approaches to adaptive natural language generation. In: Lemon O, Pietquin O (eds) Data-driven methods for adaptive spoken dialogue systems: computational learning for conversational interfaces. Springer, New York. doi:10.1007/978-1-4614-4803-7_6

López Cózar R, Araki M (2005) Spoken, multilingual and multimodal dialog systems: development and assessment. Wiley, Chichester. doi:10.1002/0470021578

López-Cózar R, Callejas Z, McTear M (2007) Testing the performance of spoken dialogue systems by means of an artificially simulated user. Artif Intell Rev 26:291–323. doi:10.1007/s10462-007-9059-9

Möller S (2005) Quality of telephone-based spoken dialogue systems. Springer Sciennce + Business Media, Heidelberg. doi:10.1007/b100796

Möller S, Smeele P, Boland H, Krebber J (2007) Evaluating spoken dialogue systems according to de-facto standards: a case study. Comput Speech Lang 21(1):26–53. doi:10.1016/j.csl.2005.11.003

Möller S, Engelbrecht K-P, Schleicher R (2008) Predicting the quality and usability of spoken dialogue services. Speech Commun 50:730–744. doi:10.1016/j.specom.2008.03.001

Raux A, Langner B, Black A, Eskenazi M (2005) Let's go public! Taking a spoken dialog system to the real world. In: Proceedings of the 9th European conference on speech communication and technology (Interspeech'2005—Eurospeech), Lisbon, Portugal, 4–8 September 2005, pp 885–888. http://www.isca-speech.org/archive/interspeech_2005/i05_0885.html

Reiter E, Belz A (2009) An investigation into the validity of some metrics for automatically evaluating natural language generation systems. Comput Linguist 35:529–558. doi:10.1162/coli.2009.35.4.35405

Reiter E, Robertson R, Osman LM (2003) Lessons from a failure: generating tailored smoking cessation letters. Artif Intell 144:41–58. doi:10.1016/S0004-3702(02)00370-3

Ruttkay Z, Pelachaud C (eds) (2004) From brows to trust. Evaluating embodied conversational agents. Springer, Netherlands. doi:10.1007/1-4020-2730-3

Schatzmann J, Georgila K, Young S (2005) Quantitative evaluation of user simulation techniques for spoken dialogue systems. In: Proceedings of the 6th SIGdial workshop on discourse and dialogue, Lisbon, Portugal, 2–3 Sept 2005, pp 45–54. http://www.isca-speech.org/archive_open/sigdial6/sgd6_045.html

Schatzmann J, Weilhammer K, Stuttle M, Young S (2006) A survey of statistical user simulation techniques for reinforcement-learning of dialogue management strategies. Knowl Eng Rev 21:97. doi:10.1017/S0269888906000944

Scheffler K, Young S (2001) Automatic learning of dialogue strategy using dialogue simulation and reinforcement learning. In: Proceedings of 49th annual meeting of the association for computational linguistics: human language technologies (HLT), Portland, Oregon USA, 19–24 June 2011, pp 12–18. http://dl.acm.org/citation.cfm?id=1289246

Schmitt A, Ultes S (2015) Interaction quality: assessing the quality of ongoing spoken dialog interaction by experts—and how it relates to user satisfaction. Speech Commun 74:12–36. doi:10.1016/j.specom.2015.06.003

Schmitt A, Ultes S, Minker W (2012) A parameterized and annotated spoken dialog corpus of the CMU let's go bus information system. In: Proceedings of the eight international conference on language resources and evaluation (LREC'12). Istanbul, Turkey. http://www.lrec-conf.org/proceedings/lrec2012/summaries/333.html

Sim DYY, Loo CK (2015) Extensive assessment and evaluation methodologies on assistive social robots for modelling human–robot interaction—a review. Inf Sci 301:305–344. doi:10.1016/j.ins.2014.12.017

Suhm B (2008) IVR Usability engineering using guidelines and analyses of end-to-end calls. In: Human factors and voice interactive systems. Springer US, Boston, MA, pp 1–41. doi:10.1007/978-0-387-68439-0_1

Tur G, de Mori R (eds) (2011) Spoken language understanding: systems for extracting semantic information from speech. Wiley, Chichester, UK. doi:10.1002/9781119992691

Walker MA, Litman DJ, Kamm CA, Abella, A (1997) PARADISE: a framework for evaluating spoken dialogue agents. In: Proceedings of the 8th conference on European chapter of the association for computational linguistics (EACL), Madrid, Spain, 7–12 July 2005, pp 271–280. https://aclweb.org/anthology/P/P97/P97-1035.pdf

Walker MA, Litman DJ, Kamm CA, Abella A (1998) Evaluating spoken dialogue agents with PARADISE: two case studies. Comput Speech Lang 12(4):317–347. doi:10.1006/csla.1998.0110

Walker M, Kamm CA, Litman DJ (2000) Towards developing general models of usability with PARADISE. Nat Lang Eng 6(3–4):363–377. doi:10.1017/s1351324900002503

Walker MA, Rudnicky A, Prasad R, Aberdeen J, Bratt EO, Garofolo J, Hastie H, Le A, Pellom B, Potamianos A, Passonneau R, Roukos S, Sanders G, Seneff S, Stallard D (2002) DARPA Communicator: cross-system results for the 2001 evaluation. In: Proceedings of the 7th international conference on spoken language processing (ICSLP2002), vol 1, Denver, Colorado, pp 273–276. http://www.isca-speech.org/archive/archive_papers/icslp_2002/i02_0269.pdf. Accessed 21 Jan 2016

White RW (2016) Interactions with search systems. Cambridge University Press, Cambridge

Williams JD (2008) Evaluating user simulations with the Cramér-von Mises divergence. Speech Commun 50(10):829–846. doi:10.1016/j.specom.2008.05.007

Williams JD, Raux A, Ramachandran D, Black A (2013) The dialog state tracking challenge. In: Proceedings of the 4th annual SIGdial meeting on discourse and dialogue (SIGDIAL), Metz, France, 22–24 Aug 2013, pp 404–413. http://www.aclweb.org/anthology/W13-4065

Yang Z, Levow G-A, Meng H (2012) Predicting user satisfaction in spoken dialog system evaluation with collaborative filtering. IEEE J Sel Top Signal Process 6:971–981. doi:10.1109/JSTSP.2012.2229965

Young S (2002) The statistical approach to the design of spoken dialogue systems. Tech Report CUED/F-INFENG/TR.433. Cambridge University Engineering Department. http://mi.eng.cam.ac.uk/~sjy/papers/youn02b.ps.gz

Chapter 18
Future Directions

Abstract As a result of advances in technology, particularly in areas such as cognitive computing and deep learning, the conversational interface is becoming a reality. Given the vast number of devices that will be connected in the so-called Internet of Things, a uniform interface will be necessary both for users and for developers. We describe current developments in technology and review a number of application areas that will benefit from conversational interfaces, including smart environments, health care, care of the elderly, and conversational toys and educational assistants for children. We also discuss the need for developers of conversational interfaces to focus on bridging the digital divide for under-resourced languages.

18.1 Introduction

In the course of this book, we have shown how conversational interfaces that allow humans to interact with machines using natural spoken language have long been a dream but are now beginning to become a reality. In this chapter, we describe future prospects for the conversational interface. We look at two aspects: advances in technology and applications that use conversational interfaces.

18.2 Advances in Technology

While speech recognition was long seen as a major stumbling block due to high error rates, more recently it has become apparent that language understanding and conversational ability are also important aspects of an intelligent conversational interface. In this section, we outline some recently emerging technologies and indicate areas for future development.

© Springer International Publishing Switzerland 2016
M. McTear et al., *The Conversational Interface*,
DOI 10.1007/978-3-319-32967-3_18

18.2.1   Cognitive Computing

Although our focus in this book has been on what is required to design and implement a conversational interface that would enable humans to communicate naturally with virtual personal assistants and other smart devices, ultimately there also has to be an intelligence behind the conversational interface that acts on the interpretation of the human's queries and commands and provides appropriate responses. As Sara Basson,[1] formerly of IBM Research and now at Google, puts it:

> The key to these systems is not their ability to communicate through speech and language, but that these systems will be intelligent, personalized, and constantly learning—about the environment, about the users, and their preferences and expectations. Speech and language technologies will not drive the era of cognitive systems, but cognitive systems may well drive the era of speech and language interaction. Perhaps it wasn't enough for speech technology to get smarter and better. The underlying systems needed to be smarter and better, in order to take adequate advantage of speech interfaces.

The technology being referred to here has been called *cognitive computing*. In cognitive computing, several technologies are combined, including natural language processing, information retrieval, machine learning, and reasoning. The first and best-known demonstration of cognitive computing was IBM's Watson. Watson is a computer system that was developed initially to compete with human champions in the American TV quiz show Jeopardy. In order to compete effectively, Watson had to scan through 200 million pages of structured and unstructured documents to find potential answers, evaluate and rank the answers to come up with a single answer with a sufficient degree of confidence, and beat the human competitors to the buzzer within a time of 3 s. In 2011, Watson competed in Jeopardy against two former winners and received the first prize.

Since the success in Jeopardy, Watson is now being used to provide intelligent question answering capabilities in specialist domains such as health care and financial services. With cognitive computing, it is predicted that computer systems will be able to learn by interacting with data, to reason about the information extracted from the data, and to interact intelligently with humans and extend human capabilities—in other words, to contribute to the ultimate virtual personal assistant.

18.2.2   Deep Learning

Deep learning involves extracting patterns from data and classifying them by learning multiple layers of representation and abstraction. One of the attractions of deep learning is that the models and algorithms that are used in one application can be applied to a diverse range of other applications. For example, applications

---

[1]http://www.speechtechmag.com/Articles/Column/The-View-from-AVIOS/Building-Smarter-Systems-with-Cognitive-Computing-94590.aspx. Accessed February 21, 2016.

involving computer vision, speech recognition, natural language understanding, audio processing, information retrieval, and robotics can all make use of the same models and algorithms, whereas with previous approaches, problem-specific methods would have been used for the different applications (LeCun et al. 2015). The success of deep learning can be attributed to the availability of vast amounts of data, more powerful processors to process this data, and new models for learning, in particular, a fast learning algorithm developed by Hinton and colleagues for learning deep belief networks (Hinton et al. 2006).

With respect to conversational interfaces, deep learning has brought about dramatic improvements in speech recognition, language understanding, and question answering. The next area for development is conversational interaction. Currently, most interaction with virtual personal assistants is not conversational. Instead, this interaction consists mainly of one-shot queries to which the assistant provides a response. There is generally no way to engage in dialog, except in a fairly limited sense as in Google's Voice Interactions, where the user can ask follow-up questions.[2] Moreover, the conversational capabilities of most current virtual personal assistants are rule-based, meaning that dialog designers predict and design a range of user inputs and system responses in the form of rules.

An alternative to this approach is to have the system learn from the interactions in which it engages. This is the approach being adopted by Steve Young, Professor of Information Engineering at the University of Cambridge. Young proposes that conversational systems should be able to learn online from their interactions with users, learn from their mistakes, and gradually become more intelligent as a result of participating in more interactions. In the approach being developed by Young and colleagues in research projects in the Dialogue Systems Group and in the spin-off company VocalIQ (VIQ), deep learning is combined with reinforcement learning (see Chap. 10). Young draws the analogy of a child gradually learning to become more conversationally competent by engaging in and learning from conversations. As Young[3] put it:

> VIQ is learning across whole dialogs. What the system is trying to do is get a reward from the user. The system's reward is to satisfy the user's need. It might take a long conversation before the user gets what they want, but as long as the system ends up with a positive reward for that interaction, it propagates the reward back amongst everything it's done over the dialog.

Apple acquired VocalIQ in October 2015, and the expectation is that VocalIQ's self-learning platform will make a major contribution to the next generation of conversational interfaces.[4]

---

[2]https://developers.google.com/voice-actions/interaction/. Accessed February 21, 2016.

[3]http://www.fastcolabs.com/3027067/this-cambridge-researcher-just-embarrassed-siri.   Accessed February 21, 2016.

[4]http://techone3.in/apple-buys-artificial-intelligence-startup-vocaliq-to-expand-siri-3734/. Accessed February 21, 2016.

18.2.3   The Internet of Things

The Internet of Things (IoT) is a term used to refer to a massive network of connected devices and sensors that can collect and exchange data. Indeed, it has been predicted that by 2020, the IoT network will consist of more than 29 billion connected devices.[5] IoT enables communication between these devices and sensors and also in some cases with virtual personal assistants as well as humans who can make use of this information. For example, a smart car not only will assist drivers with advice about traffic flow and congestions but will also be able to communicate with devices in the home to do tasks such as controlling security and heating or checking with the refrigerator what food items need to be bought when the car is in the vicinity of the user's favorite supermarket.

Applications involving the IoT will make use of technologies such as cognitive computing and the conversational interface. For example, IBM recently introduced the notion of Cognitive IoT in which systems not only will interact with data but will also learn and adapt.[6] Conversational interfaces are likely to be the only way to communicate with devices as many of them will not have conventional graphical user interfaces (GUIs) to accept input from human users and display output. Moreover, given the large number of devices that we might communicate with in our activities of daily living, a single and unique conversational interface will be preferable to a situation in which each device has its own interface. As recommended by Deborah Dahl, principal at Conversational Technologies and chair of the Multimodal Interaction Working Group at the Worldwide Web Consortium (W3C), the use of a standard for speech understanding results such as the W3C's Extensible Multimodal Annotation (EMMA) would provide an interface between VPAs and devices in IoT that would make it easier for users and developers to interact with this wide variety of connected devices.[7]

18.2.4   Platforms, SDKs, and APIs for Developers

Given the technological advances discussed in the preceding sections, it will be important for developers to have access to platforms, SDKs, and APIs that will enable them to develop powerful and effective conversational interfaces. In this section, we first review some tools that are currently available and then make some suggestions for future work.

---

[5]http://www.businesswire.com/news/home/20131003005687/en/Internet-Poised-Change-IDC. Accessed February 21, 2016.

[6]http://www.ibm.com/internet-of-things/. Accessed February 21, 2016.

[7]http://www.speechtechmag.com/Articles/Column/Standards/Talking-to-Everything-User-Interfaces-for-the-Internet-of-Things-103795.aspx. Accessed February 21, 2016.

There have been many tools that have been developed in industry and universities to enable researchers to build spoken dialog systems and voice user interfaces. Some of these tools are proprietary and some are open source. A recent addition is Sirius, an open end-to-end virtual personal assistant that was developed by Clarity Lab at the University of Michigan (Hauswald et al. 2015).[8] Sirius includes the key elements of a virtual personal assistant, such as speech recognition, image matching, natural language processing, and question answering, building on open-source projects that include technologies similar to those found in commercial systems: for example, for speech recognition, several technologies are available, including Gaussian mixture models (GMMs) and deep neural networks; for natural language processing technologies such as regular expressions, conditional random fields, and part-of-speech tagging; for question answering the OpenEphyra system, which uses techniques similar to those used in IBM's Watson; and for image processing Speeded Up Robust Features (SURF), a local feature detector and descriptor that uses state-of-the-art image matching algorithms. Developers can select from the available technologies and customize them by using their own components and algorithms.

To support the development of cognitive systems, IBM has made available the IBM Watson Developer Cloud that provides services enabling developers to create cognitive applications on the IBM Bluemix cloud platform. Language services include Dialog, Language Translation, Natural Language Classifier, and AlchemyLanguage, a collection of APIs that offer text analysis using natural language processing technologies, such as entity extraction, sentiment analysis, language detection, and parsing. Other relevant services for developers of conversational interfaces are Speech to Text and Text to Speech.[9] Similarly, Microsoft's Project Oxford offers a series of APIs, including Speech, Speaker Recognition, and Language Understanding.[10]

Other platforms for developers of conversational interfaces include: Houndify,[11] Silvia,[12] and the Teneo platform from Artificial Solutions.[13] The Virtual Human Toolkit from the Institute of Creative Technologies at the University of Southern California is a set of tools and libraries for creating virtual human conversational characters that interact using speech and also track and analyze facial expressions, body posture, and acoustic features as well as generating nonverbal behaviors.[14]

---

[8]http://sirius.clarity-lab.org/. Accessed February 21, 2016.

[9]http://www.ibm.com/smarterplanet/us/en/ibmwatson/watson-cloud.html. Accessed February 21, 2016.

[10]https://www.projectoxford.ai/. Accessed February 21, 2016.

[11]http://www.soundhound.com/houndify. Accessed February 21, 2016.

[12]http://silvia4u.info/technology/. Accessed February 21, 2016.

[13]http://www.artificial-solutions.com/natural-language-interaction-products/. Accessed February 21, 2016.

[14]https://vhtoolkit.ict.usc.edu/. Accessed February 21, 2016.

Platforms, SDKs, and APIs such as these provide a useful resource to developers, given the complexity of the components required to design and implement a conversational interface. In the future, tools will also be required that will make it easier for developers to easily integrate conversational interfaces with other types of application and device. At present, most of this integration is done on an ad hoc basis, but there is a need to have agreed standards and tools that can be used by developers wishing to adapt and tailor their applications to their specific requirements.

18.3   Applications that Use Conversational Interfaces

Until recently, interacting with applications using speech involved either stilted, system-directed dialogs over the telephone or commands using a restricted vocabulary to control objects in the environment. However, given the developments in technology outlined above as well as in earlier chapters of this book, virtual personal assistants and social robots can process conversational and multimodal input from users and retrieve and process data from sources on the Web and from personal and environmental sensors, and they can use this data to respond to and interact intelligently with users. In the following sections, we outline a number of application areas that are currently benefitting from these new developments in conversational technology and offer some suggestions for future work.

18.3.1   Enterprise Assistants

While VPAs such as Google Now, Siri, Cortana, Alexa, M, and many others are intended primarily to assist individual users to find information and accomplish tasks, there is an increasing demand for specialized enterprise assistants that will provide customer-facing support for a company's products and services as well as customer-enabling support that helps customers find information about those particular products and services that meet their current requirements. As Meisel[15] has predicted, enterprise agents such as these will be as necessary for companies as traditional company Web sites.

Several companies are heavily involved in specialized enterprise assistants, including Interactions,[16] NextIT,[17] and Nuance Nina.[18] Enterprise agents bring

---

[15]http://wp.avios.org/wp-content/uploads/2015/conference2015/Bill%20Meisel.pdf.   Accessed February 21, 2016.

[16]http://www.interactions.com/. Accessed February 21, 2016.

[17]http://www.nextit.com/. Accessed February 21, 2016.

[18]http://www.nuance.com/for-business/customer-service-solutions/nina/index.htm.   Accessed February 21, 2016.

enormous benefits to companies. For example, NextIT claim that their virtual assistant for Amtrak, Ask Julie, saved $1 million in customer service e-mail costs alone, while Nuance provide a number of case studies showing how their applications address the challenges of customer service.[19] Similarly, Kasisto show how their specialized intelligent conversational agent enhances interactions involving mobile financial applications.[20]

As the technology for conversational interfaces matures, we can expect that an ever-increasing number of companies and enterprises will want to deploy virtual personal assistants with conversational interfaces. There will be a need for tools that make use of standard processes and techniques and that can be easily mastered and tailored to the specific needs of particular businesses.

18.3.2   Ambient Intelligence and Smart Environments

Ambient intelligence (AmI) is a term used to refer to the way devices in IoT work together to support people in their activities of daily living within smart environments such as the home and the car. The conversational interface provides a natural way for humans to interact with these smart environments.

The vision of the smart home is that the devices within the home will be connected with each other and that many appliances and processes, such as controlling lighting, heating, and security, will be automated. For example, the lighting might automatically adjust to take account of the time of day and the presence of people in a particular part of the house. Likewise, in a smart car, various functions will be automated, such as adjusting the headlamps in response to ambient weather and visibility conditions.

The conversational interface is likely to be the preferred method for communicating with smart devices. For example, in the home, it would be easier to say "Turn up the heating to 20° in the living room" as opposed to going to the thermostat and carrying out the action manually. For elderly people and people with disabilities, using a conversational interface would certainly be an advantage. Similarly, in smart cars, there are many actions that would be dangerous to perform manually, such as searching for music on the vehicle's audio device.

The next years are likely to see many developments in the area of AmI and smart environments. This is where a conversational interface will be important both for users and for developers. As Deborah Dahl[21] puts it in respect of users:

---

[19]http://www.nuance.com/for-business/resource_library/index.htm. Accessed February 21, 2016.

[20]http://kasisto.com/. Accessed February 21, 2016.

[21]http://www.speechtechmag.com/Articles/Column/Standards/Talking-to-Everything-User-Interfaces-for-the-Internet-of-Things-103795.aspx. Accessed February 21, 2016.

Right now there are more than 100 apps in the iTunes Store and Google Play Store for controlling just one type of light bulb, the Philips Hue. Finding and using the right app for all of our connected objects will be very difficult. To manage this, we need a natural, generic means of communication that doesn't require installing and learning to use hundreds of apps.

And with reference to developers:

IoT interfaces will be much easier to develop if the results they produce are not only available to developers, but are in a standard format too. That way developers won't have to work with completely different API formats for all the current personal assistants and all those that may come along in the future.

18.3.3 Health care

Conversational interfaces in which users can talk with a virtual nurse or doctor have been in use since around 2000. The first systems provided telephone-based monitoring and advice, for example, the Homey system, a telemedicine service for hypertensive patients (Giorgino et al. 2005), and the DI@L-log system that monitored users with type 2 diabetes (Harper et al. 2008). The dialogs in these systems were system-directed, in that the system asked a series of questions to elicit information from the user. For example, the DI@L-log system asked users to provide values for their weight, blood pressure, and blood glucose levels and then output relevant advice or, in the case of a problem, such as an excessive rise in blood pressure or blood glucose levels, generated an alert to the user's clinician.

A new generation of conversational interface, known as relational agents, provides more natural interaction. These relational agents are being developed in the Relational Agents Group[22] at Northwestern University, Boston, by Bickmore and colleagues (Schulman et al. 2011; Bickmore et al. 2016). Relational agents build and maintain long-term relationships with people and use technologies such as those developed in the embodied conversational agent (ECA) tradition that include features of face-to-face communication, for example, hand gestures, facial expressions, body posture, and spoken interaction. The effectiveness of conversational interfaces in terms of user satisfaction has been demonstrated in a number of studies, for example, Bickmore et al. (2016).

Within the commercial world, there are several companies that specialize in conversational interfaces in the healthcare domain, for example, NextIT's ALME health coach[23] and Nuance's healthcare solutions.[24] Sense.ly provide a number of solutions in the healthcare domain, including a virtual nurse called Molly.[25] As the

---

[22]http://relationalagents.com/. Accessed February 21, 2016.

[23]http://www.nextithealthcare.com/. Accessed February 21, 2016.

[24]http://www.nuance.com/for-healthcare/index.htm. Accessed February 21, 2016.

[25]http://sense.ly/. Accessed February 21, 2016.

costs of health care escalate, we can expect that virtual assistants for specialist areas of health care will be an important supplement to care provided by healthcare professionals.

18.3.4 Companions for the Elderly

In Chap. 13, we discussed social robots such as Amazon Echo, Jibo, and Pepper that act as personal assistants in the home by answering queries, providing information, and in some cases controlling devices in a smart home. Pepper is also being used in stores in Japan to welcome and inform customers.

Social robots are increasingly being used for more specialized functions, such as acting as companions for the elderly. This is a particularly important issue given aging populations in many countries. Various European Union (EU) research projects have investigated the use of artificial companions, such as LIREC[26] and COMPANIONS.[27] Similarly, Dahl et al. (2011) describe a personal assistant that enables elderly users to perform various tasks of everyday living using an intuitive conversational interface, while McCarthy et al. (2015) discuss usability aspects of mobile devices for the elderly.

In addition to providing companionship, building relationships, and supporting the sorts of tasks that can also be performed using VPAs on a smartphone, some social robots provide physical support for activities of daily living to enable independent living at home. Recent projects involving social robots that provide physical as well as cognitive and social assistance include the EU projects ACCOMPANY[28] and Mobiserv[29] and a project at Toyota in Japan involving a household robot called Robina that in addition to conversing with people can provide medical and nursing care, perform housework, and carry and use objects.[30]

Another type of application involves activity monitoring, in which environmental sensors are used to detect the execution status of activities along with biosensors that log information about the user such as heart rate and body temperature. One example is the Planning and Execution Assistant and Trainer (PEAT) that helps elderly users and those with cognitive impairments to plan activities. The system monitors their progress and helps to replan when changes occur in the plan.[31] Another example is a system developed at the Toronto Rehabilitation

---

[26]http://lirec.eu/project. Accessed February 21, 2016.

[27]http://www.cs.ox.ac.uk/projects/companions/. Accessed February 21, 2016.

[28]http://accompanyproject.eu/. Accessed February 21, 2016.

[29]http://www.mobiserv.info/. Accessed February 21, 2016.

[30]http://www.toyota-global.com/innovation/partner_robot/family_2.html#h201. Accessed February 21, 2016.

[31]http://www.brainaid.com/. Accessed February 21, 2016.

Institute that detects when a person has fallen in the home and engages in a conversation to provide support and assistance.[32]

Conversational interfaces for the elderly pose a number of problems. The speech of an elderly person, in particular someone suffering from dementia, will be more difficult to recognize automatically. Furthermore, the content of the utterances of an elderly person may also be difficult to process semantically, particularly if that person is confused, overexcited, or anxious. Additionally, instructions spoken to an elderly person on how to carry out a sequence of actions, for example, to prepare a meal, need to be carefully delivered with the option to repeat and pause, and also, if the interface is conversational, to ask for clarifications and alternative explanations. For reasons such as these, conversational interfaces for the elderly will need to be specially adaptive and responsive to the needs of these users.

18.3.5 Conversational Toys and Educational Assistants

Social robots such as Amazon Echo and Jibo are described as family robots with which all members of the family including children can interact. However, there are also conversational toys and various types of educational assistant that are specifically intended for young children. These include toys that behave in a similar way to VPAs on smartphones and in smart robots, i.e., answering questions, carrying out commands, and engaging in chat, as well toys and gadgets that play a more educational role.

Hello Barbie from Mattel in conjunction with ToyTalk is an example of a commercially available conversational toy. Hello Barbie is operated by pressing a push-to-talk button on its belt buckle. The speech recognition technology is specially tuned to the speech of young children, and chatbot technology is used to respond to the children's inputs and to retrieve answers from data sources on the Web. The toy has some ability to learn, for example, by remembering what the child said previously and using this information in future conversations. A video and some examples of what Barbie can say are available at the Mattel Web site.[33] Some other commercially available conversational toys are reviewed in the VirtualAgentChat blog[34] (see also Stapleton 2016).

The educational benefits of conversational toys and educational assistants and issues concerning their usability are being studied in a number of research projects. MIT's Personal Robots Group has a range of projects, including Storytelling Companion, which investigates the effectiveness of a social robot for children's language learning, and DragonBot, a robot that runs on Android phones and that helps preschool children with learning language. Information about these and other

---

[32]http://www.idapt.com/. Accessed February 21, 2016.

[33]http://hellobarbiefaq.mattel.com/. Accessed February 21, 2016.

[34]http://virtualagentchat.com/category/conversational-toys/. Accessed February 21, 2016.

robots at MIT's Personal Robots Group can be found here.[35] Other research includes the following: the EU ALIZ-E project that investigates how children interact with social robots over a longer period of time[36]; the TRIK robot that helps children with severe communication disabilities to learn language (Ljunglöf et al. 2009); and the Robots4Autism project that uses robots and tablets to teach social and conversational skills to children with autism.[37]

Conversational toys for children have certain special requirements. Firstly, they have to be able to process the speech input of young children, and secondly, they have to be able to engage with children across different stages of development, including children with delayed or impaired communication.

Automatic recognition of children's speech poses particular challenges. Due to the physiological characteristics of children, for example, a shorter vocal tract and articulators that are still under development, their speech has particular characteristics such as a higher fundamental frequency compared with adults. Articulation problems include the production of consonantal clusters, which at different stages of development typically involve consonantal cluster reduction (e.g., "stop" pronounced as "top") and substitutions (e.g., "frog" pronounced as "fwog"). Moreover, these characteristics change over time as the child matures. In addition to these acoustic characteristics, there are also issues with language models for children's speech as children's vocabulary and word combinations differ from those of adults and also change over time. Various studies have addressed the problems associated with automatic recognition of children's speech, for example, Hagen et al. (2007) and Russell and D'Arcy (2007); see also the deliverable from the EU project iTalktoLearn.[38]

Another problem for conversational interfaces involving children concerns the sorts of topics that interest children across different ages, as well as children's ability to engage in dialog using conventional principles of conversation, such as turn-taking and topic management (Narayanan and Potamianos 2002; Kruijff-Korbayová et al. 2012; Gray et al. 2014).

In addition to the research issues discussed here, there is a need for sophisticated conversational interfaces that will enable children to take roles and interact in various games and adventures. ToyTalk has produced a number of apps in which children can have conversations with characters in a range of games and adventures.[39] Further work is required to make it easier for developers to integrate conversational interfaces with similar apps and also with the wide range of serious games that are becoming available and that can be used to educate children in a novel way. For an overview of the use of natural language processing techniques in serious games, see Picca et al. (2015).

---

[35]http://robotic.media.mit.edu/project-portfolio/. Accessed February 21, 2016.

[36]http://www.aliz-e.org/. Accessed February 21, 2016.

[37]http://www.robokindrobots.com/robots4autism-home/. Accessed February 21, 2016.

[38]http://www.italk2learn.eu/wp-content/uploads/2014/09/D3.1.pdf. Accessed February 21, 2016.

[39]https://www.toytalk.com/. Accessed February 21, 2016.

18.3.6 *Bridging the Digital Divide for Under-Resourced Languages*

We have shown the importance of conversational interfaces for many different aspects of technological advance and the many benefits they bring for their users. However, only a small fraction of the languages in the world have the resources required to implement conversational interfaces (e.g., there are currently 40 languages supported by Google Voice Search and more than 6000 languages in the world).

It is necessary to distinguish between minority and under-resourced languages. Minority languages are spoken by a minority of the population of a territory or by a small number of people in the world, while under-resourced languages are those with a limited presence on the Web and a lack of electronic resources for speech and language processing (e.g., recordings, vocabularies, dictionaries, and transcriptions). Minority languages are not necessarily under-resourced. For example, Iberian languages other than Spanish, such as Galician, Basque, and Catalan, have many resources provided by the main technological companies (e.g., Google, iOS) and active research groups in speech and natural language processing technologies, while languages with a high number of speakers of languages from developing countries, such as Swahili or Bengali, are under-resourced.

META, the Network of Excellence forging the Multilingual Europe Technology Alliance, conducted a large study in 2012 on European languages to assess the level of support they receive through language technologies.[40] The results showed that 21 European languages (most of them official) were facing "digital extinction" as there were no relevant updated online contents and services in those languages. Currently, EU research efforts are directed toward covering all European languages and generating pan-European multilingual digital services. For example, the CITIA Baselayer for Multilingual Speech Technology[41] emphasizes the need to cover this gap: there are over 50 million speakers of the 25 languages comprising the twenty sixth to the fiftieth most used in Europe (Finnish to Montenegrin). To address the challenge, they propose the construction of an open multilingual infrastructure and in 2014 and 2015 have built a technology road map that will enable this vision.[42]

Besacier et al. (2014) present a comprehensive survey of the challenges and methods that are currently being used to tackle them. The methods mainly involve generating new resources in the target language, or trying to bootstrap systems and resources developed for other languages into the target language, or a combination of both. On the one hand, there is the possibility of adopting new acquisition methodologies so that new resources can be compiled using crowdsourcing and

---

[40]http://www.meta-net.eu/whitepapers/overview. Accessed February 21, 2016.

[41]http://www.lt-innovate.org/citia/citia-baselayer-multilingual-speech-technology.   Accessed February 21, 2016.

[42]https://directory.sharpcloud.com/html/#/story/e505f7dc-2b77-41cf-9646-28b752e600b7. Accessed February 21, 2016.

other approaches. This poses difficulties, for example, in finding and reaching native speakers and experts with technical skills and a good knowledge of the target language. On the other hand, different authors are investigating how to use already available resources for other languages. This can be done either by porting resources from better covered languages (the cross-lingual approach), or by using the source and target languages together (the multilingual approach) (Schultz and Kirchhoff 2006). The Workshop on Spoken Language Technologies for Under-resourced languages (SLTU)[43] brings together researchers working on these topics.

18.4 Summary

Looking at future prospects for the conversational interface, there are two aspects: developments in technology that will enable conversational interfaces to become more usable and effective and a range of appropriate applications where a conversational interface can be usefully employed.

Cognitive computing combined with deep learning enables virtual personal assistants to access and harness the vast range of data and knowledge on the Internet and use it to provide information, advice, and services to human users. By linking together millions of smart devices, IoT provides the infrastructure for intelligent interaction with these devices. New development environments are required to make it easier for developers to create systems that include these advanced technologies. As more interfaces are developed, there will be more users of these interfaces and this will in turn provide more opportunities for the next generation of developers.

There are many applications of the conversational interface. For businesses, specialized assistants provide a friendly and natural face to customers. For individual users, virtual personal assistants are being developed for a variety of contexts and users: to enable people to interact with devices in smart environments such as smart homes and smart cars; to provide more effective health care; applications for elderly people support independent living; and conversational toys and educational assistants provide entertainment and learning opportunities for young children.

There are still many problems to be resolved. Technology is advancing at incredible rates, and some of the techniques described in this book will soon be overtaken by new developments. Currently, there is a wide variety of solutions. What will be important in the future is to focus more on aspects such as usability. People will not want to have a multitude of interfaces and assistants and are likely to prefer a scenario where there is one main assistant that interacts seamlessly with other specialist assistants through a single common conversational interface. Realizing this scenario remains a significant challenge for developers in the coming years.

---

[43]http://www.mica.edu.vn/sltu/. Accessed 21 February 2016.

Further Reading

A search of the Internet will return many documents on recent developments in conversational interfaces. Here are two particularly interesting articles: "The future of voice: what's next after Siri, Alexa, and OK Google,"[44] and "Advanced AI to power a new generation of intelligent voice interfaces,"[45] which also includes a useful diagram showing how voice interfaces require much more than just speech recognition and what other components are required to intelligently process a user's query.

The techniques used to build Watson are described in a collection of papers published in the IBM Journal of Research and Development (Ferrucci 2012). Baker (2012) provides a popular account of the building of Watson and its participation in the Jeopardy quiz show. Various IBM white papers provide further information about Watson and its use as a tool for the development of cognitive applications.[46,47,48]

There are numerous references on deep learning. We can point to a forthcoming book that is available online.[49] For conversational interfaces using deep learning and reinforcement learning, there is a lecture available on iTunes by Steve Young entitled "Towards open-domain spoken dialogue systems" in the MIT-Apple Human Language Technology Lecture Series.[50] A paper by Vinyals and Le, researchers at Google, describes a neural conversational model that learns how to generate fluent and accurate responses in conversation from a dataset of movie transcripts.[51]

Regarding applications, in addition to the references within the main text of this chapter, we can point to the following: the Journal of Ambient intelligence and Smart Environments[52]; the Web page of the Workshop on Child Computer Interaction (WOCCI)[53]; and an article on digital assistants for smart cars.[54]

---

[44]http://recode.net/2015/10/27/the-future-of-voice-whats-next-after-siri-alexa-and-ok-google/. Accessed February 21, 2016.

[45]https://mindmeld.com/. Accessed 21 February 2016.

[46]http://www.ibm.com/smarterplanet/us/en/ibmwatson/what-is-watson.html. Accessed February 21, 2016.

[47]http://www.research.ibm.com/software/IBMResearch/multimedia/Computing_Cognition_WhitePaper.pdf. Accessed February 21, 2016.

[48]http://www.ibm.com/developerworks/cloud/library/cl-watson-films-bluemix-app/. Accessed February 21, 2016.

[49]http://www.deeplearningbook.org/. Accessed 21 February 2016.

[50]https://itunes.apple.com/us/itunes-u/human-language-technology/id787393959?mt=10. Accessed February 21, 2016.

[51]http://arxiv.org/abs/1506.05869. Accessed 21 February 2016.

[52]http://www.iospress.nl/journal/journal-of-ambient-intelligence-and-smart-environments/. Accessed February 21, 2016.

[53]http://www.wocci.org/2016/home.html. Accessed February 21, 2016.

[54]http://www.patentlyapple.com/patently-apple/2015/10/apple-has-acquired-vocal-iq-a-company-with-amazing-focus-on-a-digital-assistant-for-the-autonomous-car-beyond.html. Accessed February 21, 2016.

References

Baker S (2012) Final Jeopardy: man vs machine and the quest to know everything. Houghton Mifflin, Boston

Besacier L, Barnard E, Karpov A, Schultz T (2014) Automatic speech recognition for under-resourced languages: a survey. Speech Commun 56:85–100. doi:10.1016/j.specom. 2013.07.008

Bickmore TW, Utami D, Matsuyama R, Paasche-Orlow M (2016) Improving access to online health information with conversational agents: a randomized controlled experiment. J Med Internet Res 18(1):e1

Dahl D, Coin E, Greene M, Mandelbaum P (2011) A conversational personal assistant for senior users. In: Perez-Martin D, Pascual-Nieto I (eds) Conversational agents and natural language interaction: techniques and effective practices. IGI Global, Hershey, pp 282–301

Ferrucci DA (ed) (2012) IBM Watson: the science behind an answer. IBM J Res Dev 3(4), May–June 2012 (Baker S)

Giorgino T, Azzini I, Rognoni C, Quaglini S, Stefanelli M, Gretter R, Falavigna D (2005) Automated spoken dialogue system for hypertensive patient home management. Int J Med Inform 74(2–4):159–167

Gray SS, Willett D, Lu J, Pinto J, Maergner P, Bodenstab N (2014) Child automatic speech recognition for US English: child interaction with living-room-electronic-devices. In: Proceedings of workshop on child computer interaction (WOCCI) 2014, Interspeech 2014 Satellite Event, Singapore, 19 Sept 2014. http://www.wocci.org/proceedings/2014/wocci2014_proceedings.pdf

Hagen A, Pellom B, Cole R (2007) Highly accurate children's speech recognition for interactive reading tutors using subword units. Speech Commun 49(12):861–873

Harper R, Nicholl P, McTear M, Wallace J, Black LA, Kearney P (2008) Automated phone capture of diabetes patients readings with consultant monitoring via the web. In: Proceedings of engineering of computer based systems (ECBS) 2008, 15th annual IEEE international conference, Belfast, UK, 31 Mar–4 Apr 2008, pp 219–226

Hauswald J, Laurenzano MA, Zhang Y, Li C, Rovinski A, Khurana A, Dreslinski R, Mudge T, Petrucci V, Tang L, Mars J. (2015) Sirius: an open end-to-end voice and vision personal assistant and its implications for future warehouse scale computers. In: Proceedings of the twentieth international conference on architectural support for programming languages and operating systems (ASPLOS), ASPLOS '15, New York, NY, USA, 2015:223–228

Hinton GE, Osindero S, Teh YW (2006) A fast learning algorithm for deep belief nets. Neural Comput 18:1527–1554

Kruijff-Korbayová I, Cuayáhuitl H, Kiefer B, Schröder M, Cosi P, Paci P, Sommavilla G, Tesser F, Sahli H, Athanasopoulos G, Wang W, Enescu V, Verhelst W (2012) Spoken language processing in a conversational system for child-robot interaction. In: Proceedings of workshop on child computer interaction (WOCCI) 2012, Interspeech 2012 Satellite Event, Portland, Oregon, 14 Sept 2012

LeCun Y, Bengio Y, Hinton G (2015) Deep learning. Nature 521:436–444. doi:10.1038/nature14539

Ljunglöf P, Larsson S, Mühlenbock K, Thunberg G (2009) TRIK: a talking and drawing robot for children with communication disabilities. In: Jokinen K, Bick P (eds) Proceedings of the 17th Nordic conference of computational linguistics (NODALIDA) 2009, vol 4, pp 275–278

McCarthy S, Sayers H, McKevitt P, McTear M, Coyle K (2015) Intelligently adapted mobile interfaces for older users. In: Xhafa F, Moore P, Tadros G (eds) Advanced technological solutions for e-health and dementia patient monitoring. Medical Information Science Reference (IGI Global), Hershey PA, USA, pp 36–61

Narayanan S, Potamianos A (2002) Creating conversational interfaces for children. IEEE T Speech Audi P 10(2):65–78

Picca D, Jaccard D, Eberlé G (2015) Natural language processing in serious games: a state of the art. Int J Serious Games 2(3):77–97

Russell M, D'Arcy S (2007) Challenges for computer recognition of children's speech. In: Proceedings of the ISCA special interest group on speech and language technology in education (SLaTE) 2007, Farmington, Pennsylvania USA, 1–3 Oct 2007, pp 108–111

Schultz T, Kirchhoff K (eds) (2006) Multilingual speech processing. Elsevier Academic Press, Amsterdam

Schulman D, Bickmore TW, Sidner CL (2011) An intelligent conversational agent for promoting long-term health behaviour change using motivational interviewing. In: AI and health communication, papers from the 2011 AAAI spring symposium, Technical Report SS-11–01, AAAI Press, Menlo Park, California, pp 61–64

Stapleton A (2016) Conversational toys and devices. In: Proceedings of mobile voice conference 2016, San Jose, CA, 11–12 Apr 2016

Index

© Springer International Publishing Switzerland 2016
M. McTear et al., *The Conversational Interface*,
DOI 10.1007/978-3-319-32967-3

CPSIA information can be obtained
at www.ICGtesting.com
Printed in the USA
LVOW05*2039110617

537726LV00001B/90/P

9 783319 329659